Tooth
and
Claw

"Tho' Nature, red in tooth and claw"
Alfred, Lord Tennyson, 1850

Tooth
and
Claw

TOP PREDATORS OF THE WORLD

Robert M. Johnson III, Sharon L. Gilman,
and Daniel C. Abel
Illustrated by Elise Pullen

PRINCETON UNIVERSITY PRESS

PRINCETON AND OXFORD

TO ALL THE MAGNIFICENT PREDATORS AND OTHER LIVING THINGS OF OUR SMALL, MUTUAL PLANET. MAY WE LEARN TO BETTER SHARE AND CARE FOR THEIR SPACE.

Published by Princeton University Press
41 William Street, Princeton, New Jersey 08540
99 Banbury Road, Oxford OX2 6JX

press.princeton.edu

Library of Congress Cataloging-in-Publication Data
Names: Johnson, Robert M., III, author.
Title: Tooth and claw : top predators of the world / Robert M. Johnson III,
    Sharon L. Gilman, and Daniel C. Abel ; illustrated by Elise Pullen.
Description: Princeton : Princeton University Press, [2023] | Includes
    bibliographical references and index.
Identifiers: LCCN 2022026585 (print) | LCCN 2022026586 (ebook) | ISBN
    9780691240282 (hardback) | ISBN 9780691245379 (ebook)
Subjects: LCSH: Predatory animals. | Predation (Biology) | Predatory
    animals—Conservation.
Classification: LCC QL758 .J64 2023 (print) | LCC QL758 (ebook) | DDC
    591.5/3—dc23/eng/20220622
LC record available at https://lccn.loc.gov/2022026585
LC ebook record available at https://lccn.loc.gov/2022026586

British Library Cataloging-in-Publication Data is available

Editorial: Robert Kirk and Megan Mendonça
Production Editorial: Kathleen Cioffi
Jacket Design: Katie Osborne
Production: Steven Sears
Publicity: Matthew Taylor and Caitlyn Robson
Copyeditor: Laurel Anderton
Typeset and designed by D & N Publishing, Wiltshire, UK

Jacket images: (Clockwise) Clouded leopard by Adalbert Dragon/
Shutterstock. Sand shark by blickwinkel/Alamy. *Canis lupus* photograph
by Robert Johnson. American alligator by Joe Austin Photography/Alamy

This book has been composed in DIN 2014

Printed on acid-free paper. ∞

Printed in Italy

10 9 8 7 6 5 4 3 2 1

CONTENTS

PREFACE

If you are reading this book, odds are good that you are part of the large and growing number of people who have a fascination with predators. Looking fondly on these beasts of prey, and wanting to conserve them, is a recent occurrence in human history. This is largely because most of us are under no immediate threat from them. This was not true for most of the thousands of generations that preceded us. Bears, sharks, and Tigers posed a real danger to life and limb, as well as to the livestock our ancestors relied on to make a living or put food on the table. While the danger predators pose is an issue for some people around the world, the dramatic decline in populations of large predators makes these encounters increasingly rare. Our peaked interest is leading us to better understand the importance of top predators in the environment and our need to ensure their continued existence.

This book was originally planned to focus exclusively on *apex predators*, those at the very pinnacle of the food chain. While we still discuss the importance of apex predators and highlight a number of them, we decided to broaden the scope of the book to include predators that may not be at the very top of predatory hierarchy but are within whispering distance—that is, the larger group of *top predators*. As you may know, nature does not fit into nice, neat little boxes. While we scientists do our best to define and put labels on the animal kingdom, there are a lot of blurred lines. An apex predator in a particular environment may not be at the top of the food chain where its territory overlaps with that of an even bigger predator. A Great White Shark, for example, would definitely count as an apex predator, until it is killed and eaten by an

Orca. Similarly, apex predators and their very close cousins have much in common, from their biology, to their predatory behavior, to their relative importance in the ecosystem. It makes sense to discuss these groups as a whole and highlight their interconnectedness.

The title of this book, *Tooth and Claw*, is in homage to Alfred, Lord Tennyson's 1850 poem *In Memoriam*. In his grief over a lost friend, Tennyson describes nature as "red in tooth and claw," referring to the seeming callousness with which predators dispatch their prey. While there seem to be as many interpretations of this work as there are cantos in the poem itself, there appears to be an underlying struggle to reconcile a caring, loving world with the brutality of nature. In this book we explore top predators not as "fiends of hell" but rather as an integral part of a fully functioning biological world. Their importance in nature cannot be overstated.

Collectively, the authors have nearly a century of experience working with, studying, conserving, and teaching about the world's predators. Indeed, these have been a focus in our lives. Our passion led us to write this book to share our knowledge and enthusiasm with a wider audience. A greater understanding of predators will foster in us all a heightened and educated appreciation of their significance. While labeling predators as "killing machines" makes a good headline, it ignores the subtler impact they have on the natural world. Predators act to reduce disease and promote biodiversity in natural ecosystems. A predator-rich environment is a healthier environment. Since we do not yet fully recognize all the roles predators play in an ecosystem, it behooves us to protect them as well as the ecosystems in which they reside, or at the very least ensure that we are not the cause of their extinction. The fate of predators will also likely be our fate. Conservation and proper stewardship should loom large in our thoughts. We have written this book not only *about* the top predators of the world, but in many ways *for* them.

ACKNOWLEDGMENTS

ROBERT JOHNSON: First, although it may be a bit cliché, I'd like to thank my fellow authors for their mentoring and patience in this epic project, as it has been several years in the making. This is undoubtedly a better book because of your persistence and dedication. I'd like to thank Dr. Brian Davis for thoroughly reviewing, vetting, and contributing to the chapters on cats and dogs. Although I try to stay current, Brian's relevance in current predator DNA analysis has kept me both honest and up to date. To my paramour Mari Gent, I express my deepest gratitude for your understanding of why I spent so many late nights (and early mornings) writing at the computer and for your unwavering support as my unpaid research assistant (much to your chagrin). Our sleep schedule will return to normal one of these days. I'd like to express my appreciation to Myrtle Beach Safari not only for access to knowledge but for allowing me to meet and learn about many of the predator species featured in this book. While I could go on into perpetuity, I'll keep it short. I want to thank the many friends and family members who did not get as much of my attention as you deserved while I wrote the following pages. Most importantly, I'd like to acknowledge the many animals I have encountered in the wild spaces that remain in this world. It is because of you that I've written your stories and tried to share my passion for them and for your importance in this fragile world.

SHARON L. GILMAN: Thank you to my coauthors, first, for coming up with the idea to teach a course on apex predators and include me, and then to think to write a book! And our immensely talented illustrator came along at just the right time. It continues to be a pleasure and honor to work with you all. Coastal Carolina University (CCU) has provided me a place to hone my skill at conveying the wonder of the natural world to students while continuously learning new things myself. I don't really want to thank *all* my students, but a lot of them have enriched my life and buoyed my spirit. The kids are all right, mostly. Fabulous Pennsylvania public school teachers and then professors at McDaniel College and the University of Rhode Island's Graduate School of Oceanography have contributed more to what I know about science, biology, and communicating than I could ever explain. I thank them all. A lot of what I know about vertebrates I have learned indirectly from Dr. F. Harvey Pough, by virtue of using his wonderful textbooks for my Vertebrate Zoology course since its inception in 2006. I had the good fortune to contribute some student activities to the most recent edition and got to know Harvey in the process. I took advantage of his vast store of vertebrate knowledge by having him review several of our chapters. I also had three CCU colleagues, Drs. Rob Young, Chris Hill, and Derek Crane, review chapters or sections for me. I thank them all for their kindness and expertise. This is a more accurate book thanks to their efforts. Please blame any remaining errors on me! (Except in the shark chapter; those are Dan's.) Finally, I want to thank my mom and dad for instilling in me a love for animals and the natural world, the understanding that writing well matters, and the belief that I could do anything. Jeff and Nicole have allowed me to see the world through new, young eyes, and their mom is very proud that they both love biology and travel:

kindred spirits! And thanks to Craig for putting up with my great enthusiasm for all living things, even those he does not much like, and partnering in the interesting adventure that is our life.

DANIEL ABEL: My inspirations for coauthoring this book are indistinguishable from those that motivate me and am thankful for on a daily basis: Mary (my wife), Juliana (my daughter), and Louis (my son); Billy and Alan (my brothers), Sara (my sister), and Ellen, Tamara, and Paul (their spouses); the enduring legacy of my parents, Harris and Ruth; other family members; my mentors and role models throughout my life; my colleagues; my students; anyone who challenges me to be a better thinker, writer, and person; my friends; kind strangers who greet you in passing with a smile; people who strive to conserve our planet; and especially, the natural world. I remain deeply indebted to my two coauthors, Rob and Sharon, and our illustrator (and my former student), Elise. Their vibrant voices and creativity illuminate the book from beginning to end. And let me not forget the photographers who selflessly allowed us to use their spectacular photos, continuing the tragedy of not remunerating them properly.

I have intentionally avoided naming most of those whom I have acknowledged because I always forget someone whose contribution merited their inclusion. I hope they already know the depth of my appreciation for their impact on my life and how much I value them.

ELISE PULLEN: Thank you to our extremely knowledgeable authors for bringing me to the team and for giving me the opportunity to play a part in their creative vision. I feel very fortunate to have been able to see this project grow and to learn so much along the way. A special thank-you to Dr. Daniel Abel, who as a professor and graduate adviser has continued to encourage both my academic and artistic pursuits. I am very grateful for my professors from Florida Southern College and Coastal Carolina University, especially those who let me get away with constantly drawing on my notes during class! My family has always fostered a love of the outdoors and wildlife. Our phone calls almost always include stories about neat animal encounters on any given day. Many thanks to my loving parents and brother for always being there to cheer me on. I also count myself lucky to have such amazing friends; thank you for all the encouragement and support.

Finally, we'd like to thank the folks at Princeton University Press. In the predatory world of book publishing, we were fortunate to work with Robert Kirk, one of the world's most formidable "Apex Editors." Robert gave us the liberty to write a manuscript that was not only educational but also included personal anecdotes, our own observations, and our own distinct humor—in short, the book we wanted to create. Perhaps equally impressive from a writer's perspective, his responses to all our questions and comments were thorough and immediate. There was no "I'll get back to you next week." This man should be a global model for efficiency.

To Laurel Anderton, your copyediting is really the glue that holds this book together. Your attention to detail was spot-on. We particularly love that you corrected our capitalization errors for a rule that we made up. You are "jaw-some."

Thanks also go to Megan Mendonça, David M. Campbell, Kathleen Cioffi, and Sydney Bartlett for helping produce such an excellent book and getting it into the hands of fellow nature lovers. We know that they could not have done this work alone. For those of you we have not mentioned or have not directly communicated with, please know that you have both our sincere thanks and our gratitude for your tireless efforts in doing such a "Buteo-ful" job. (Look it up. It's in this book.)

AUTHORS' NOTE

Google, Amazon, Walmart, Target, and Apple are all *proper nouns*, and since proper nouns are typically capitalized, these are as well. Groups of organisms, such as monkeys, sharks, dogs, cats, and so forth, are considered *common nouns* and are not capitalized. Common names of specific species are proper nouns that are not usually capitalized unless a part of that name is a capitalized proper noun, such as Bryde's whale or American eel. While the explanation may be a tad confusing, you are probably used to this convention.

In writing this book, we were aghast that the capitalized names of corporations conferred respect and the lowercase names of living organisms were more dismissive. Our solution in this book is to capitalize the common names of all organisms, a trend that some taxonomic institutions have begun. At the same time, we will conform to the practice of using lower case for groups of organisms. So, Sandbar Shark, but sharks; African Wild Dog, but dogs; Spotted Hyena, but hyenas. (Note, however, that the varieties of Tigers and Wolves are subspecies.)

There is one additional complication with common names: in many cases they are ambiguous and one common name may refer to several different species. In the southeastern United States, local residents refer to several different shark species—Sandbar Sharks, Atlantic Sharpnose Sharks, Finetooth Sharks, Spiny Dogfish, and others—as "sand sharks." Much of the time this is harmless, but if two or more different species have the same common name, it can spell trouble if one species is in decline but the others are not. To avoid this potential problem, every organism has its own unique two-part name. The Sandbar Shark is *Carcharhinus plumbeus*. The Atlantic Sharpnose Shark is *Rhizoprionodon terraenovae*. The former is recovering from decades of overfishing whereas the latter is stable. Calling both species sand sharks jeopardizes the Sandbar Shark.

In this book, we use a two-pronged approach. Since virtually all the predators we describe have unique and unambiguous common names, and since common names are less intimidating, we use (and duly capitalize) these throughout the book. At the same time, we include the unique two-part scientific name when each animal is first introduced in the chapter discussing that group.

Finally, we note some good-quality videos you might want to check out, and we've made some of our own. To make this easy for you, we've created a YouTube page specifically for this book where our videos are all featured. Here is the link: https://youtube.com/@toothandclaw-toppredators.

INTRODUCTION TO TOP PREDATORS

Predators, whether a White Shark, Tiger, Nile Crocodile, or even the insect known as an Antlion (two-thirds of the millions of species of insects are predatory!), fascinate us. But this book is not about *all* predators, although such a publication would be immensely fun to read (and very, very long!). Here, we focus on the *top* predators of the animal world. With apologies to the insects and other invertebrates, the top predators we include in this book are those iconic and charismatic beasts featured on the cover, and some lesser-known but equally fascinating sharks, cats (big and small), nonavian reptiles, raptors, wild dogs (Wolves and their kin), bears, marine mammals, and others.

Top, as we use the term, refers to the top of the food chain. In most cases, the top predators in an ecosystem are also its *apex predators*, but not always. All apex predators are top predators, but not the converse. Dining on a diverse menu of prey items lower on the ecological pyramid, the apex predator is often so big and fierce that it need not worry about being dinner for a bigger, fiercer beast. What distinguishes an apex predator from other top predators is simply that nothing eats an adult apex predator. Before reading the rest of this chapter, examine the animals in figure 1.1. Which of these do you think are apex predators?

Tiger? Check. Gray Wolf? Check. Biology, however, resists neat categorizations. What about a Leopard Seal? Big canine teeth. A reputation for ferocity. Rules the Antarctic Ocean. Ask a Crabeater Seal, Weddell Seal, or Antarctic Fur Seal and its answer will be an unambiguous *yes*. Unless an Orca shows up. White Shark? Big check! Unless an Orca shows up. Orca? No doubt. What about Eurasian Eagle Owls? Yes, but truly so only at night, since they are inactive in daylight. What about a Coyote? Although a pack of Coyotes is a fearsome sight, these midsize canids (members of the dog family), which are increasingly common throughout North America, are classic *mesopredators*—the predators typically one level beneath those at the apex, but above virtually everything else. Here is where it gets tricky. In the case of Coyotes, the apex position is often no longer occupied in much of their range. Pumas and Wolves could eat, as well as compete with, Coyotes for food and they would win, but those animals have mostly been eradicated where the Coyote is common. Ecologists call this phenomenon

IMAGES NOT DRAWN TO SCALE

FIGURE 1.1 Apex predator or not? Look at the line drawings of these animals and decide whether you would categorize each one as an apex predator or not.

mesopredator or *trophic release*,[1] resulting from the loss of the apex predator, which ecologists call *trophic downgrading* or a *trophic cascade*. With the apex cat (Puma) and dog (Gray Wolf) gone, the Coyote takes their place. Is the Coyote then an apex predator? Do you achieve apex predator status in your ecosystem just because the actual ones have been wiped out? Maybe not.

What started out as an easily defined term becomes muddled and complicated with further consideration. We thus decided to be inclusive in our coverage of predators in this book. We highlight a few incontrovertible apex predators, but we include an animal if it is ecologically at the top of the heap most of the time, regardless of how or at what time of day or year it is there. Hence, Lions, mako sharks, and Komodo Dragons, certainly, but also Coyotes, otters, lynx, and owls. And we throw in some extras that do not fit in our taxonomically grouped chapters because, well, they are interesting top predators, too. We will point out where each fits in the grand scheme of things. You can disagree with our categorizations, but we hope you'll still like the animals and stories.

Now look in the mirror. Is this beast you see in the reflection an apex predator? Clever how we have avoided including humans in this discussion so far, no? Rest assured that this topic has not escaped our consideration. We have devoted the last chapter of this book to the question of whether humans are apex predators. Thus, you can contemplate the answer while you read the chapters preceding our take on the issue.

WHY PREDATORS MATTER

You probably think predators are fascinating and important, or you wouldn't have picked up this book. You are fortunate to have the luxury of thinking that, because if you had been obliged to live outside in North America when the continent had its full original complement of Pumas and Grizzly Bears, "fascinating" might not be the word that came to mind when you considered these animals. "Terrifying" is more likely how you might describe these and other predators, with their huge teeth and scary claws. These animals are easily capable of killing and eating you.

Fast-forward to the present. Suppose you are a Namibian farmer and you and your family live safely indoors, but Cheetahs sometimes eat some of your few, precious calves. They take food and money away from your family, which barely ekes out an existence in the best of times. You might loathe instead of respect those Cheetahs. Or perhaps you are taking your family on an annual vacation to an expansive Florida beach with incredible white sand, where a Bull Shark bit a bather the previous year. You just might choose to skip a dip in the water and instead work on your tan in what you perceive as the safer confines of the beach because of your fear of what toothy beast could lurk beneath the waves.

But a predator, even if it can eat you—maybe *because* it can eat you—warrants a certain level of admiration. Humans are smart, pretty big, and strong. We tend to dominate when we move to a place. Predators are also strong and smart and tend to dominate when they move to a place, although in very different ways. They are, with some exceptions, our counterparts in the natural world. They run the show if we do not. Until recently in our evolution, we couldn't reliably kill them. In a one-on-one battle, minus a spear or a gun, we lose. Just like we respect (or fear) a powerful person, we respect (or fear) these powerful animals. Most are attractive, too. Some people like *alligator* boots, *bear* skin rugs, *leopard* print clothing. History suggests that even where humans stand a reasonable chance of losing their lives or resources to a predator, most of us retain a certain respect for these beasts. Some of us are even crazy about them.

The earliest known humans lived in tribes, and very early on they developed religions based on *animism*. Animism sounds like a belief system based on animals, but the Latin word *anima* means "soul" or "life." A Tiger acts like a Tiger because it has a tiger-spirit. You can call on the tiger-spirit to protect you, to avenge you, or to give you power, like a Tiger. Cave paintings by very early humans in the Chauvet Cave near Vallon-Pont-d'Arc, France, a UNESCO World Heritage Site, date back more than 30,000 years and artistically depict owls, bears, horses, mammoths, rhinos, Leopards, and other large animals, perhaps most famously the Panel of the Lions. (Lions and Leopards in Europe! Times have changed.) Rock carvings at Twyfelfontein, Namibia, were created by San Bushmen (one of the oldest living races of humans, who have been around for 20,000 years) to depict watering holes in the desert and animals that might be encountered there, including Lions (fig. 1.2). Our ancestors clearly spent a lot of time thinking about these big beasts, and for good

FIGURE 1.2 These rock carvings by San Bushmen at Namibia's Twyfelfontein are approximately 20,000 years old and depict native animals, including Lions.

reason. Big animals represented life and death, as food and predators, respectively. By about 1400 BCE, the Sumerians had developed an alphabet, so we have written records of how they viewed nature in the Tigris-Euphrates valley, an area known as the cradle of civilization, at that time.[2]

Divine forces governed all aspects of life of these early humans, and these forces were very often represented by animals. This phenomenon was common worldwide. You are probably familiar with the totem poles of the indigenous peoples of the Americas. These carvings are representations of the animal spirits that people looked to for guidance and protection. Indigenous peoples hunted and ate animals, but in a respectful way so as not to upset the animal spirits and face retribution. With animals abundant and people less so, this worked fine.

With increasing numbers of people, and associated development of new methods to exploit natural resources, there was a bit of a shift in the view of Earth, reflecting humanity's increasing ability to "manage" nature. Some interpretations of more recent religious traditions give humans dominion over nature. Humans still depend on nature (perceptions and actions to the contrary notwithstanding), but the situation reverses from nature taking care of humans to humans exploiting and managing nature. Hinduism stresses harmony with nature, encouraging what we might now call "wise use," but it is still use.[3]

It is entirely possible that all this use would have worked out fine when humans lived within the planet's carrying capacity. *Carrying capacity* is an ecological concept defined as the population size of a species, in this case humans, that can live sustainably on the planet (or in any of its component ecosystems). Living *sustainably* means using resources like food, habitat, energy, and water at rates no faster than they can be replenished, thus leaving adequate and abundant resources for future generations. Arguably the last time we truly lived sustainably on the planet was when we were hunter-gatherers about 12,000 years ago, before we transitioned to agriculture. At the end of 2022, the human population numbered about 8 billion. Our ever-increasing population is giving the planet, and its predators, fits. Scholarly estimates of the Earth's human carrying capacity range from 2 billion to 10 billion, depending on lifestyle and resource consumption.[4] To maintain optimal biodiversity, it is likely closer to 2 than 10, and maybe even lower than that!

When early humans faced a choice between building a home in a bottomland hardwood forest or leaving the habitat pristine for a Wolf pack, they built, and killed or scared off the Wolf pack. This single act may not noticeably or irreversibly alter nature, as long as there are other habitats around that are also good for Wolves, and enough prey for both Wolves and humans. If it is a choice between your family or the Wolf pack getting enough to eat, the ones with the bigger brain win.

Okay, so the burgeoning human population needs space and food. Even if humans are not exactly predators (a topic we will revisit later), we are competing with predators, and we are winning. In fact, we are slaughtering them (sad pun intended)! So, like them or not, predators' numbers are dwindling. Does this matter? Let's set aside the ethics of this for now (but *only* for now) and look at the topic as a practical matter.

Consider big sharks. Sharks are not furry and cute like, say, a bear (although who could resist a Horn Shark or Nurse Shark pup, really?). They eat crustaceans, fish, or even mammals, so in theory

they are competing with us for food. We can eat them, and some people do (the market for shark meat and fins drives the current fishery for them), but there are lots of fish we can eat instead. Some sharks, like many of the predators in this book, occasionally kill people. Why should we care whether these animals continue to exist? Many of us do not. Indeed, some beachgoers claim they would feel much safer if sharks disappeared, at least from their beach. Four categories of arguments can be made in favor of caring about the plight of sharks and other predators: ecological, economic, ethical, and spiritual.

First, ecology. To understand the ecology argument, you need to know what an ecosystem is and how it works. An *ecosystem* (fig. 1.3) is a community of organisms interacting with each other and their environment. Energy comes into an ecosystem mainly from the sun in the form of light and heat. Organisms that can photosynthesize convert the radiant energy in sunlight to carbohydrates, which represent food energy. These photosynthesizing plants, algae, and bacteria, known collectively as *primary producers*, are the base of the food web in an ecosystem. They make food but also use energy to live and grow and reproduce themselves. In fact, they use most incoming energy themselves, or it is lost as heat, because the chemical reactions that occur in living organisms are not very efficient, leaving just 10%–20% useful energy to pass along. This varies a bit depending on what organism and what ecosystem you're looking at, but overall, ecologists use the 10% value (to the point that the phenomenon is called the "Rule of 10"). This is where the idea of ecosystem *trophic*, or feeding, levels comes from. Primary producers are the base, or first trophic level, in an ecosystem, and if you've ever seen a diagram of trophic levels, it's a pyramid (fig. 1.4). That base of primary producers is wide because nearly all the incoming energy goes through them and only 10% of that energy is left over for the

FIGURE 1.3 Animals found in a pelagic marine ecosystem (though not as tightly compressed as pictured). Plankton, the lower level of the food web, are not included.

organisms the next level up. So, there are a lot of primary producers, and the combined mass of them averaged over, say, a year eclipses the mass of every trophic level above them.

The next level is the secondary producers (fig. 1.4). We also call them *primary consumers*. If we stick with the marine ecosystem in which sharks live, that second producer level would include zooplankton, filter feeders like oysters and clams, and small fish, as well as larger animals like Green Sea Turtles and Marine Iguanas, which both feed on algae. The secondary producers also use (or lose as heat) about 90% of the energy they consume to maintain themselves and grow and reproduce, leaving just 10% of the 10% (which equals 1% of the energy of the primary producers) they originally received to pass along to the next trophic level. And so it goes, like a pyramid, with the biomass of

organisms (numbers and overall mass) diminishing with each level as less energy becomes available.

So, who eats the secondary producers? Bigger animals prefer these larger packages of food. So bigger fish eat smaller fish. Loggerhead Sea Turtles and Walruses eat filter-feeding shellfish. So do Horn Sharks and Nurse Sharks. Surprisingly, giant baleen whales and Whale Sharks, the largest fish in the sea, do not. Instead, they eat zooplankton. Why? Because that's where the energy is, at the lowest level of consumers, since it has not been lost as it has at higher trophic levels.

And here is a major consequence of being a big animal: there cannot be that many of you because the organisms of each trophic level use up about 90% of their energy on themselves or lose it as heat, leaving less and less as you climb the trophic pyramid. So, there can't be very many big

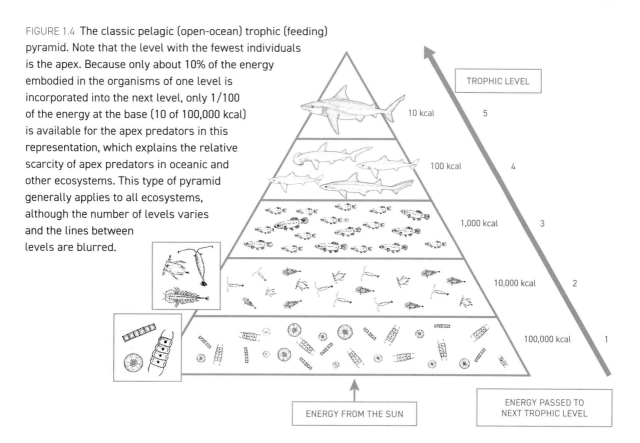

FIGURE 1.4 The classic pelagic (open-ocean) trophic (feeding) pyramid. Note that the level with the fewest individuals is the apex. Because only about 10% of the energy embodied in the organisms of one level is incorporated into the next level, only 1/100 of the energy at the base (10 of 100,000 kcal) is available for the apex predators in this representation, which explains the relative scarcity of apex predators in oceanic and other ecosystems. This type of pyramid generally applies to all ecosystems, although the number of levels varies and the lines between levels are blurred.

TROPHIC LEVEL

10 kcal — 5

100 kcal — 4

1,000 kcal — 3

10,000 kcal — 2

100,000 kcal — 1

ENERGY FROM THE SUN

ENERGY PASSED TO NEXT TROPHIC LEVEL

predators, in this case big sharks and dolphins. Similarly, a rain forest ecosystem can't support too many Jaguars. An African plain ecosystem can't support too many Lions. There is just not enough energy for very many of these big animals. That is why neither the world nor any of its ecosystems will ever be overrun with top predators.

Back to the marine ecosystem. Since a big shark *could* eat you, you might be happy that when you go for a swim in the ocean, the place is not choked with big sharks. As we just explained, the ecosystem cannot support that many big sharks, plus we humans are particularly effective at catching and killing off the ones that are there. The 2021 assessment by the International Union for the Conservation of Nature (IUCN) found that 37% of shark, ray, and chimaera (closely related cartilaginous fish) species are threatened with

extinction. In some coastal regions, some of the biggest shark species have disappeared or experienced population decreases.

If you are not a fan of sharks, so what? Isn't this a cause for celebration? Emphatically, no! Sharks play important and irreplaceable roles in marine ecosystems. They, like top predators in other ecosystems, serve two principal functions: controlling the population and evolution of organisms below them in the food chain. Some shark species are considered *keystone species* in their ecosystems. In an architectural arch, the keystone is the block at the top of the arch. If you pull it out, the arch collapses. Likewise, keystone species have disproportionate importance in their environment. They are sentinels of ecosystem stability and health—that is, they are ecosystem regulators and protectors. Often a top predator plays this role, though not always. Ecologists have understood this since a fundamental study in the 1960s showed that one species of starfish—a starfish!—was found to be essential for maintaining species diversity in a Pacific Northwest rocky intertidal zone.[5] More relevant here is that in Hawai'i, for example, Tiger Sharks are keystone predators in seagrass ecosystems, regulating sea turtle populations and thus preventing them from overconsuming seagrass.

Scientists have discovered several other instances where predators are keystones in ecosystems. Wolves may play a big part in maintaining the forest ecosystem by removing Elk, which otherwise gobble up all the young trees, leaving the resident birds and small mammals without homes and food.[6] Coyotes cull those smaller prey and in doing so prevent overpopulation.[7] Grizzly Bears similarly help control Moose and Elk populations, carry and deposit seeds in their droppings, and move the carcasses of salmon they have eaten from stream to forest, providing nutrients to the latter.[8] Pumas and Jaguars cover large areas and have diverse diets, affecting prey populations and the behavior of both prey and scavengers.[9] American Alligators excavate burrows that become shallow pools, necessary habitat for birds and fish.[10] As we learn more about the habits of more animals, we will undoubtedly find other keystones

From an ecological standpoint, the world needs these keystone predators, but do we have to care whether there are intact seagrass beds, rain forests, or other ecosystems? There is a strong economic argument in favor of caring about— and preserving—*ecosystem services*, which are resources that ecosystems provide for free, such as food, clean air, clean water, flood and erosion control, building materials, pollination, and so forth, which we depend on as a species and which are the linchpin of our economy (fig. 1.5). More significantly, there are no technological replacements for most ecosystem services. Intact wetlands store and filter water, preventing flooding and providing clean drinking water. Intact forests sequester carbon dioxide, helping mitigate climate change. Both produce oxygen for us to breathe. Filter feeders in aquatic systems remove pollutants from the water as they eat. Many plants filter pollutants out of the air. Healthy populations of pollinators like bees are necessary to pollinate the plants that produce almost all the fruits and vegetables we eat. These ecosystem services are provided on a large scale and free of charge, and we lose them at our peril. For example, according to the Ecological Society of America, our free pollinators are worth $4–6 billion a year in the United States alone.[11] From an economic perspective, we need to conserve the ecosystems that provide services like pollination, oxygen, and so forth. To do that, we need to conserve the animals that function as keystone species. In many cases, these are the top predators.

Important on a more localized economic scale, the human hunter who likes to stalk Elk with gun or camera, or the birder looking to check off species on his or her life list, similarly requires intact habitats. And in the process of stocking up on gear and permits, and traveling to and then staying at these places, these humans support a surprisingly extensive and lucrative economy. According to the National Wildlife Federation, hunters and anglers contribute $200 billion or more annually to the US economy. [12] You may question the ethics of conserving animals in order to kill them (we will talk more about that later), but what about hunters with cameras and binoculars instead of guns? The economic input of bird-watchers in the United States alone is more than $180 billion according to the US Fish and Wildlife Service.[13] This is difficult to quantify for African countries, but a recent paper in *Biological Conservation* estimates that trophy hunting contributes around US$200 million annually to African governments and conservation organizations.[14] This is not a lot when you consider that general tourism to Africa contributes US$36 billion to the continent, according to the World Bank.[15] But surely a high percentage of those tourists are there hoping to see iconic wildlife. Whether you agree with it philosophically or not, both the hunters with guns and those with cameras provide incentive for the local people to conserve large mammals, other predators, and their ecosystems.

The final part of the economic argument is that products from the natural world contribute directly to our economy. Most of us don't farm or go out and hunt our own food or collect our own plant material from which to make cloth and clothing. We don't individually chop down the trees that build our houses. When the human population was smaller we did such things, but now we do them on an industrial scale, at great cost to the environment. Given our numbers, there is not much choice. But there is something important but less obvious that we still get directly from intact habitats and that provides, or in some cases protects, products we need: biodiversity. A current example of how this works involves bananas, which we recognize are

not predators—please excuse this as an example. Bananas are grown commercially, and almost every commercial banana worldwide is the Cavendish variety. They grow well, ship well, and keep well, and people like to eat them. But in the early 2010s the fungal pathogen "tropical race 4" started killing off Cavendish banana trees. Sixty years ago, the same thing happened, killing off the Gros Michel strain of bananas we were all eating, and driving the switch to the apparently less tasty Cavendish.[16] Taste aside, so what? Well, in an industrial world, all bananas are one kind, and if this kind turns out to be susceptible to a new pathogen, no more bananas. Biodiversity to the rescue! There are almost 1,000 varieties of bananas in the world. About half are edible. All are found in the wild. So, if something kills all your Cavendish bananas, what do you do? Try out some different types or do some crossbreeding and see what works to replace the Cavendish. But you can do that only if there are still other bananas out in the wild to try. Also, most of our medicines have come from plants and other natural products. Of course, this was true when

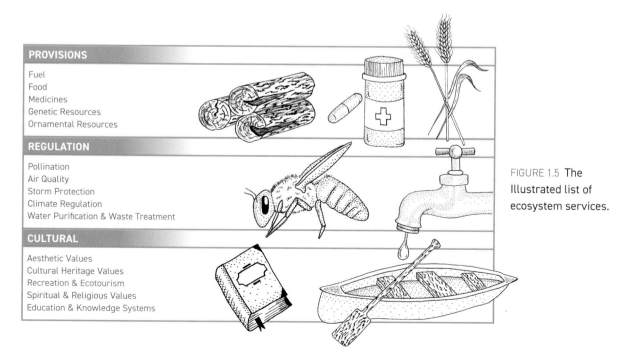

PROVISIONS

Fuel
Food
Medicines
Genetic Resources
Ornamental Resources

REGULATION

Pollination
Air Quality
Storm Protection
Climate Regulation
Water Purification & Waste Treatment

CULTURAL

Aesthetic Values
Cultural Heritage Values
Recreation & Ecotourism
Spiritual & Religious Values
Education & Knowledge Systems

FIGURE 1.5 **The Illustrated list of ecosystem services.**

we were all living directly off the land, but even in the last 30 years, about 50% of the approved drugs in the world have come from natural products, mostly plants.[17] You still need to have the plants to discover their medicinal value. And a diversity of plants requires a diversity of intact habitat.

Global estimates of the economic value of ecosystem services like those above and others (such as flood control, climate stabilization, clean air, etc.) range as high as US$54 *trillion*—that's 54 followed by 12 zeroes—*every year*.[18] And consider this: there are no effective, large-scale technological substitutes for these ecosystem services, and there likely never will be. So, ethics aside, biodiversity matters to you.

And predators help preserve biodiversity. Some scientists refer to animals like big predators as *charismatic megafauna*. What this means is that people like these toothy beasts. They're sexy. Arguably, not many people want to save the rain forest for Assassin Beetles, amazing predators in their own right. But Jaguars? Yes, for all the reasons you are reading this book. If you want people to care about climate change, you don't go on about oak trees, lovely though they may be. You tell them about starving Polar Bears, often with graphic pictures of emaciated bears. These charismatic megafauna are advocates for themselves, and in order to save them we are obliged to try to save the ecosystems on which they depend and the biodiversity those contain. Predators can help save biodiversity, but biodiversity can also save predators, as we'll see in examples in subsequent chapters. As John Muir, founder of the Sierra Club, famously put it: "When we try to pick out anything by itself, we find it hitched to everything else in the universe."[19]

These animals thus help us survive, but we are smart and maybe we could figure out innovative technology to perhaps replace a small number of the services their ecosystems provide, although it would certainly cost money. We can design drugs. We can filter water. If we continue to replace forests with agriculture, we will lose more animals, but we might carry on, at least for the short term. Are we morally obliged to be good stewards of our planet and our fellow creatures? Do we want to live in a world of pavement and endless fields of genetically modified soybeans? Does wildness matter? These are ethical questions you must answer for yourself, but humans seem to have an intrinsic need for nature, something physical and spiritual. In his book *Last Child in the Woods*, Richard Louv makes the case that experiences in nature are vital for the physical and emotional health of children and adults.[20] Likewise, ecologist E. O. Wilson defines *biophilia* (in a book by that name) as "the urge to affiliate with other forms of life."[21] He argues that the need for nature is innate in humans because we evolved with nature. More people visit zoos and aquaria annually in the United States than go to all sporting events combined. People prefer to live near parks and open land. People still like to hike and camp, forgoing modern conveniences. We bring animals and plants into our homes. We tend gardens. Studies have shown that viewing natural landscapes in pictures or through windows reduces stress. Even as we are increasingly isolated from the natural world by the modern one, we seem to still crave contact. It's good for us.

In the end, there is simply this: nature inspires us. A rainbow. A sunset. A starry night. A snow-capped mountain. The ocean in a storm, or on a cloudless summer day. A waterfall. A snowflake. A Tiger. Yes, you can see beautiful pictures of these things, but you can also go outside and see them for yourself, for real. Maybe. And even if you can't, isn't it worth something knowing they are out there? Being Wolves, Cheetahs, or Tiger Sharks. Our planet home is theirs too. And let us not forget: it was theirs first. We should be able to share, for our own good, but also just because. Predators matter.

2
SHARKS

We begin our survey of the magnificent top predators of the world with the group that is most often misunderstood, the sharks. There are at least three aspects of this misunderstanding. First is the perception that all sharks are at the very top of the food chain—that is, they are all apex predators. Second, the idea that sharks are "living fossils" is broadly accepted. Finally, the most widely accepted misconception is that sharks are apt to attack you when you are in their presence.

MISCONCEPTION 1:
ALL SHARKS ARE APEX PREDATORS

If you close your eyes and imagine a shark, chances are you will envision an iconic species like a White (*Carcharodon carcharias*), Tiger (*Galeocerdo cuvier*), or Bull Shark (*Carcharhinus leucas*), or perhaps a Sand Tiger[1] (*Carcharias taurus*), Lemon Shark (*Negaprion brevirostris*), or Grey Reef Shark (*Carcharhinus amblyrhynchos*) (fig. 2.1A and B). In anyone's book (including the one you are reading), these are considered formidable top predators. But, while all of the 541 species of sharks currently recognized—ranging from the 8 in (20 cm) Dwarf Lantern Shark (*Etmopterus perryi*) to the 40 ft (12 m) Whale Shark (*Rhincodon typus*)—are predators, only a handful are considered true apex predators. The issue is that the apex predators among sharks (White, Tiger, Bull, etc.) are not *typical*

sharks. The maximum size of almost two-thirds of the living shark species is no larger than 3.3 ft (1 m), and about half live in the deep sea (> 650 ft, or 200 m). Moreover, across all the ecosystems inhabited by sharks, different species occupy one of three, and in some cases four, different feeding (or trophic) levels, which means that many sharks are in fact prey for other organisms, often other sharks (fig. 2.2A and B). Thus, a typical shark is not a large, gray, fast-swimming, coastal apex predator but rather is a small, brown, slow-moving, deep-sea, midlevel predator (fig. 2.1C and D).

MISCONCEPTION 2:
SHARKS ARE LIVING FOSSILS

To be a living fossil, an organism must look like its ancestors—that is, its anatomy must have evolved minimally over geological time. Horseshoe crabs,

FIGURE 2.1 You probably consider the (A) Lemon Shark (*Negaprion brevirostris*) and (B) Sand Tiger (*Carcharias taurus*) to be typical sharks. In fact, the (C) Cuban Dogfish (*Squalus cubensis*) and (D) Gulper Shark (*Centrophorus granulosus*) are more representative of sharks as a group. A, Lesley Rochat; B, Tanya Houppermans; C, Lance Jordan

crocodiles, and even ferns are considered living fossils. To the extent that sharks and their relatives settled on a body shape and predatory lifestyle that was successful early in their approximately 450-million-year evolutionary history, and that some aspects of these have persisted to the present, then yes, sharks conform to the definition of a living fossil.

But other characteristics of sharks have changed significantly over evolutionary time, challenging that perception. One of the most significant of these is that the jaws of modern sharks have become more protrusible and have a larger gape than those of their ancestors (fig. 2.3), endowing them with the ability to grasp, shear, and manipulate

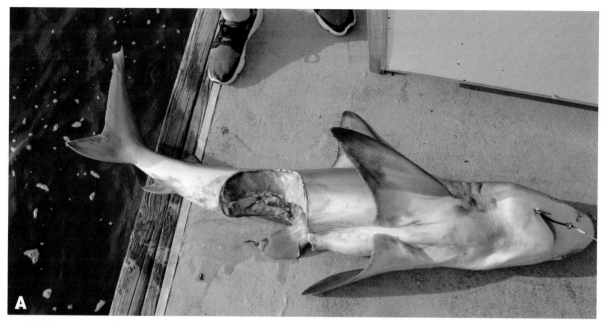

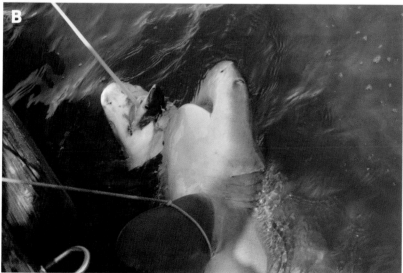

FIGURE 2.2 (A) Trophic levels played out on a hook: a shark on a scientific longline killed by another shark. (B) More evidence from an experimental longline: a large Bull Shark (*Carcharhinus leucas*) ate a Sandbar Shark (*Carcharhinus plumbeus*) that ate a Finetooth Shark (*Carcharhinus isodon*) used as bait.

their prey in ways that would be impossible with a fixed jaw like yours.

A second major evolutionary change in sharks is that their fins have become slightly more movable, and they differ from the more rigid fins of early sharks. Modern sharks are thus more maneuverable than many of their ancestors and are better at catching prey or avoiding being preyed on.

Finally, bony fish like tuna, mullet, perch, and others trace their origins back to the same geological period as sharks (the Devonian, nearly 400 million years ago), so if sharks are living fossils, so are bony fish! Thus, modern sharks are not primitive creatures but are advanced and well-adapted beasts. Plus, most sharks look nothing like we think their ancestors looked.

FIGURE 2.3 Protruding jaws of an Atlantic Sharpnose Shark (*Rhizoprionodon terraenovae*), tagged and about to be released.

MISCONCEPTION 3:
SHARKS ARE APT TO ATTACK YOU

Refuting this last misconception is easy in principle, yet this is the most difficult one to refute. Yes, shark bites and attacks do occur, but the number of shark-related fatalities and serious but non-life-threatening injuries is negligibly low compared to the threats we face every day and consider routine. We shorten our lives by smoking, drinking sugary sodas, eating too many carbs, and texting while driving. In the United States, air pollution causes as many as 200,000 premature deaths annually. Breathing the dust emitted by our tires and running shoes as they erode is a bigger risk to our health than sharks are. The number of documented, unprovoked shark bites and attacks hovers around 100 annually, according to the Florida Museum of Natural History's International Shark Attack File, although this is likely an underestimate because in remote or sparsely populated areas, bites and attacks are almost certainly underreported. Still, the total is low.

Note that we have used both *bite* and *attack* in describing shark-human interactions. The former typically causes only minor injuries after bite-and-release behavior associated with mistaken identity.[2] A spate of these often occur in the summer months along the southeastern coast of the United States, particularly in naturally murky inshore waters where Blacktip Sharks (*Carcharhinus limbatus*)—streamlined, swift-swimming eaters of small schooling fish—are thought to confuse human hands and feet with these fish and may bite. However, within fractions of a second, they recognize the mistake and release (*Blech! I wasn't expecting that!*), often leaving only a series of shallow puncture marks. The term *attack* is reserved for cases involving more significant injury and may involve repeated bites after the first, a behavior associated with Bull Sharks, for example. Unfortunately, in the case of White Sharks, a bite may be fatal even if it is the result of mistaken identity, because it is typically an awfully big bite.

You have about a 0.00125% chance of being killed by a lightning strike and an even smaller chance of being killed by a shark while in the water. We know far more people who have experienced direct lightning strikes than people who have been attacked by a shark. (As field biologists who at times work at sea in dangerous weather, we are aware that the odds of being struck by lightning are increased from those while ashore. We thus try to avoid thunderstorms while on the water, in part because we fear the headline that might result: "Shark Biologists Prove Adage: Struck by Lightning!"). One of our students wrote in an essay that you are more likely to be struck by "lighting" than bitten by a shark. We are quite sure she meant to write "light*n*ing," but the statement was correct as written: more people are injured by falling light fixtures than by sharks annually.

Beachgoers are more threatened by errant surfboards, Jet Skis, rip currents, jellyfish, bacteria in the water, SUVs in the beach parking lot, skin cancer, and toasters (yes, *toasters*; see below) as they prepare breakfast before heading to the shore than they are by sharks. But foremost in their heads as their toes sink into the hot sand is the foreboding, ominous *Jaws* theme.

Speaking of the theme from *Jaws*, a very cool study titled "The Effect of Background Music in Shark Documentaries on Viewers' Perceptions of Sharks"[3] concluded that attitudes toward sharks

are influenced or reinforced by how ominous the background music is (apparently if it plays only in your head, like it probably is now). Don't you just love science?

In one of the most iconic scenes from the 1975 blockbuster movie *Jaws*, with the sounds of seagulls in the background and a gentle wind blowing under blue skies, a group of carefree adults and kids are frolicking close to shore when one of them spots something obviously sinister in the water. She shrieks in terror and the lifeguard blows his whistle, leading everyone to chaotically run from the water amid a chorus of screams. Then silence, as the stunned people on the beach stare seaward and a deflated raft drifts ashore, with an arc of fabric removed by the White Shark where a child had been. Filmmaker Lesley Rochat,[4] an award-winning shark conservationist, has produced a variation of this scene to highlight our misconception about the probability of shark attacks. In her version, the focus is not the remnants of the raft but instead a toaster bobbing at the surface, followed by the superimposed message "Last year 791 people were killed by defective toasters, 9 by sharks. Rethink the shark." The filmmaker's conservation message is that we should be afraid *for* sharks rather than *of* them.

In the new millennium, the years 2001 and 2015 have stood apart, at least in the eyes of the American media, as Years of the Shark because of the perception of a significant increase in the number of shark bites and attacks. In 2001, beginning on that summer's Fourth of July weekend, several gruesome shark attacks occurred, including serious injuries to an eight-year-old boy and others, and fatalities in North Carolina and Virginia. News footage of migrating Blacktip Sharks, a common annual occurrence, fueled the hysteria (fig. 2.4). Although the summer shark-bite season started out furiously in 2001, by its end this was not a particularly unusual year.

The year 2015 started out similarly to 2001 along the US East Coast, although with no fatalities, and ended as a slightly atypical year. Explanations for the uptick include more beach visitors, and two events that could be associated with global climate change: slightly elevated water temperatures that persisted longer than usual, and wind and ocean current patterns that may have pushed shark prey and hence sharks closer to shore.

So, it boils down to these conclusions: first, your daily routine activities are more dangerous to you than sharks are. Second, given enough time, even extremely unlikely events, including shark bites, will happen. Third, no species of shark bites humans with enough frequency for us to be considered its prey.

As biologists and conservationists, we wish we could say that people are interested in sharks, indeed in all the predators in this book, because of the vital ecological role they play in maintaining biodiversity. Or because of their fascinating adaptations—for example, the streamlined shape and high-performance metabolic machinery of a Shortfin Mako (*Isurus oxyrinchus*) that make it one of the speediest beasts in the sea. However, it is clear that for many, the fascination with sharks owes more to fear (which is called *galeophobia* or *selachophobia*) and the perception of the danger they pose than to their ecological roles, biological attributes, or threats to their survival. Same with many of this book's other predators, unfortunately.

FIGURE 2.4 **Blacktip Shark (*Carcharhinus limbatus*) aggregations off the east coast of Florida are annual occurrences.** Stephen Kajiura

WHAT IS A SHARK?

A shark is a *fish*—that is, an aquatic vertebrate with fins for locomotion and gills for breathing. As you saw above, biologists distinguish between two major groups of fish, *bony* and *cartilaginous* fish. The former group (including tuna, mullet, catfish, etc.) numbers approximately 35,000 species and is the most species-rich and diverse assemblage of all vertebrates. We will briefly discuss these

as predators in chapter 9. In contrast, there are only about 1,300 different kinds of sharks and rays (collectively known as *elasmobranchs*), as well as about 53 species of more obscure, largely deep-sea relatives known as *chimaeras*.

A critical difference between sharks and bony fish is the material that makes up their internal skeleton. Bony fish have a skeleton composed of dense, rigid, calcified (that is, impregnated with calcium and phosphorus compounds), hard, living tissue called

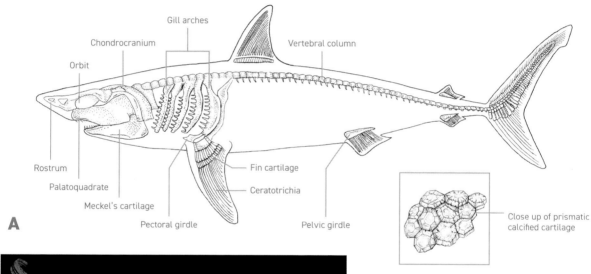

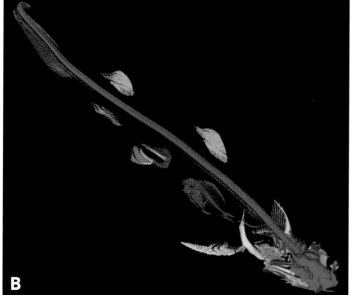

FIGURE 2.5. (A) Cartilaginous skeleton of a shark. Palatoquadrate and Meckel's cartilage refer to the cartilaginous upper and lower jaws, respectively. The rostrum refers to the snout. Ceratotrichia are the unbranched fin rays that are the essential ingredient in shark fin soup. (B) Computed tomography (CT) scan of a Lemon Shark (*Negaprion brevirostris*) showing the cartilaginous skeleton. Note the bony skeleton of a prey item in the Lemon Shark's stomach. How many of the parts labeled in A can you identify in B?

(A), adapted in part from Whitely, 2015, http://www.fossilsofnj.com/shark/cartilage.htm, accessed 12/4/21; (B), Gavin Naylor

bone, like yours. Sharks and their relatives have a skeleton made of cartilage (fig. 2.5).

Cartilage is a tough but lighter, generally more flexible material than bone. Chemically, cartilage is composed principally of protein and sugar molecules. Humans and other mammals possess three different kinds of cartilage in the nose, ears, bronchial tubes, trachea, ribs, ends of long bones, and disks between vertebrae. Cartilage makes up about 0.6% of a typical mammal's body weight, but from 6% to 8% of a shark's total mass.

A major benefit of cartilage to sharks is that its lightness in part compensates for the heaviness of their muscle-bound body as well as their lack of a swim bladder, a gas-filled internal "balloon" found in most bony fish that allows them to adjust their buoyancy.

Cartilage is not as hard as bone because it lacks bone's high levels of calcium and phosphorus salts. However, in the jaws and vertebrae of sharks, where strength is needed, the cartilage is fortified

by incorporating calcium in the form of the mineral apatite in a pattern resembling a prism, known as *prismatic calcified cartilage*. In some shark species, the vertebrae are as strong as bone. The hardness of bone thus does not equal mechanical superiority, just a different pathway to evolutionary success.

In addition to differences in skeletal composition, there are other differences between the sharks and rays and the bony fish (fig. 2.6). One of the most obvious distinguishing characteristics of sharks is their five to seven bilateral (that is, on either side) gill slits, whereas bony fish have a single bilateral flap, the operculum, covering their gills.

Sharks have an asymmetrical tail (caudal) fin in which the upper lobe is longer than the lower lobe. The lobes of the tail fin in bony fish are typically symmetrical. This *heterocercal* (meaning "different tail") caudal fin of sharks is characterized by the vertebral column extending almost to the tip of the upper lobe, which stiffens it, in contrast to stopping at the base of the tail in bony fish. This stiffening of

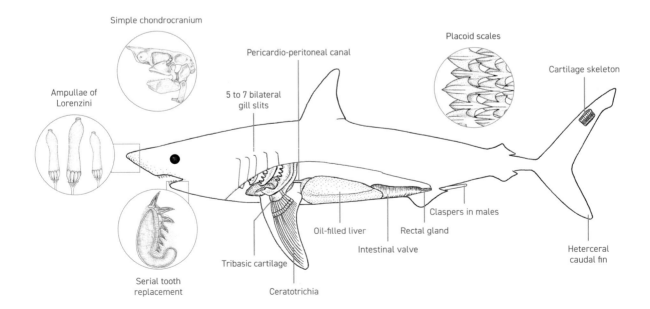

FIGURE 2.6 Selected features that distinguish a typical shark from a typical bony fish. See text for explanation.

the upper lobe of the caudal fin in combination with the upper lobe sloping rearward provides upward and forward thrust, which is needed to overcome the negative buoyancy of the shark's heavily muscled body.

Sharks all use internal fertilization—that is, the male transmits sperm directly into the female—whereas the overwhelming majority of bony fish fertilize externally. Bony fish typically invest fewer resources into a large number of eggs (several million in the case of pollack, for example), and the likelihood of survival is small for every hatchling.

Internal fertilization produces a limited number of offspring, but more resources are invested in each one. The advantage for sharks is that each individual has a greater chance of survival than a single bony fish hatchling. Early development of the shark embryo occurs inside the female for sharks that lay eggs, like Port Jackson Sharks (*Heterodontus portusjacksoni*) and Small-spotted Cat Sharks (*Scyliorhinus canicula*), and complete development occurs internally for sharks like White Sharks and Sandbar Sharks (*Carcharhinus plumbeus*).

To transmit sperm, male sharks all have special external, paired organs known as *claspers*, one of which is inserted into the female. The term *claspers* dates back to Aristotle, who incorrectly thought that males used these exclusively as a way to clasp females, not as sex organs, and that sharks used external fertilization. Although Aristotle's interpretation was incorrect, he was accurate in concluding that a coupling mechanism was required for males to transfer semen to females while both were swimming, and claspers indeed play a role in keeping a pair of reproducing sharks from separating. During reproduction, the end of the clasper splays out to reveal miniature grappling hooks for anchoring in the female during sex. For the limited number of sharks for which mating has been observed (e.g., Nurse

Sharks, *Ginglymostoma cirratum*; and Whitetip Reef Sharks, *Triaenodon obesus*), it may involve ritualized behaviors prior to coupling, to position and align the clasper before inserting it. Males may bite females and grasp their pectoral fins to keep them in place during copulation, and in some species females have thicker skin than males to defend themselves during courtship.

Sharks also replace their teeth regularly. Their teeth are not firmly embedded in bone like yours and those of other vertebrates. They are instead loosely attached by collagen, an abundant fibrous protein common in animals that is used to connect and support parts of the body. In sharks there are one or more rows of functional teeth, with additional rows in varying states of maturity lying in wait on a very slow-moving collagen conveyor belt. This setup ensures that a row of fresh, sharp teeth is always available. If one or more teeth on the front line are lost, their replacements will soon be in transit. A shark may produce and lose thousands of teeth over its lifetime, and the replacement time is measured in weeks. (Useful tip: if you are at an aquarium that displays sharks, ask one of the attendants whether there are any spare shark teeth or ray tooth plates to give you.) Surely, it is wasteful to make and discard teeth, no? See box 2.1 for sharp insight on this issue.

A shark's skin is covered in scales, specifically *placoid* scales. These are not like bony fish scales but rather are like miniature teeth, with an inner pulp cavity surrounded by a layer of dentin and one of hard enamel. Like teeth, they are shed and replaced, although more slowly. The scales have various functions among sharks. They protect against struggling prey and parasites, safeguard females from male bites during mating season, and reduce the frictional drag that may slow the shark while swimming. In fact, beginning in 2000, Speedo marketed swimsuits attempting to mimic the drag-reducing structure of sharkskin. One study showed

BOX 2.1 **"OH THE SHARK, BABE, HAS SUCH TEETH, DEAR ..."**

Is it extravagant and wasteful to continually make and discard teeth instead of producing one permanent set? One of the central tenets of evolution is that the energy supply is often limiting, so conserving energy is key, right? Evolution often entails compromises, though, and natural selection can be thought of as a continuous experiment that tests what works and what does not. The individual with the characteristics that work best, enabling it to live to reproduce the most offspring that can also reproduce, is the one that passes the evolutionary test. The individual with less successful characteristics likely does not live to pass its genes to the next generation. You may know this concept as *survival of the fittest.*

So yes, making and shedding teeth may waste mineral and energy resources, but in combination with other shark characteristics, it is the winning strategy. One of the biggest threats to predators, particularly terrestrial ones but also some marine animals, is starvation. If a Lion loses one of its long canine teeth, its odds of surviving are significantly decreased. If a Tiger Shark loses several teeth, no problem! Replacements are on the way.

that the skin, perhaps not surprisingly, worked best on actual sharks, but Speedo's 2020 Olympic line still featured a layer of sharkskin-inspired, drag-reducing fabric.[5]

Members of a large and well-known family of sharks, the carcharhinids, or requiem sharks, possess a nictitating eyelid, or membrane, in the lower portion of each eye (fig. 2.7). Some birds, lizards, frogs, and even a few mammals (e.g., seals, Polar Bears, camels, and Aardvarks) have functional nictitating membranes as well. The structure, which is covered in scales in sharks, rises to cover and protect the eye as the shark is about to eat, or when it is threatened. White Sharks and cow sharks have different methods of protecting their eyes. White Sharks roll their eyes and expose the tough layer around the eyeball known as the *sclera*, while cow sharks suck their eyes well back into their sockets.

Several other anatomical features distinguish sharks from bony fish, including an intestine that spirals or scrolls, electricity sensors called *ampullae of Lorenzini* (described below), a simple skull, and a large, oil-filled liver, which, among other things, provides buoyancy since sharks lack the swim bladder found in most bony fish.

FIGURE 2.7 Nictitating membrane covering the eye of a Tiger Shark (*Galeocerdo cuvier*).

BIOLOGY AND ECOLOGY

The Devonian period (about 419–359 million years ago) is known as the Age of Fishes, because bony fish dominated in the warm Devonian seas. At about the same time, cartilaginous fish, specifically sharks, also became abundant, but the following period, the Carboniferous (358–299 million years ago), was the Golden Age of Sharks. Sharks dominated not only in diversity (there were as many as *3,000* distinct species then, compared to about 541 now), but also in size. They were the main predators in oceans as well as in rivers and lakes, where few sharks live now. Among those sharks swimming today, the most primitive—that

is, those least changed from their ancestors—are in the group known as cow sharks (figs. 2.8 and 2.31). We know this primarily from the presence of teeth in sediment 180 to 200 million years old.

Based largely on teeth, we also know about the extinct beast known as Megalodon, or Megatooth (*Otodus megalodon*), the largest predatory fish to ever swim the world's oceans, whose maximum length is conservatively estimated to be from 50 ft (15 m) to as long as 59 ft (18 m) (fig. 2.9).

Megalodon was not a direct ancestor of the White Shark, as you might have mistakenly been led to conclude, but was an evolutionary dead end. White Sharks evolved from the related lineage of mako sharks. Megalodon lived from around 23

FIGURE 2.8 Broadnose Sevengill Shark (*Notorynchus cepedianus*) in a kelp forest near Cape Town, South Africa. Members of this family (cow sharks) are the most primitive of the living sharks. Lesley Rochat

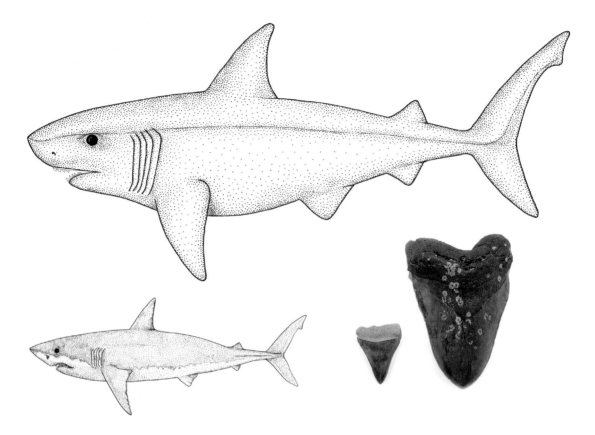

FIGURE 2.9 How Megalodon (*Otodus megalodon*) (*top*) might have looked compared to a White Shark (*Carcharodon carcharias*) (*bottom*). Fossilized Megalodon tooth (*far right*) next to a White Shark tooth. Sharks are drawn approximately to scale to each other, as are teeth.

million to 2.6 million years ago and preyed on the abundant smaller marine mammals of the time. Why did Megalodon go extinct? There are several hypotheses, including cooling of the oceans about five million years ago, which it may have been unable to tolerate, disappearance of its marine mammal prey, or even competition with White Sharks. Is there even the teeniest possibility that a Megalodon still lurks somewhere in the unexplored deep, feasting on whales? No. The deep sea is too cold and there is simply not enough food down there to support a beast this large.

There are currently 541 named species of sharks, with doubtless more to be added as we explore the deep sea and other poorly explored regions, and as we more widely employ modern molecular techniques to distinguish cryptic species (i.e., those that may be indistinguishable externally). There are also 651 rays and 53 holocephalans (chimaeras, or ghost sharks), both close relatives of sharks. All are classified as chondrichthyans. Rays, in fact, are sometimes called "pancake" or "flat" sharks because of this close relationship. (Not to be outdone, biologists who study rays refer to sharks as "sausage rays.") Although the taxonomy of sharks and their relatives is still the subject of debate, one currently accepted construct recognizes 1,245 species organized into 14 orders and 64 families. All are magnificent beasts with fascinating adaptations, which we will explore in this chapter.

Organisms employ one of two main life history strategies, or exist along the continuum between these extremes, in their quest for evolutionary success. On the one hand are organisms like fruit flies, which grow quickly, mature early, and lay lots of small, quickly hatching eggs. These organisms can prosper in a wide variety of environments. They are called *r-selected*, a reference to their speedy growth rates (the variable *r* in the classic growth equation). A cod that releases millions of eggs at spawning is a fishy example.

Most sharks, as well as most of the other top predators in this book, are *K-selected*. *K* refers to the first letter of the German word for the *carrying capacity* of their habitat. Carrying capacity is the maximum number of individuals of a given species that an area's resources, such as food, can support in the long term without significantly depleting or degrading those resources. The life history characteristics of sharks have suited them well throughout their evolutionary history for over 400 million years. Unfortunately, these same characteristics make them particularly vulnerable to modern human pressures, such as overfishing, habitat destruction and degradation, and climate change, which can reduce their populations and make recovery difficult. Let's look at some examples of these K-selected characteristics.

Sharks are slow growing. It takes female Sandbar Sharks (see fig. 2.28) in the Atlantic Ocean about 15 years to grow from 16–20 in (45–50 cm) at birth to sexual maturity at about 53.5 in (136 cm), an average growth rate of just 2.4 in (6 cm) per year. Sharks are generally long-lived and reach sexual maturity late in life. Sandbar Sharks, for example, live 30–35 years, and Greenland Sharks (*Somniosus microcephalus*) may live 400 or more years! Spiny Dogfish (*Squalus acanthias*) take as long as 20 years to reach sexual maturity. Gestation (period of pregnancy) is also slow going; about a year for Sandbar Sharks, as

long as two years for a Spiny Dogfish, and three and a half years for frilled sharks! And fecundity (number of offspring) is typically low. Sandbar Sharks produce about eight pups every two years.

Why do these life history characteristics matter? For one thing, they make it difficult for shark populations to quickly recover from perturbations like overfishing. And while the life history characteristics we list above generally apply to sharks as a group, there are differences in individual species, and it is critical for scientists to learn these (e.g., growth rate, age of maturity, reproductive characteristics, nursery grounds, position in food web, migration patterns, etc.) to manage and conserve populations.

Sharks inhabit nearly all marine environments, from tropical coral reefs to cold Arctic waters and from estuaries and coasts to the deep sea and open ocean. There are also a small number of sharks that live in fresh water. Juveniles often use different habitats than adults, and males and females of some species may segregate for much of their lives.

About 40% of shark species live in coastal estuaries and along the continental shelves of warm temperate and tropical environments. Here, food is diverse and abundant, and there is a high variety of habitat types. Fewer than 3% of shark species (e.g., Blue Sharks, *Prionace glauca*; Oceanic Whitetips, *Carcharhinus longimanus*; and Silky Sharks, *Carcharhinus falciformis*) are truly pelagic (open ocean), primarily because the open ocean is food poor and there are no barriers to separate the sharks long enough for them to diversify into more species. Thus, more pelagic shark species are found *circumglobally* (spanning the globe) than in isolated areas. Only about 7% of species live in the Arctic or in fresh water.

It may have surprised you earlier to learn that over half of all shark species live in the deep sea (> 650 ft, or 200 m). Most deep-sea shark species are small (maximum total length < 3.3 ft, or 1 m), and this high diversity is in large part a response to the relatively low competition for resources and the lower predation risk compared to shallower environments, where many other competitors live, too. Sharks do not, however, live in the *very* deep sea. Sharks rarely occur deeper than 9,800 ft (3,000 m; the average depth of the world's oceans is 13,100 ft, or 4,000 m), likely because food is very limited and the pressure is high down deep. Water weighs a lot, and every 32.8 ft (10 m) you descend adds the equivalent of about one atmosphere of pressure compared to sea level. We will discuss this a bit more when we get to marine mammals. Not many organisms have the adaptations required to manage the deepest seas.

Why does a shark live where it does? First, it must be capable of adapting to factors such as temperature, salinity, dissolved oxygen, pH (acidity), pressure (as a function of depth), and light. Consider Sandbar Sharks (if you have not guessed by now, this gorgeous beast of a shark is one of our favorites) in Winyah Bay in northeastern South Carolina, one of the US Atlantic coast's largest estuaries. We recently discovered that juveniles have internal adaptations that enable them, but not adult Sandbar Sharks or most other sharks, to live in the middle bay region. Salinities here (as low as 7 parts per thousand, or ppt) are typically much lower than in the lower bay (about 25–30 ppt) and open ocean (35 ppt). This juvenile adaptation, tolerance of low salinities, allows them to use the area as a refuge from predation from most larger sharks.

Why don't more sharks live in fresh water? Among sharks, only the Bull Shark and three species of river sharks, plus several species of rays, can survive in truly freshwater ecosystems like rivers and lakes. Since modern sharks evolved in the marine environment, moving to fresh water or even brackish water exposes them to

environmental conditions that are foreign, even deadly. In salt water, which is slightly saltier than the interior of the shark, a shark neither bloats nor dehydrates. To accomplish this balance, sharks have a special organ called the rectal gland, which takes the salt that moves into the body from the seawater and deposits it right back into the seawater, a revolving door of sorts.

The situation changes drastically for a saltwater shark in fresh water. There, water floods into the shark through the gills and digestive tract as it breathes and eats, and the salts and other compounds critical for the shark's survival diffuse *from* the shark *into* the freshwater environment. For a typical shark living in the ocean, being plonked in fresh water would soon prove fatal. Bull Sharks and other river sharks have adapted to become more like carp, Largemouth Bass, and other freshwater bony fish by lowering the total salt concentration of their internal fluids so that there is less tendency to gain water and lose salts. Relative lack of prey and loss of function of some of their senses are also problems for sharks in fresh water and likely account for the inability of more species to adapt over evolutionary time. For sharks the advantages of living in fresh water, such as abundant prey, were insufficient to allow them to overcome the problems.

Ecological constraints also influence where a shark lives, as well as how many an ecosystem can support. For a shark to be successful in a particular ecosystem, it must achieve a balance between competing for resources like food while avoiding being eaten. It also needs to find mates. At any one time in Winyah Bay, there may be adult and/or juvenile Sandbar Sharks, Atlantic

OPPOSITE PAGE:

FIGURE 2.10 The sharks of Winyah Bay, a large estuary in northeastern South Carolina. Note the diversity of species whose habitats may overlap in this bay.

Sharks of Winyah Bay
GEORGETOWN, SC

Atlantic Sharpnose Shark

Lemon Shark

Blacknose Shark

Nurse Shark

Blacktip Shark

Sandbar Shark

Bonnethead

Scalloped Hammerhead

Bull Shark

Spinner Shark

Dusky Smoothhound

Finetooth Shark

Spiny Dogfish

ILLUSTRATED BY ELISE PULLEN
www.artbyestudio.com
@Art_By_E_

Sharpnose Sharks (*Rhizoprionodon terraenovae*), Blacktip Sharks, Blacknose Sharks (*Carcharhinus acronotus*), Finetooth Sharks (*Carcharhinus isodon*), Bull Sharks, Lemon Sharks, Spinner Sharks (*Carcharhinus brevipinna*), Bonnetheads (*Sphyrna tiburo*), Scalloped Hammerheads (*Sphyrna lewini*), Spiny Dogfish, Dusky Smoothhounds (*Mustelus canis*), and others (not to mention dolphins, alligators, and large bony fish) in an area of about 25 mi^2 (16,000 ac, or 65 km^2) (fig. 2.10).

Some shark species live in home ranges (areas occupied daily) smaller than the average grocery store. Others migrate across entire ocean basins. *Migration* refers to predictable movements between areas. Sharks and other animals migrate different distances or have home ranges of different sizes for a variety of reasons: to gain access to prey, to mate, to avoid potential predators, and/or to live in an environment that suits their internal function. These movements occur over short timescales (e.g., tidal cycles, day-night, lunar) and longer ones (seasonal). Many sharks are said to be highly migratory.

Some sharks migrate to and from nursery areas, which provide food, reduce competition between different age classes, and minimize risk of predation. Within these nursery areas, some sharks, like juvenile Lemon Sharks around Bimini, Bahamas, establish small and well-defined home ranges. Others, such as juvenile Sandbar Sharks in Chesapeake Bay, are nomadic, ranging widely each day within estuaries. In both cases, the sharks move to adjacent, more expansive habitats as they grow, just like you. Most coastal sharks exhibit some form of seasonal migratory behavior, although there can be migratory and nonmigratory populations within a species. Of the deep-sea species we know about, many do not migrate, except for some that move vertically daily. Why? Because of the constancy of the deep-sea environment, one place looks like any other and has the same characteristics.

Some sharks migrate to avoid dangerous, even lethal, climate extremes or to take advantage of more favorable climates. Juvenile Sandbar Sharks, for example, can survive temperatures down to only about 59°F (15°C), but Chesapeake Bay drops to 41°F (5°C) in winter. So, these juvenile sharks must migrate or die.

SHARKS AS PREDATORS

What do sharks eat? We have already established that it is not people, with rare exceptions. Most shark species are thought to have diverse diets. They are opportunists capable of switching prey based on what is available. Others have more specialized diets. Horn Sharks (*Heterodontus francisci*) feed primarily on hard-shelled prey that live on the seafloor, and Bonnetheads eat mostly crabs. The frequency of eating varies with the species, life stage, and type and availability of prey. Some species (e.g., deep-sea sharks) may eat one big meal that lasts for weeks, whereas others (e.g., Shortfin Makos) feed more frequently.

On average, sharks consume about 2%–3% of their body weight per day, their *daily ration*. This varies depending on the energy demands of the specific shark species or life stage and the energy content of the prey. Dramatic shifts in the diet occur with growth in species like Tiger and White Sharks. Both feed mostly on bony fish as juveniles. But as adults, Tiger Sharks commonly eat other sharks, rays, sea turtles, birds, and mammals, and coastal White Sharks eat largely marine mammals. The shape of the teeth of White Sharks even changes as they grow, reflecting this dietary shift.

As shown in figure 2.2, juveniles or smaller shark species are regularly prey themselves. Being large minimizes the threat of being preyed on, but even adult White Sharks fall prey to larger beasts like Orcas. White Sharks have also been spotted with scars that clearly demonstrate encounters with large squid, but these are certainly the result of defensive actions by the squid to avoid being eaten. The Love-Heart Squid (we are not making this up), which reaches a meter in length and occurs in the South Pacific in deep water, has been shown to eat deep-sea sharks, including the Birdbeak Dogfish (*Deania calcea*). Videos of sharks, most notably smaller species or juveniles, being eaten by groupers, moray eels, Sea Otters, and even seabirds have been posted to various websites. Still, the most common predators of sharks are, you guessed it, other sharks.

We think of sharks as the ultimate predators, but they have also evolved several successful *anti*predatory adaptations. We talked about their tough skin and mechanisms to protect their eyes. Some species school, which affords them the safety of numbers. Others have fins with stout, sharp spines (e.g., bullhead sharks, gulper sharks, dogfish) and camouflaging. Many smaller species or juveniles of larger species (e.g., Lemon Sharks,

whose main predators are adult Lemon Sharks) occupy protective habitats like mangrove-fringed shallow creeks. Angel Sharks bury themselves in the substrate, and many benthic sharks (e.g., bamboo sharks) hide in crevices. Swell Sharks (*Cephaloscyllium ventriosum*) inflate with water when threatened, in the hope of becoming too big to eat or just startling the heck out of a predator! But they are also predators themselves.

If you were asked to design the prototype perfect marine apex predator, you might design a Shortfin Mako (figs. 2.11 and 2.25). Sleek, beautiful, and packing a high-performance metabolic engine, these sharks excel at finding and catching prey. Makos possess a suite of adaptations (see below) that are more for preying on other animals than avoiding being preyed on, since not many animals prey on them as adults.

An important aspect of sharks that contributes to their effectiveness as predators is their external anatomy. Consider the fins (fig. 2.12), which play various roles in forward thrust, balance,

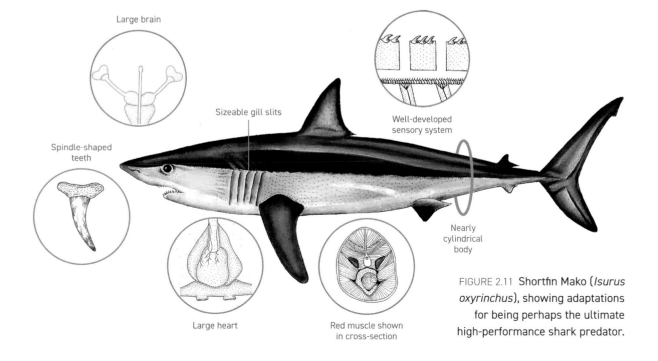

Large brain

Well-developed sensory system

Sizeable gill slits

Spindle-shaped teeth

Nearly cylindrical body

Large heart

Red muscle shown in cross-section

FIGURE 2.11 Shortfin Mako (*Isurus oxyrinchus*), showing adaptations for being perhaps the ultimate high-performance shark predator.

and maneuverability. We already mentioned the distinctive heterocercal tail fin on sharks. In addition, most sharks have three median fins (that is, along the centerline): two dorsals and a single anal fin. Some groups have lost a dorsal fin through evolution and have only one, for example the cow sharks (which include the Bluntnose Sixgill Shark, *Hexanchus griseus*, fig. 2.31). All of the approximately 160 species of dogfish sharks have lost their anal fin.

Sharks have two sets of paired, bilateral (on both sides) fins: the more forward pectorals, and the pelvics toward the rear. These fins could also technically be called appendages or limbs, or even arms and legs, although we do not call them that in fish. All vertebrates have these appendages except where they are lost or vestigial, as in snakes, reflecting our common ancestry.

To move, sharks use mainly the tail fin. Moving provides lift, which is necessary in almost all sharks when they swim (see box 2.2 for an exception) since, as we established earlier, their muscle-bound bodies make them heavier than water and they lack a swim bladder, the gas-filled inner balloon of bony fish. This lift is generated by the uneven shape of the tail as it sweeps back and forth; by the broad pectoral fins (not coincidentally looking like airplane wings); by other fins; and by the somewhat flattened body and head in some species (e.g., hammerheads).

Also, most sharks have lost the ability to pump water over their gills, so movement is necessary to irrigate, and thereby oxygenate, the gills. Contrary to a common misconception, however, not *all* sharks need to keep moving. Bottom-associated sharks, like the nurse sharks, bullhead sharks, bamboo sharks, and others, spend all or most of their time on the seafloor, so sinking is not really an issue. In addition, their low-activity lifestyle reduces their metabolism, so they also need less oxygen, which they can acquire by slowly pumping seawater over their gills. See box 2.2 for an alternative strategy.

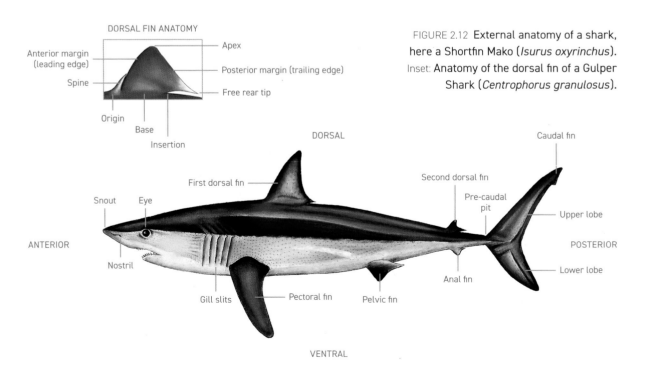

FIGURE 2.12 External anatomy of a shark, here a Shortfin Mako (*Isurus oxyrinchus*). Inset: Anatomy of the dorsal fin of a Gulper Shark (*Centrophorus granulosus*).

DORSAL FIN ANATOMY

Anterior margin (leading edge)
Apex
Spine
Posterior margin (trailing edge)
Free rear tip
Origin
Base
Insertion

DORSAL

Caudal fin
First dorsal fin
Second dorsal fin
Pre-caudal pit
Snout
Eye
Upper lobe
ANTERIOR
POSTERIOR
Nostril
Lower lobe
Gill slits
Pectoral fin
Pelvic fin
Anal fin

VENTRAL

Big, predatory sharks must often catch fast-swimming prey. How do they do this? Most top predatory sharks are successful at catching prey because of a suite of highly developed senses to detect the prey, jaws that can protrude outward and are full of shearing or grasping teeth, body shading and behaviors that make them hard to detect, and adaptations for fast swimming, to name a few. One group of sharks, however, stands out as the master of high-performance, elite ocean predators: the mackerel sharks.

Members of the family Lamnidae (also called the lamnid sharks), the mackerel sharks consist of five species, the White Shark (fig. 2.23), Shortfin Mako (fig. 2.25), Longfin Mako (*Isurus paucus*), Salmon Shark (*Lamna ditropis*), and Porbeagle (*Lamna nasus*). These animals just look like they are built for speed, with muscular, highly streamlined,

BOX 2.2 FLOATIES FOR SAND TIGERS

You may have heard of a hovercraft, but did you know that there is a hover *shark*? Well, not exactly, but close. The Sand Tiger (fig. 2.1B), also called the Grey Nurse Shark, Ragged Tooth, or Raggie, occurs worldwide in temperate and subtropical waters. It is also a common shark in bigger marine aquaria where its large size (up to 10.5 ft, or 3.2 m) and toothiness (it has the unmistakable look of needing corrective braces) make it popular. As Sandbar and Blacktip Sharks and others in the main aquarium tank swim deliberately past you, the Sand Tiger Shark inches forward at a snail's pace, nearly hovering, giving you a chance to examine it closely. How, you might marvel, can such a hefty beast maintain its position in the water column without moving more speedily? Air bubbles, that is how. Both in captivity and in the wild, Sand Tigers will gulp air at the surface, and the lining of the stomach prevents it from diffusing out. In so doing, Sand Tigers create what other sharks cannot: they make their own ersatz swim bladder. Using a floatation device like the swim bladder common in bony fish saves energy, since it is much more energetically costly to move under your own power than to float (hence, pool floats). Energy efficiency, as we note throughout this book, is critical to evolutionary success.

Sand Tiger Shark (*Carcharias taurus*) inside the shipwreck *Aeolus* off the Outer Banks of North Carolina.

almost cylindrical bodies and high-efficiency tails. A glimpse inside would reveal a relatively large brain and a high-performance cardiovascular system, specifically a bigger heart and larger blood volume than in other sharks, to meet the demands of the robust swimming muscles for oxygen and energy, and for removing toxic waste products. The additional oxygen required by these Porsches of the pelagic is supplied by huge gills and large gill slits, and by ram ventilation, keeping the mouth partially open when swimming. And, to add insult to injury (literally!) to the prey, lamnids have striking countershading (dark colored from above, light colored from below) that hinders the ability of prey to detect them.

As if the above adaptations were not enough, the mackerel sharks have evolved a game-changing strategy used by only a handful of other fish, as well as mammals and birds; they elevate their body temperature above that of the environment, known as *endothermy* (the opposite condition, *ectothermy*, refers to the temperature of an organism being the same as that of its environment). The overwhelming majority of bony fish and sharks are ectotherms, since elevating and maintaining a body temperature above that of the heat-sapping water that bathes these animals is a difficult problem to overcome. Members of only seven fish families, including 14 species of tuna, some billfish, five species of mackerel sharks, three species of thresher sharks, and at least two species of manta rays, have evolved ways of trapping heat and elevating their body temperatures above that of the environment.

Why would a marine organism want to raise its body temperature? Temperature is one of the most important environmental factors for all living organisms. For our discussion, let us consider only fish. Temperature affects not only their diversity, abundance, and distribution but also their behavior, activity, growth, development, metabolism, heart rate, digestive rate, and so on. Within the range of temperatures where fish occur, other variables being equal (e.g., prey availability), there are advantages to living at higher temperatures. Within limits, as the environmental temperature increases, there is an increase in the rate of digestion, processing of sensory information, and muscle power.

Consider the effect of temperature on just one factor, metabolic rate, or how fast an organism uses energy. The metabolic rate of most fish, including sharks, as well as amphibians and reptiles, decreases in colder temperatures and increases in warmer ones. Deep-sea sharks, specifically those that do not migrate vertically daily, live in an environment whose temperature is about 39°F (4°C) day and night, across all seasons. These animals have a lower metabolic rate than, say, a Blue Shark at the warmer surface. Sluggish deep-sea sharks thus expend less energy and require less energy from prey than counterparts that inhabit shallow water, which is convenient since there is much less food down deep.

Here then are the advantages of endothermy. First, it enables elevated cruising speeds. All endothermic fish, which include tuna, are fast-swimming, highly mobile predators. In addition, warming the brain increases sensory acuity and processing. Thus, endothermic animals are also better at finding prey and avoiding predators. Endothermy allows mackerel sharks and tuna to move independently of temperature, both latitudinally and vertically within the water column. Cold water, in other words, does not slow them down much. This expands the habitats where they can live.

If being an endotherm is so advantageous, why are only a handful of marine organisms endotherms? First, it is hard to be warm in anything but tropical water. The heat generated by a shark, bony fish, squid, or crab is rapidly lost to the water, about 75,000 times faster than heat is lost in air. Second, trapping body heat requires the evolution of specialized internal systems in the case of fish, or blubber or fur in the case of marine mammals. Although there are big advantages to being an endotherm, ectothermy works well for most marine organisms. Finally, there is a cost to being endothermic: the metabolic furnace that provides the heat, specifically the muscles, requires loads of energy, and thus endothermic marine organisms must be superior predators to supply the calories required to heat the body. And they are!

SENSING PREY

Like you, sharks have a suite of senses that allow them to interpret their surroundings and respond. These include those that we humans and other vertebrates possess as well as one or two that are foreign to us. These sensors enable a shark to detect prey, predators, conspecifics (members of the same species), other organisms, and structures. They also allow a shark to orient itself.

A *sense* can be thought of as a group of specialized cells that work as part of a system to detect some form of physical energy or substance in the environment and transmit information about it to the central nervous system, which may then initiate a response. The five traditional human

senses are sight, hearing, touch, taste, and smell. These can also be described by the form of energy they detect. Vision is a form of *photoreception*, detecting energy in the form of light waves. Smell and taste represent types of *chemoreception*, responses to chemical stimuli. And touch and hearing represent *mechanoreception*, a means of detecting pressure or distortion. Sharks possess mechanoreceptors, chemoreceptors, and photoreceptors. They also possess *electroreceptors*, capable of detecting an electrical field, and possibly also *magnetoreceptors*, capable of detecting magnetic fields. Although the entire suite of senses integrates to paint a complete sensory picture of the environment, different senses in sharks come into play at various distances and for specific functions (fig. 2.13).

In 2015, a Dutch Caribbean Coast Guard helicopter hovered over a shipwreck survivor in the water near Aruba and was about to rescue him when, according to the *Guardian* newspaper,[6] a shark bit and killed him. Immediately, speculation surfaced that the low-frequency *whomp-whomp-whomp* of the helicopter's rotors played a role in attracting the shark. Is that possible? What sounds attract sharks?

FIGURE 2.13 Hierarchy of shark senses as a function of detection distance.

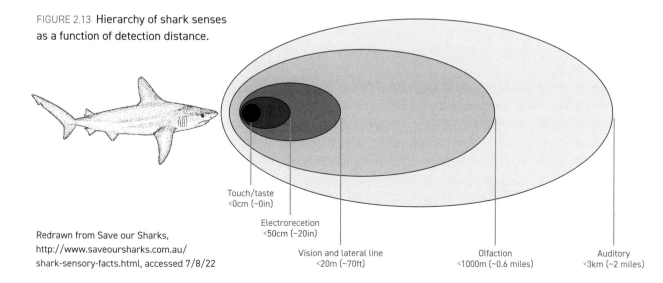

Touch/taste <0cm (~0in)

Electrorecetion <50cm (~20in)

Vision and lateral line <20m (~70ft)

Olfaction <1000m (~0.6 miles)

Auditory <3km (~2 miles)

To answer that question, we must consider mechanoreception, the sense associated with touch, sound, and posture. Water conveys sensory information from mechanical disturbances (sounds and movements) at significantly greater distances than air, so it is no surprise that sharks have adapted to take advantage of this information. Let us consider the ears and lateral line. Yes, sharks have ears, though you would see them only upon dissection, the only external sign being small, paired *endolymphatic pores* (fig. 2.14). These pores open into ducts that lead to the inner ears in the shark's head. Sharks hear in the range of 40–1,500 Hz, with highest sensitivity around 200–400 Hz. The range for humans is reported as 20–20,000 Hz, so sharks hear better in the low-frequency range. Irregular, low-frequency sounds, like those emitted by injured fish, attract sharks. Other sounds, particularly loud ones, repel sharks. Sharks live in an environment that can be relatively noisy. Sounds under a frequency of 1,000 Hz that a shark might encounter and perceive in the marine environment include those associated with swimming fish schools, fish sounds, and waves.

Back to the helicopter and the unfortunate swimmer. While the shark that killed the shipwreck victim could have been attracted to the swimmer using any of its senses, or was simply in the vicinity, the sound of the hovering helicopter unfortunately may have played a role. It certainly did not scare it away, in any case.

The lateral line (fig. 2.15), along with other mechanoreceptors, allows the shark to detect predators, prey, other organisms, and others of its species. It also detects water flow and other physical characteristics of the surroundings and allows for perception of the shark's own body. The lateral lines are canals that run along both sides of the shark from the front of its head to the base of its tail. You can see this line on just about any fish. Along the lateral line, the sensory cells are lined up in a water-filled tube, or canal, under the skin.

Lateral lines are believed to sense *distant touch*, an experience terrestrial organisms lack. You cannot feel a change in air pressure as another human moves close to you, but a shark can feel a pressure change when another shark swims close. And "distant" in this case is an exaggeration, since the lateral line is thought to detect objects only a few body lengths away. If you have ever tried to catch a small fish in a dip net, you know how well this system works. That fish seems to anticipate where your net is coming from. It can do that because it feels the pressure as the net pushes the water ahead of it.

In author Dan's annual Biology of Sharks course at the Bimini Biological Field Station in the Bahamas, one of the students' favorite activities is attracting and hand-feeding juvenile Lemon Sharks[7] in an isolated tidal lagoon. Students spread out along the mangroves, and a *chum bag*, a mesh sack containing minced fish, is positioned on a stake in the sediment where the incoming tide slowly carries pieces of the fish deeper into the lagoon. After perhaps 30 minutes, we begin to see ripples created by the dorsal and caudal fins of the juvenile sharks breaking the water's surface as they wend their way toward us. After conditioning the sharks to take the squid or herring, the students each hand-feed one or two sharks. This exercise is a nice demonstration of how a shark senses its environment. Since the lagoon is less than 330 ft (100 m) long, the sharks likely hear us and initially move away because we are quite noisy. Other senses, including vision, electroreception, and other forms of mechanoreception, cannot come into play at such a distance. What remains is chemoreception, specifically smelling (olfaction) and tasting (gustation).

The ability to detect environmental chemicals like blood is the most ancient of the senses, having evolved over 500 million years ago. Chemoreception in water is quite different than in

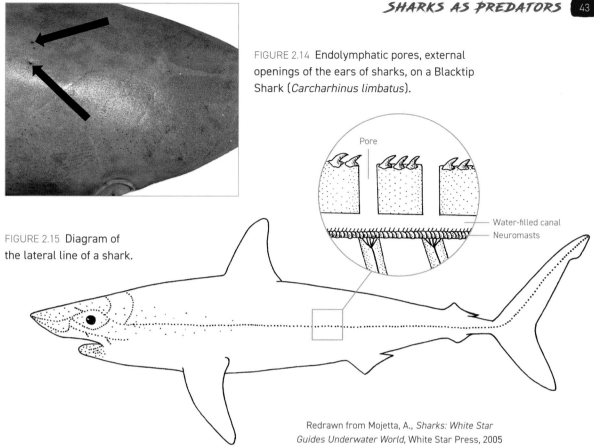

FIGURE 2.14 Endolymphatic pores, external openings of the ears of sharks, on a Blacktip Shark (*Carcharhinus limbatus*).

FIGURE 2.15 Diagram of the lateral line of a shark.

Pore

Water-filled canal

Neuromasts

Redrawn from Mojetta, A., *Sharks: White Star Guides Underwater World*, White Star Press, 2005

air since chemicals in water need to be dissolved as opposed to being in the gaseous state in air. Because water is heavier and more resistant to flowing than air, it often moves more slowly, and odors diffuse more slowly as well. Odors also dissipate more quickly in air than in water. Chemoreception is most important for feeding and reproduction.

You may have heard that sharks are "swimming noses" capable of detecting a drop of blood from several miles. This statement is not wrong, but it requires qualification. Yes, sharks have an extremely well-developed sense of smell, but they are not the only bloodhounds of the ocean, since many bony fish have similar sensitivities. For a shark to detect a drop of blood, specific molecules in that blood capable of stimulating the shark's smell receptors must physically encounter these receptors. In other

words, the molecules must actually reach the shark's nose, which takes some time.

Back to the lagoon. The chemosensory system of these juvenile Lemon Sharks brings the sharks in. They smell the odorous chum and follow their noses. They then use their other senses in the final stages of locating the food source, which we hope is not our thighs! (And in all the years we have done this, not one shark has made that mistake.)

The organs of smell are paired *olfactory sacs* just above the mouth, in cartilaginous nasal capsules. Water is typically channeled into the nares (you might call these nostrils), the channels that lead into the olfactory sac by flaps on the outermost opening, and then the water moves out through the innermost opening (fig. 2.16). The small distance between the outer and inner opening might mislead you to think that the water that enters travels only a short way before it exits. In reality, the water takes a circuitous path through a heavily folded sensory surface, the *olfactory rosette*, which greatly increases the surface area for sensing environmental chemicals. Think of a spiral playground slide and the extra time it takes to reach the bottom compared to a straight one. The slow transit time ensures that compounds in the water contact the sensors in the rosette. Most sharks must be moving, or there must be a water current, for the system to work.[8] When a shark perceives a chemical stimulus of appropriate strength, it will generally swim toward it while continuously sampling the water, sort of "sniffing" as it goes.

Another form of chemoreception is taste, or gustation. Perhaps surprisingly to you, taste is assumed to have a role secondary to smell and is more specialized in its function. Taste is a way to assess food quality; however, since it is among the most poorly studied senses, its function is not well understood. Taste receptors in sharks are located on taste buds inside the mouth and on the gills. Unlike in mammals, taste receptors are not concentrated on the tongue (yes, sharks have tongues, but theirs are immobile pieces of cartilage, not at all like yours). The location and density of taste receptors vary by species. In bottom-dwelling sharks, these receptors are more evenly distributed throughout the mouth than in open-water species. These different distributions of taste receptors make good sense. Benthic species, such as Nurse Sharks and Brownbanded Bamboo Sharks (*Chiloscyllium punctatum*), often manipulate their prey in their mouths prior to swallowing, during which time they assess the palatability of the ingested item. In pelagic sharks, the highest densities of taste receptors are in areas immediately adjacent to the teeth and at the front of the mouth, the first point of contact when biting a food item. At this critical point, the predator must instantaneously assess whether the item represents food on its menu and is worth

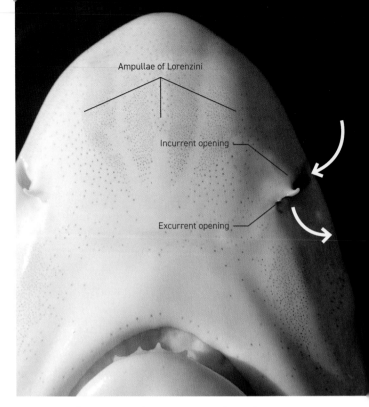

FIGURE 2.16 Incurrent and excurrent openings of the nares (i.e., nostrils) of a shark. White arrows depict direction of water flow. Ampullae of Lorenzini (small pores) are also visible on the underside of the head of this Atlantic Sharpnose Shark (*Rhizoprionodon terraenovae*).

continuing to consume. The concentration of taste receptors near the mouth also explains the *bite-and-release* shark bites we discussed earlier.

The main organ of vision in sharks is, of course, the eye. Sharks have a typical vertebrate, image-forming eye, with some modifications. Sharks possess a *tapetum lucidum*, a tissue behind the retina composed of reflective crystals, which reduces internal glare and scattering of light and improves vision in low-light conditions. This is the same tissue that is responsible for the "deer in the headlights" reflection from the eyes of cats, raccoons, and others. We will see this in other predators, but the tapetum lucidum of a shark is twice as efficient as that of a cat. Light levels can be extremely low underwater.

The lens in sharks is a transparent structure responsible for all the refraction (bending of light) necessary to focus light on the retina. The lens in most species is large and powerful. Unlike the focusing of terrestrial vertebrates, which is achieved by changing the shape of the lens, focusing in sharks is achieved by moving the whole lens backward for far vision or forward for near vision.

The retina is responsible for converting the image that the lens focuses on it into nerve impulses. Like you and most vertebrates, sharks have a duplex retina—that is, the retina houses both rods and cones, the cells responsible for vision in low-light and bright-light conditions, respectively. The presence of cones containing multiple visual pigments implies a basis for color vision, which has been experimentally verified in several shark species.

Another favorite activity of students in our annual Biology of Sharks course in Bimini, Bahamas, is snorkeling with Caribbean Reef Sharks. Typically, within minutes of our arrival at the snorkeling spot sharks appear, having been attracted by the sounds of the boat engine and the water slapping at the hull. They know from experience that this means "snack time." When a piece of bait (not a student!) hits the water, the sharks use their hearing and other mechanosenses to approach the source of the sound. Some detect the odor in the eddies of the water current and move upstream, in the direction of the source. Then, in these clear topical waters, the sharks zero in on the bait using their vision.

This suite of senses often works exquisitely for the first shark at the bait, but sometimes in the dance of the feeding sharks, two sharks will converge on the bait from different locations simultaneously. Both sharks will then trust the more than 400 million years of evolution leading up to that moment, and when they are less than 3 ft (1 m) away from the bait, they will open their mouths almost in unison and deploy their protective nictitating membrane, since their prey in natural circumstances might fight back. This leaves them temporarily blind. If the bait were a live fish, the sharks would detect the minute electrical current that all living organisms emit, and this signal would guide them to the prey with surgical precision, a phenomenon that may also come into play with the dead bait. At this point their systems fool them, because the stronger electrical signal they detect is emanating not from the bait, but from each other. As the bait drifts safely away, at least until the next shark senses it, the two sharks will attempt to bite each other. Fortunately, we have never witnessed any damage done, perhaps because of the gustatory receptors near the teeth that inform each shark of its mistake before the bite is completed.

The electroreceptive system that guides a shark to its destination consists of receptors called ampullae of Lorenzini (fig. 2.16). The ampullae, gel-filled tubes with surface pores, are concentrated on the head of sharks and can detect extremely weak electric fields of other organisms and even inanimate objects. Sharks have been known to sometimes chomp down on boat engines that continually produce a minute electrical current rather than bait for this reason.

TEETH

Sharks have a variety of methods of ingesting food, and these are associated with variation in the form of their teeth. Look at the teeth shown in figure 2.17. Serrated teeth, found in Tiger Sharks, for example, are for shearing. The shark extends its protrusible jaws (see below) into its large prey and swings from side to side, removing a large chunk of the prey, perhaps a dolphin, seal, turtle, or even another shark.

The prey of some sharks are smaller fish that can be swallowed whole, in which case grasping is more important than shearing, and those teeth tend to be long, slender, and nonserrated, like those of a Shortfin Mako or Lemon Shark. In the Shortfin

Mako (fig. 2.25), the teeth are directed rearward, preventing any prey unlucky enough to be captured from getting away. Both species of frilled sharks (Frilled and African Frilled) have recurved teeth like those of a python, and there are about 300 of these rearward-pointing, interlocking teeth in about 25 rows! If a prey item is unfortunate enough to be grasped by a frilled shark, there is no way it can escape. Noted shark biologist Dean Grubbs once required assistance to remove his hand from the mouth of a Frilled Shark (*Chlamydoselachus anguineus*), a *dead* Frilled Shark.

Some sharks and rays, such as the Tawny Nurse Shark (*Nebrius ferrugineus*), Mexican Horn Shark (*Heterodontus mexicanus*), and Cownose Ray (*Rhinoptera bonasus*), include shelled prey like scallops, clams, snails, crabs, sea urchins, and others in their diet. Feeding on hard-bodied prey like these is known as *durophagy* (*dur* = hard; *phag* = eating). It is not surprising that durophagous sharks and rays do not require teeth that tear or snag, but rather teeth that are broader and smaller, and in some cases these teeth are organized into crushing plates.

In many sharks, the teeth in the upper and lower jaws differ. In the Caribbean Reef Shark (*Carcharhinus perezi*) and related species, including the Galapagos (*Carcharhinus galapagensis*), Sandbar, and Dusky (*Carcharhinus*

obscurus) Sharks, among others, the teeth in the lower jaw grasp while those in the upper jaw slice.

In addition to formidable teeth, the way a mouth can move impacts biting ability. *Jaw suspension* refers to how the upper and lower jaws connect to the skull and other supporting structures. To get an idea of the importance and role of jaw suspension, place an apple in a big bucket of water and try to pick it up using only your mouth (no hands!), an activity called bobbing for apples. Bobbing for apples is not easy for humans, or indeed any terrestrial vertebrate, because our upper jaw is firmly affixed to our skull. This condition works for you (except when bobbing for apples), given your evolution as a consumer of food on land. But would this type of jaw-skull connection work for an aquatic predator like a shark? No. The restricted mobility of your human jaw—that is, its inability to protrude—combined with a small gape (opening), limits the scope of your diet. Can you imagine a shark with a jaw like yours, needing to continuously reposition itself so that its very small mouth was in the right place to catch and bite a prey item in the water? If that were the case, we would very likely be discussing sharks as a minor group or even in the past tense, as evolutionary experiments gone bad—dead ends.

How then is the jaw suspension of sharks different from yours? Although there is variation among distinct groups of sharks, the most

FIGURE 2.17 Teeth diversity in sharks. (A) Bluntnose Sixgill Shark (*Hexanchus griseus*), (B) Tiger Shark (*Galeocerdo cuvier*), (C) Blue Shark (*Prionace glauca*), (D) Bull Shark (*Carcharhinus leucas*).
R. Dean Grubbs

evolutionarily advanced types, like Grey Reef Sharks, Bull Sharks, and Tiger Sharks, feature a loosening of the connection between the upper jaw and the skull. This allows the jaw to protrude to varying degrees (fig. 2.3), and the shark can thus bite from a slightly greater distance. This large gape also facilitates grasping, shearing, manipulation, and ingestion of prey and has contributed immensely to the success of modern sharks.

Shark bites are among the strongest in the animal kingdom, right? Actually, while the bite force of a White or Bull Shark is high, it is in the same range as that of crocodiles, alligators, big cats, and others. The force of shark bites closely

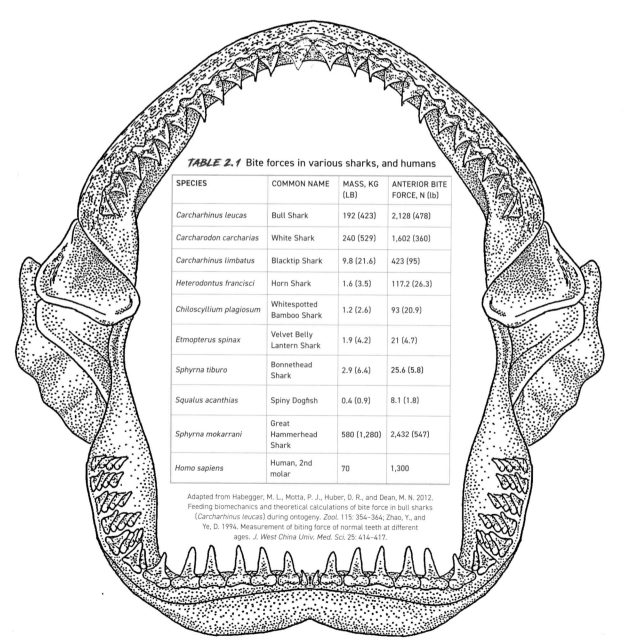

TABLE 2.1 Bite forces in various sharks, and humans

SPECIES	COMMON NAME	MASS, KG (LB)	ANTERIOR BITE FORCE, N (lb)
Carcharhinus leucas	Bull Shark	192 (423)	2,128 (478)
Carcharodon carcharias	White Shark	240 (529)	1,602 (360)
Carcharhinus limbatus	Blacktip Shark	9.8 (21.6)	423 (95)
Heterodontus francisci	Horn Shark	1.6 (3.5)	117.2 (26.3)
Chiloscyllium plagiosum	Whitespotted Bamboo Shark	1.2 (2.6)	93 (20.9)
Etmopterus spinax	Velvet Belly Lantern Shark	1.9 (4.2)	21 (4.7)
Sphyrna tiburo	Bonnethead Shark	2.9 (6.4)	25.6 (5.8)
Squalus acanthias	Spiny Dogfish	0.4 (0.9)	8.1 (1.8)
Sphyrna mokarrani	Great Hammerhead Shark	580 (1,280)	2,432 (547)
Homo sapiens	Human, 2nd molar	70	1,300

Adapted from Habegger, M. L., Motta, P. J., Huber, D. R., and Dean, M. N. 2012. Feeding biomechanics and theoretical calculations of bite force in bull sharks (*Carcharhinus leucas*) during ontogeny. *Zool.* 115: 354–364; Zhao, Y., and Ye, D. 1994. Measurement of biting force of normal teeth at different ages. *J. West China Univ. Med. Sci.* 25: 414–417.

matches what they need in order to grasp and hold on to their prey, or to take out a chunk of flesh. Our expectation of extremely high bite forces is based on continuous media reinforcement of that perception, or our belief that sharks simply look like they should have high bite forces. Recent studies have provided estimates of anterior bite force (at the front of the jaws) for several species, shown in table 2.1

Surprisingly, in some species such as the Tiger Shark and cow sharks, the jaw cartilages are weakly calcified and bend easily, and thus they do not generate the expected high bite force. Therefore, the dried jaw of a Tiger Shark is usually deformed compared to that of, say, a Bull Shark. The weak calcification permits the jaws to bend across the surface of prey (e.g., sea turtles in the case of Tiger Sharks), thus allowing most of the functional teeth to make contact. Shaking the head or rolling the body removes large chunks of flesh from the prey.

How high are bite forces of durophagous species, those that eat hard prey and typically possess large jaw musculature (fig. 2.18) and teeth designed for grinding? Anterior bite force for a 3.5 lb (1.6 kg)

Horn Shark is only 26.3 lb (117.2 N), but maximum bite force on the posterior molars is 76 lb (338 N). These numbers seem low compared to measurements for Bull and White Sharks, but relative to the shark's size, they represent one of the highest bite forces among sharks.

HOW ARE SHARKS DOING?

Such fearsome predators must rule the sea, right? So how are sharks doing? There is both good and bad news about the status of shark populations. The good news is that more people and governments are beginning to appreciate the intrinsic, ecological, and even financial value of sharks in their habitats, more than as food or products. This has led to effective, science-based management of some species, including designation of some areas (e.g., the Bahamas and Palau) as shark sanctuaries.

But there is plenty of bad news. First, policy makers do not know as much as they need to about the life history characteristics of most

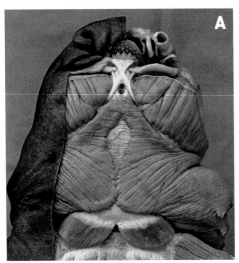

FIGURE 2.18 (A) Ventral view of a Horn Shark (*Heterodontus francisci*) showing hypertrophied jaw musculature used to crush hard prey. (B) Side view of Horn Shark. Physiological Research Lab, Scripps Institution of Oceanography

sharks in order to effectively conserve them, especially those in the deep sea. Nor do scientists understand with any degree of precision the health, behavioral, and ecological impacts on sharks of many environmental threats. What we do know is that removing them from an ecosystem causes changes, although accurately understanding those changes is not easy. But a reef with a healthy shark population is an ecologically healthier place than one without sharks.

Sharks, even those whose populations are stable, live in an ocean imperiled by human-caused climate change that is warming and acidifying their environment, with repercussions for every aspect of their biology and ecology. Also, even if sharks are protected, in many areas enforcement of regulations may be limited or absent. One of the biggest current threats to sharks is overfishing. Sharks are caught for their meat, fins, liver, cartilage, skin, teeth, and jaws, as well as other body parts. Live sharks are captured for use in aquaria. They also serve as ecotourism draws.

Since the 1980s, shark fins (fig. 2.19) have been the most economically valuable part of the shark and are still one of the most profitable, after shark meat. They are used mostly for shark fin soup, which is considered a delicacy and status symbol in countries in East and Southeast Asia. All fins on a shark are used except the upper lobe of the caudal, where the vertebral column extends all the way to the tip.

It is important to distinguish between illegal finning and the legal shark fin trade. The former involves removal of the fins from a shark, most often immediately upon capture when the animal is still alive. The now less valuable, finless, dying shark is then thrown back into the water. The legal shark fin trade involves fins from legally caught sharks that are typically brought back to port whole and are used for meat and other products. Numerous countries[9] have various kinds of restrictions or bans on finning or on shark fin soup.

Peak global value for the shark fin trade was about US$300 million in the early 2000s. Hong Kong is still the world's biggest trader of shark fins, with about 40%–50% of the global total, followed by Trinidad and Tobago. Commercial fishers from Spain and Indonesia are responsible for the largest shark catches for the shark fin industry. In the

FIGURE 2.19 **(A)** Shortfin Mako (*Isurus oxyrinchus*) and **(B)** Blue Sharks (*Prionace glauca*) with fins removed in a market in Cádiz, Spain.

United States, there is a legal shark fin trade. If shark fisheries are allowed, it seems likely that fins will be traded so that as much of the carcass as possible is used. How do you feel about this policy?

Shark cartilage is widely used as a health supplement, although the health benefits are not supported by sound, science-based evidence. The claims include protecting against or curing eczema, ulcers, hemorrhoids, arthritis, and other diseases and disorders, most notably cancer. Sharks can develop cancer, and no reputable studies have shown that eating their cartilage will protect you from it.

Shark jaws and teeth, as well as preserved shark embryos and juveniles, are bought and sold at trade shows and markets and online. All are legal to purchase, except those from species like sawfish and others that are on the US endangered species list or are present in Appendix I of CITES (the Convention on International Trade in Endangered Species of Wild Fauna and Flora), which prohibits their trade.

As a group, shark and ray fisheries represent less than 1% of the total marine capture fishery. The top five countries with the highest landings of sharks and rays are India, Indonesia, Mexico, Spain, and Taiwan. These five plus Argentina, the United States, Pakistan, Malaysia, and Japan are responsible for about 60% of the shark and ray landings worldwide. The remaining 40% are from small countries, mostly island nations or poorer countries in Africa, and mostly from artisanal fisheries, which, as we stated above, are difficult to get good data from. This remains a vexing problem.

Bycatch—that is, catch of nontargeted species—impacts shark populations and complicates managing them (fig. 2.20). Sharks constitute bycatch on most fishing gear, including longlines, trawls, and gill nets. High-seas drift gill net fisheries, such as those for squid and salmon in the North Pacific, have large shark bycatch. Purse seining for tuna also entails considerable bycatch. In New South Wales, Australia, declines of more than 90% were seen in a large suite of deep-sea sharks between the late 1970s and the late 1990s as the result of bycatch in a deep-sea trawl fishery. Ghost fishing gear—that is, hooks and nets that are lost or abandoned—also causes mortality in sharks (and other endangered taxa like marine mammals, seabirds, and sea turtles).

FIGURE 2.20 **Dead Greenland Shark (***Somniosus microcephalus***) on the deck of a bottom trawler in the North Atlantic.** Juan Vilata/Shutterstock

Mortality rates for shark bycatch vary with the species and method and can be substantial. On pelagic longlines, some sharks, such as Smooth (*Sphyrna zygaena*), Scalloped, and Great Hammerheads (*Sphyrna mokarran*), as well as all the threshers, suffer mortality rates of at least 25% and sometimes more than 50% before being boated. It is likely that most of the sharks released alive in this fishery do not survive. In contrast, the mortality rate of Blue, Silky, and Oceanic Whitetip Sharks—the three species most often caught on pelagic longlines—as well as the makos, is only 5%–20% when landed, and the few data available suggest that survival after release may be quite high. We know little about the survival rates of deep-sea sharks released after capture, but the evidence suggests, particularly for small species, that very few survive, even if they are released relatively unharmed.

Globally, sharks are protected by several agreements, including CITES, an agreement currently between 183 governments. Also, the IUCN (International Union for the Conservation of Nature) Shark Specialist Group prioritizes species at risk, monitors threats, and evaluates conservation action. The Memorandum of Understanding on the Conservation of Migratory Sharks, an environmental treaty of the United Nations, includes 40 species of sharks and rays.

In federal waters of the United States—that is, from 3 to 200 mi (5–300 km) from shore—the primary law governing fisheries management is the Magnuson-Stevens Fishery Conservation and Management Act. Within three miles of shore, management of marine resources is in the jurisdiction of states, which typically work closely with federal managers. The major means by which fisheries are regulated are Fishery Management Plans (FMPs), which assess the health of the stock in question and then establish scientifically sound regulatory measures. Currently, US Atlantic sharks

(classified as *large coastal sharks, small coastal sharks,* or *pelagics*) are under management. By 2016, large coastal sharks were managed at a quota that was 8% of historical peak landings, and the most vulnerable species remained prohibited. Because the Spiny Dogfish stock apparently recovered, landings in this fishery increased, making up 39% of the 2016 commercial landings. They are now often commercially called Cape Sharks, apparently a more a palatable term for consumers than "dogfish." (Although we eat catfish, no?) Sandbar Sharks are on a trajectory to recover in the second half of the twenty-first century, several decades away. Dusky Sharks remain one of the most overfished sharks on the US East Coast, and their full recovery is not expected until after 2100. 2100! Do not eat sharks unless you know they are part of a sustainable fishery!

Along the US West Coast, sharks are managed under the Fishery Management Plan for West Coast Fisheries for Highly Migratory Species. Managed species include all three species of thresher sharks, Silky Sharks, Oceanic Whitetip Sharks, Blue Sharks, both species of makos, and the Salmon Shark (*Lamna ditropis*).

Outside the United States, sharks are managed by a variety of methods, although in the United States as well as in other countries, enforcement remains an issue. A growing number of countries (16 as of 2022) have declared their territorial waters to be shark sanctuaries to reduce shark mortality and aid in their conservation.

While targeted fisheries and bycatch are considered the principal current dangers for sharks, climate change, habitat degradation and destruction, pollution (nutrient pollution from agricultural runoff and undertreated human sewage, plastic pollution, and other chemicals), exotic introductions, and aggregated human disturbance (persecution, noise pollution, etc.) are also major threats. The short- and long-term

FIGURE 2.21 (A) Sandbar Shark (*Carcharhinus plumbeus*) with plastic strap encircling its body, caught on an experimental longline. (B) Same shark being successfully released after the strap was removed. It swam away strongly. George Boneillo

impacts of these threats on sharks are not all known. They may be minimal, or they could range from sublethal impacts that affect an organism's overall fitness (e.g., its internal functions and behavior) to mortalities that cause declines in populations. Here, we briefly consider plastics, habitat degradation, and climate change as examples of the threats sharks face.

Figure 2.21 shows a Sandbar Shark we caught on an experimental longline in 2016 in Winyah Bay, South Carolina, with a plastic packaging strap fully encircling the shark near the gills. This shark likely survived after we removed the strap, since it appeared otherwise healthy, was actively feeding (it took the bait on our longline), and swam off strongly once released, but many sharks and other marine life do not survive similar entanglements.

Plastic is so commonplace that we are said to be in the Age of Plastics. Estimates of the amount of plastic that finds its way into the marine environment annually range from 1.8% to 10% of annual global plastic production. Once in the marine environment, plastics cause a suite of problems at the organismal and ecosystem level, although much remains to be understood about these impacts. Broadly, sharks are affected by plastics through entanglement or ingestion. Entanglement of sharks in plastics has been reported anecdotally (as in our example above), especially from abandoned drift nets. However, the global extent of entanglement and its impacts on local populations and ecosystems are difficult to assess.

Ingestion of plastics has been studied more rigorously than entanglement and is a major line of inquiry as we write this book. Direct ingestion of large plastic by sharks seems rare, but smaller pieces are routinely swallowed. As of 2021, sharks found with ingested microplastics included 15 species, but the number will assuredly expand as more sharks are examined.

Microplastic (< 5–10 mm, or 0.2–0.4 in) toxicity may result either from adsorption of harmful chemical pollutants to the surface of the microplastics, such as PCBs, DDE, and DDT, or from additional harmful chemicals in the plastics, such as flame retardants and plasticizers. However, studies on the effects of toxic microplastics on sharks, as well as other species, are lacking. Additionally, although the subject has not yet been thoroughly studied, evidence suggests that plastics can move up the food chain.

Let us look at mangroves as an illustration of the problem of habitat degradation. Mangroves, one of the most biologically productive ecosystems on the planet, are a broad group of salt-tolerant trees that grow at the water's edge in the tropics. They play important roles in the life histories of numerous sharks as well as critically endangered rays, like the sawfish. Stilt roots of mangroves at the water's edge provide myriad hiding places for a great diversity of marine life, especially the juvenile and larval forms of fish, crustaceans, and mollusks, any of which constitute food for sharks. They also provide nursery habitat for some shark species. Mangrove communities are at high risk from development, especially in countries with few if any land-use controls. Mangrove environments are readily cleared for aquaculture, resort developments, and housing. They are also threatened globally by rapid sea-level rise and locally by agricultural runoff, oil, deforestation for biomass fuel, and gas exploration and production.

Already, as much as 50% of the world's mangroves have been lost, and more have been degraded by pollution. And mangroves also serve as important buffers against coastal storms for humans in the vicinity, so it is not just sharks we are putting at risk. One particularly well-researched species is the Lemon Shark of Bimini, Bahamas (fig. 2.22). Mangroves and the lagoons fringed by the mangroves are critical to the growth and survival of juvenile Lemon Sharks during their first few years of life, after which they move to more nearshore or coastal habitats. Young Lemon Sharks take advantage of the protection from predation offered by the mangroves (mainly from larger Lemon Sharks), as well as the abundant food supply in the lagoon.

By 2010, development of a resort in Bimini, Bahamas, had involved dredging and removal of about 166 ac (67 ha) of mangroves, representing 39% of the mangrove habitat surrounding the

FIGURE 2.22 A juvenile Lemon shark (*Negaprion brevirostris*) patrols the mangroves in Bimini, Bahamas. Annie Guttridge

system. This habitat was one of the most important Lemon Shark nurseries in the northwestern Bahamas, which also serves to recruit adult Lemon Sharks to southeastern US habitats. After the development started, survival rates and growth rates of Lemon Sharks both decreased, and sharks remaining in the area were less healthy than comparable sharks in undisturbed areas. Clearly the sharks, and undoubtedly their marine neighbors, need minimally degraded coastal habitats.

What are some of the impacts of climate change on sharks? Let us start with acidification and temperature, although since climate change will affect both these variables, separating acidification from temperature changes may produce conflicting and confounding results. The scale of impacts of ocean acidification has the potential to be enormous, causing changes in ocean chemistry that have not been seen in 65 million years, which will affect the vitality and survival of all taxonomic groups. Impacts of acidification on sharks may be direct, such as affecting internal function (by acidifying blood and tissues) and behavior of sharks, or indirect, such as changing the community structure or prey. As of 2022, few studies of the effects of acidification on sharks have been published. Like many other shark studies, these have focused on smaller species, particularly benthic sharks, that are most readily maintained in captivity and are thus not applicable to the group as a whole. They suggest decreases in resilience and fitness as acidification increases. In a study of behavioral responses to acidification, Port Jackson Sharks took about four times longer than control sharks to detect their prey in acidified water. However, that time was reduced by a third when the study was conducted in water that was warmer than normal. Dusky Smoothhounds in a different study exhibited an impaired ability to track prey using odors in acidified water.

Temperature is one of the most significant environmental influences on organisms. In contrast to studies of acidification effects, there are many studies on the effects of elevated temperature on this group, such as signals for movements, as well as impacts on behavior, respiration and metabolism, growth, swimming, reproduction and embryonic development, and foraging. However, fewer of these have focused on warming in the range associated with human-caused climate change, and thus it is too early to conclude with confidence what the specific impacts of temperature changes, especially increases, are on sharks.

Some additional climate change impacts that could affect sharks include changes in precipitation patterns that alter the salinity structure of nearshore and oceanic systems, increased intensity and frequency of tropical storms, and rising sea levels that coastal wetland communities may not be able to keep up with. Also, larger and more severe oceanic dead zones occur because of oxygen depletion triggered by excess nutrient runoff, often adjacent to the mouths of major rivers. Some of these impacts may occur on an even larger scale, such as changes in ocean circulation that could include a critical slowing of the Gulf Stream, a current that moderates temperatures along the entire coast of North America over to Europe.

What would be the ecological impacts of these changes? Many sharks and rays would be forced to migrate to higher latitudes or deeper water because of temperature increases. For example, a 2018 study[10] showed the presence of juvenile Bull Sharks in North Carolina estuaries, which had not previously been a frequently used habitat, and correlated their presence with the early arrival of summer temperatures. In moving to higher latitudes, sharks and rays may encounter ecosystems novel to them. These may cause problems like changes in the abundance and size structure of shark populations; changes in the food chain; changes in behavior; and possibly mortalities, extirpations, and even extinctions of species for which migrations may be difficult or improbable or which may already be depleted or threatened by other stressors.

Sharks have an evolutionary history reaching over 400 million years, but are they capable of surviving the current human-dominated era, the Anthropocene, and the sixth mass extinction we are causing? The most recent assessment by the IUCN is not optimistic. A comprehensive evaluation of sharks, rays, and chimaeras (all closely related cartilaginous fish) published in 2021 found that 37% of species for which sufficient data exist are threatened with extinction.

A FEW OF THE APEX PREDATORS AMONG SHARKS

WHITE SHARK
(Carcharodon carcharias)

What can we say about one of the most iconic species on the planet (fig. 2.23)? Is it overrated? Well, if overrated means that the species receives

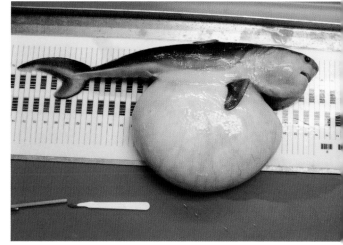

▲ FIGURE 2.23 White Shark (*Carcharodon carcharias*).
Tanya Houppermans

▶ FIGURE 2.24 Embryo of a Salmon Shark (*Lamna ditropis*), a close relative of the White Shark and makos, exhibiting its yolk stomach filled with ova it has consumed while developing in the uterus.
Kenneth J. Goldman, PhD

attention disproportionate to its ecological importance or conservation status, often to the exclusion of more interesting and endangered sharks, then yes. On the other hand, if caring about this species leads to awareness of the plight of other sharks and inhabitants of the planet in general, and thus valuing our natural environment, then no.

White Sharks and Shortfin Makos are among the five species of mackerel sharks (family Lamnidae), all of which are high-performance predators with a large, fusiform (spindle shaped, or tapering at both ends) body as long as 13–20 ft (4–6 m), a crescent-shaped caudal fin, and a pointed snout. They also maintain their body temperature above that of the water in which they reside, as we described earlier. All are pelagic, typically at 490–3,280 ft (150–1,000 m) in most temperate and some

tropical seas. As embryos, mackerel sharks grow and mature in their mother's uterus by eating the unfertilized ova the mother continues to produce, and they develop very cute pot bellies, more commonly called egg, or yolk, stomachs (fig. 2.24).

Characteristics of White Sharks include their large size (up to 20 ft, or 6.1 m, and 4,200 lb, or 1,900 kg), a jaw full of triangular serrated teeth for cutting, long gill slits, vivid color changes on the sides, and black tips under the pectoral fins. The species is globally distributed in both coastal and oceanic waters where the temperature is 54–75°F (12–24°C). Contrary to the public's perception, the overall population is not declining and in fact has been increasing in many regions during the last 20 or 30 years, although optimism must be tempered against the reality of a critically imperiled ocean, as well as population decreases in some locations.

Their diet consists of marine mammals (seals, sea lions, elephant seals, and dolphins) as well as bony fish, sharks, and rays. They will also feed opportunistically on whale carcasses. White Sharks may also occasionally eat sea turtles. Neonates and smaller juveniles consume mostly fish. We would be remiss not to mention the acrobatic, aerial predatory behavior of White Sharks at Seal Island, South Africa. While breathtaking complete breaching may occur, the repertoire of predatory activities of these White Sharks includes a suite of behaviors (e.g., broaching and lunging) that are employed in preying on Cape Fur Seals.

The White Shark is perpetually in the news, but recent reports merit particular attention in a book about predators. First near Southeast Farallon Island, south of San Francisco, and more recently off South Africa, are reports of White Sharks being killed by Orcas, which apparently removed the liver and possibly other internal organs. Even if this activity is novel, that Orcas may have enlarged their range of prey to include White Sharks would not be an unusual feat for so intelligent and, well, predatory an animal.

SHORTFIN MAKO
(Isurus oxyrinchus)

This species is perhaps the most magnificent fish in the sea (so says a shark biologist), with its beautiful coloration (brilliant blue or purple on top, white on the bottom) (fig. 2.25) and its near perfect streamlining. It is found globally in temperate and tropical waters.

Distinguishing features include a conical snout, moderately short pectoral fins, crescent-shaped tail, dagger teeth, and huge gill slits. They grow to at least 14.6 ft (4.45 m). They eat mainly bony fish and squid, often swallowing them whole, and are considered opportunistic apex predators. As they age, their teeth become broader and flatter, enabling them to widen their prey options to include organisms too large to swallow whole but from which they can remove a chunk of flesh, such as swordfish, tuna, sharks, sea turtles, and marine mammals. Specimens have been captured or observed with swordfish bills impaled in their head region and even their vertebral column. Sometimes dinner fights back.

The Shortfin Mako is considered Vulnerable in the Atlantic and Indo-West Pacific, and Near Threatened in the eastern North Pacific.

TIGER SHARK
(Galeocerdo cuvier)

Found worldwide in tropical and temperate coastal waters, this iconic, large shark (16.5 ft, or 5.0 m) is easily identified by its markings, which are most vivid in juveniles; its long caudal fin; and its wide, multicusped teeth (fig. 2.26). Like the cow sharks, Tiger Sharks have surprisingly relatively weak jaws that bend across the body of large prey (e.g., sea turtles, dead whales). They then twist or spin their bodies to carve out huge chunks of flesh.

Tiger Sharks may produce 60 or more pups every three years. Because of their high reproductive potential, Tiger Shark populations are considered healthy globally. However, a recent study showed a 71% decline in Tiger Sharks along the east coast of Australia.[11]

While Tiger Sharks have virtually no predators as adults, in some locations they eat lower in the food web, since the food item they frequently consume, sea turtles, eats primarily plants.

▲ FIGURE 2.25 **Shortfin Mako** (*Isurus oxyrinchus*).
Wildestanimal/Shutterstock.com

▼ FIGURE 2.26 **Tiger Shark** (*Galeocerdo cuvier*).
Ken Kiefer

BULL SHARK
(Carcharhinus leucas)

This is a stout shark with a robust, blunt, rounded snout (fig. 2.27). It has a large first dorsal fin situated far forward on its body. Its eyes are relatively small, but its teeth are broad and heavily serrated for shearing.

The Bull Shark is found predominantly in shallow tropical and temperate waters less than 100 ft (30 m) deep, but it can be found shallower and as deep as 538 ft (164 m). It is considered

FIGURE 2.27 **Bull Shark**
(*Carcharhinus leucas*).
Ken Kiefer

dangerous, especially in the developing world where the daily lives of inhabitants find them in Bull Shark habitat, particularly brackish and fresh waters. When Bull Sharks encounter a boat in shallow, clear tropical waters, they may go into a threat display, lowering their fins and hunching their back. When they do that, they may charge and strike the boat. Try not to fall in right then!

Bull Sharks eat a variety of bony fish as well as sharks and rays (see fig. 2.2B). Their occasional prey includes sea turtles, dolphins, seabirds, crustaceans, and squid.

SANDBAR SHARK, BROWN SHARK
(*Carcharhinus plumbeus*)

The Sandbar Shark is distributed worldwide and is an ecologically important, bottom-dwelling species of coastal temperate waters shallower than 330 ft (100 m) (fig. 2.28). In many ecosystems it shares the top spot in the food web with other sharks or is one level lower; in other words, it is not always an apex predator. The Sandbar Shark reaches a maximum length of 8 ft (2.4 m). It has an oversized first dorsal fin far forward on its body. The Sandbar Shark is the dominant shark along the

US East Coast as well as in Hawai'i, but it occupies deeper water in the latter. The biggest Sandbar Shark nursery in the world—that is, the area where they are born and/or spend their early years—is Chesapeake Bay. In large part because of their large first dorsal fin's value to the shark fin soup industry, Sandbar Sharks drove the US East Coast shark fishery until they became overfished. They are no longer overfished, but full recovery is not expected for decades.

The Sandbar Shark is a generalist when it comes to feeding. Its diet includes bony fish, crustaceans, mollusks, and other invertebrates.

OCEANIC WHITETIP SHARK
(*Carcharhinus longimanus*)

The common name describes the prominent edges of the fins, which appear to have been dipped in white paint (fig. 2.29). The Oceanic Whitetip Shark is a stocky, large (to at least 11.5 ft, or 3.5 m), pelagic shark found in temperate and tropical oceans. It has a reputation of being aggressive and dangerous. This reputation is exaggerated, but the Oceanic Whitetip Shark is inquisitive; it will likely bump people it encounters and may bite, and thus caution is always advised when diving with this species. It is considered Vulnerable by the IUCN.

Oceanic Whitetip Sharks eat mainly bony fish (e.g., marlin, tuna, mahi mahi, mackerel), sea turtles, seabirds, squid, and crustaceans. Like other pelagic sharks, they will also opportunistically feed on whale carcasses.

GREAT HAMMERHEAD
(Sphyrna mokarran)

The Great Hammerhead is one of nine species of hammerheads. The first dorsal fin is enormous, and there are large pelvic fins as well as a huge upper caudal fin (fig. 2.30).

◀ FIGURE 2.28
Sandbar Shark
(*Carcharhinus plumbeus*).
Brandon B / Shutterstock.com

▼ FIGURE 2.29
Oceanic Whitetip Shark
(*Carcharhinus longimanus*),
a ridgeback species. Ken Kiefer

Great Hammerheads are found in all tropical and warm temperate seas in both inshore and pelagic environments, from the surface to 987 ft (300 m). They reach a length of 20 ft (6 m). They feed on fish, including rays. Their weird head, called a *cephalofoil*, may play roles in both maneuverability and stability. It may also be used for prey handling by pinning rays to the bottom before eating them.

The Great Hammerhead is considered Critically Endangered. Capture mortality is very high in this species; capture prohibitions are therefore ineffective in curbing mortality in a multispecies fishery. A recent study showed that 50% of hammerheads were dead after three hours on the hook.[12]

Great Hammerheads eat a variety of bony fish, as well as other sharks and marine invertebrates. They also prey on rays.

BLUNTNOSE SIXGILL
(Hexanchus griseus)

The Bluntnose Sixgill is a member of a primitive group of five species of cow sharks (fig. 2.31). These are widely distributed, big (to 16 ft, or 5 m), stout-bodied predators with surprisingly weakly calcified jaws. This may be explained by problems depositing calcium salts at the depths at which they spend much of their time. Their odd, cockscomb-shaped teeth are like miniature saw blades, and the additional flexibility of the weakly calcified jaws allows the jaws to bend as they encounter prey, which includes small and large bony fish and sharks, bringing more of the serrated teeth in contact with the flesh, which is then more easily sliced and removed: death by a thousand cuts!

Most live in deep water (1,000–3,300 ft, or 300–1,000 m), a section known as the oceanic *twilight*

OPPOSITE PAGE:

FIGURE 2.30 **Great Hammerhead** (*Sphyrna mokarran*) in Bimini, Bahamas. Note the tight turning radius, enabled in part by the use of the laterally expanded head as a rudder. Annie Guttridge

TIHS PAGE:

FIGURE 2.31 **Bluntnose Sixgill Shark** (*Hexanchus griseus*) in Exuma Sound, Bahamas, photographed from a submersible. Spear guns used to tag sharks from the submersible are shown. R. Dean Grubbs

zone, but they may be shallow in some locations. They have a single, spineless dorsal fin set far back along the body, which reduces friction and allows them to spin more easily when sawing chunks of flesh from prey. They also have retractable eyes; four muscles contract and basically suck the eyeball back into its socket for protection from struggling prey.

SMALLTOOTH SAWFISH
(Pristis pectinata)

The Smalltooth Sawfish is a ray, but recall that the rays are the closest relatives of the sharks, and both are cartilaginous (fig. 2.32). Rays split off from the shark lineage about 270 million years ago. The Smalltooth Sawfish looks like a shark, and in its habitat, where it eats bony fish and other sharks, it occupies a high trophic level, which is why we include it here. Rays can be distinguished from sharks in that the pectoral fins of the former are connected to the body above the gills, which are on the ray's underside.

Smalltooth Sawfish can reach lengths of more than 16 ft (5 m). The most recognizable characteristic is the saw, or rostrum, which is basically the snout projected as a stout, thin blade, with a series of pointed teeth on both sides. The ampullae of Lorenzini, organs sensitive to minute electrical currents, extend all way to the end of the rostrum. The saw is used to probe the bottom for benthic prey and to slash and disable schooling fish.

This critically endangered species is found in tropical and subtropical coastal areas on both sides of the North Atlantic, but it has been extirpated from several areas, especially in the eastern North Atlantic. It was placed on the US endangered species list in 2003. Continued threats include coastal development, dredging, mangrove removal, seawall construction, alteration of freshwater flow, habitat fragmentation, climate change, and especially commercial fishing as bycatch. Extensive efforts to reverse population declines in southwestern Florida are beginning to work.

FIGURE 2.32 (A) Smalltooth Sawfish (*Pristis pectinata*) at an aquarium. (B) Scientists implanting an acoustic tag through a small surgical opening in a Smalltooth Sawfish. This will enable them to track the animal to determine its preferred habitat and movements. (A), Nick Fox / Shutterstock; (B), R. Dean Grubbs

3 NONAVIAN REPTILES

You probably learned what reptiles are in elementary school: animals that live mostly on land, have scales, and, for turtles, possess shells. Well, things change. Powerful methods for understanding relationships of all organisms using the structure of their genetic material, specifically their mitochondrial and/or nuclear DNA or RNA, have revolutionized the fields of systematics and taxonomy. Earlier taxonomists (and your teachers) weren't wrong about reptiles, but there is a somewhat surprising complication. If you look at the family tree that includes both traditional reptiles and birds (fig. 3.1), you see that the classical interpretation of what is designated as a reptile doesn't make much sense.

Either crocodylians (alligators, crocodiles, caimans, and the Gharial, as opposed to the crocodilians, which are just the crocodiles), which resemble lizards but share a more recent common ancestor with birds than with other reptiles, should be considered "birds" too, or birds should be considered "reptiles." In fact, paleontologist Robert Bakker wrote in *The Dinosaur Heresies* in 1986, "When the Canada geese honk their way northward we can say: 'The dinosaurs are migrating, it must be spring!'" We like that interpretation a lot, but it hasn't caught on yet. One solution is to continue to call birds "birds," or even "avian reptiles." Nobody we know commonly calls a hummingbird or eagle an avian reptile, but if taxonomists adopted that nomenclature, the remaining reptiles could then be referred to as "nonavian reptiles." That is where our

chapter title comes from, but for convenience we will refer to these as just reptiles, as long as you, faithful reader, understand our basis for doing so.

You may have noticed we skipped from sharks to reptiles, ignoring amphibians, including the Cane Toad (*Rhinella marina*), a predator worthy of our attention, if not one that inspires night sweats. Amphibians descended from a Paleozoic group of semiaquatic ancestors. Most species still live part of their lives in the water and part on land (*amphi* = both, *bio* = life), and they all need to stay at least damp. If you value necks, lungs, distinct feet with toes, mobile tongues, inner ears, and a skeleton robust enough to stand up without the support of water, you owe gratitude to the amphibians. All adult amphibians are carnivores, too, so like the sharks in the previous chapter, they are definitely predators!

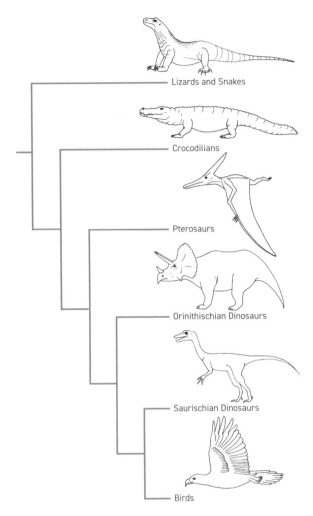

Lizards and Snakes

Crocodilians

Pterosaurs

Orinithischian Dinosaurs

Saurischian Dinosaurs

Birds

FIGURE 3.1 Reptile phylogeny. This evolutionary tree (also known as a cladogram) shows the relationships of reptiles, both avian (birds) and nonavian (classical).

The toxic Cane Toad we referenced above can weigh almost 4 lb (1.8 kg). After its ill-advised introduction to Australia in 1935, it wreaked havoc on other amphibians, reptiles (including birds), and even small mammals there. The Chinese and Japanese Giant Salamanders (*Andrias davidanus* and *japonicus*) can grow to nearly 6 ft (1.8 m) long. But these are the biggest of the amphibians and they are outliers. Most are smaller than 1 ft

(30 cm). Their teeth are tiny. They are voracious consumers of insects, but with few defenses, they are prey, not top predators. As much as we love frogs, toads, salamanders, and the weird, legless caecilians, we are leaving them out.

While some of the earliest tetrapods (*tetra* = four, *pod* = foot) evolved into amphibians, others took a more terrestrial path, which meant surmounting the biggest impediment to life on land: drying out. Overcoming desiccation was a major evolutionary advancement, equivalent to the development of jaws in vertebrates. This evolutionary leap onto land required the development of a waterproof egg, "waterproofing" in this case meaning preventing fluid from leaving rather than entering. Thus, the *amniotic* egg was hatched, so to speak, with a shell that resisted water loss, which allowed it to be deposited on land and not dehydrate. Terrestrial eggs can be bigger, producing larger offspring and therefore bigger adults. Some reptiles, like many sharks, have evolved the ability to retain the eggs inside them and give birth to live young, skipping external eggs altogether. Live birth required internal fertilization, which was also required for egg layers since they could not rely on water to transport sperm to eggs. Life on land also led to the evolution of thicker skin that contained water-repelling lipids (fats) and provided protection (from parasites, predators, etc.)

The colonization of land was accompanied by problems other than desiccation. Metabolism produces ammonia, a toxic waste product that cannot be stored in the body for long. For aquatic or amphibious animals, ammonia can diffuse out or be excreted as urine. On land it must exclusively be excreted, but in such a way that conserves one of a terrestrial animal's most valuable commodities, water. Reptiles accomplish this via a complicated physiological pathway that ends with the solid *uric acid*, which can be excreted without much water. If you or your windshield have ever been pooped on by a bird, you are familiar with this white substance. Mammals produce a water-soluble relative of uric acid, *urea*, and their kidneys excrete a concentrated urine. Either way, water is conserved.

Being bigger and more active demands faster exchange of oxygen and carbon dioxide to enable the metabolic machinery to operate, and so more efficient lungs developed. The crocodylians, some lizards, and the birds have a flow-through—that is, *unidirectional*—respiratory system, which is more efficient than the tidal flow in the lungs of mammals, including you, since a fresh supply of oxygen continually courses through the lungs.

The crocodylians are the most obvious top predators among the reptiles. In fact, their ancestors were the most diverse group of meso- to top predators in the Triassic period, 200–250 million years ago (mya). They are now the largest reptiles, with large gapes and impressive teeth. Some species are less than 6.5 ft (2 m) long, but a Saltwater Crocodile (*Crocodylus porosus*) might grow to more than 20 ft (6 m) long and weigh 2,200 lb (1,000 kg). Enormous, to be sure, but not as gargantuan as *Sarcosuchus imperator* (*sarco* = flesh, *suchus* = crocodile, *imperator* = ruler) (fig. 3.2) from the Cretaceous (145–66 mya). This beast, tipping the scales at 17,500 lb (8,000 kg) and measuring 30–40 ft (11–12 m) long, lurked in African rivers waiting to ambush, well, whatever it wanted.[1] This supercroc was the size of a bus. Luckily for us, *Homo sapiens* was not around yet. However, you have probably heard of modern crocodylians eating humans or their pets, and you may have seen videos of them going after Wildebeests and zebras. Definitely top predators.

Most lizards are carnivores, but because they are not very large, they are more likely to be prey than predators. The exceptions are the Gila Monster (*Heloderma suspectum*), Mexican Beaded Lizard (*Heloderma horridum*), the tegus, and the big monitor lizards, such as the Komodo Dragon (*Varanus komodoensis*). Monsters and dragons indeed! We will talk more about these later.

FIGURE 3.2 *Sarcosuchus imperator.* This prehistoric crocodylian was as long as a school bus and weighed nearly 4 tons. Shadowgate from Novara, Italy—Museum of Natural History, CC BY 2.0, via Wikimedia Commons

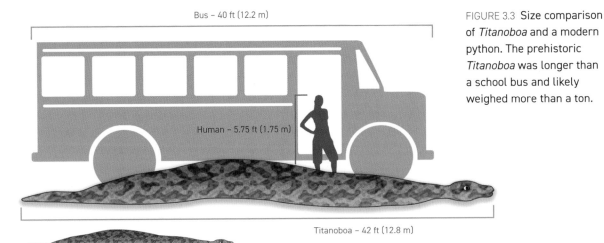

Bus – 40 ft (12.2 m)

Human – 5.75 ft (1.75 m)

Titanoboa – 42 ft (12.8 m)

Reticulated Python – 20 ft (6.1 m)

FIGURE 3.3 Size comparison of *Titanoboa* and a modern python. The prehistoric *Titanoboa* was longer than a school bus and likely weighed more than a ton.

The next reptiles that might inspire nightmares are the snakes. Snakes branched off from lizards about 128.5 mya.[2] Evolutionarily, snakes started with legs and lost them over time. For you, with legs, losing such useful appendages probably seems like a nonsensical choice, but as you will see, natural selection did not get this wrong. All snakes are predators, but few are considered top predators, although as with crocodylians there are giants in their ancestry. For example, the South American *Titanoboa* (fig. 3.3) grew to 43 ft (13 m) and weighed an estimated 2,500 lb (1,135 kg).[3] If you stood next to it, its back would reach your waist! Again, that was before the time of *Homo sapiens*. Today's closest counterpart to *Titanoboa* would be a nearly 30 ft (9 m) Green Anaconda or Reticulated Python. We will look at a couple of venomous snakes too.

Turtles typically do not inspire nightmares (except for *chelonaphobics*, people who fear turtles), but one group reaches the rarefied air of top predator, the snapping turtles. An encounter with a large snapping turtle is an unforgettable event. We will also take a brief look at those.

REPTILES AND US

Like the other beasts in this book, the crocodylians, lizards, snakes, and turtles have been and continue to be common parts of our culture. Sobek is an ancient crocodile-headed deity of ancient Egypt, representing power, protection, and fertility. Mummified crocodiles were found inside the ancient Egyptian temple of Kom Ombo. Author Roald Dahl's *Enormous Crocodile* wandered the jungle promising to eat children. If you are a fan of the University of Florida, you cheer on your Gators with an enthusiastic clapping of your arms like a snapping alligator. (Although lately you may also do this to portray a shark chomping in tune to the "Baby Shark" children's song, now the theme for the Washington Nationals baseball team in the United States—our apologies for bringing this tune to mind.)

In movies there is Ramon, from the 1980 movie *Alligator*, a mutant gator lurking in New York City sewers. The legend of gators in the sewers of New York has been around at least since a *New York Times* article describing an account of just that, published in 1935. The story really took off in 1963 with the publication of Thomas Pynchon's novel *V.*, which included a description of children all over the city buying alligators for 50 cents from Macy's, only to tire of them and flush them away, leaving alleged armies of albino alligators terrorizing sewer workers and apartment dwellers for years to come.[4]

Famous crocodylians from real life exist as well. Two-Toed Tom was a 14 ft (4 m) alligator on the border between Florida and Alabama in the 1920s. He lost all but two of his toes on one foot in a steel trap, so he was easily identified by his footprints. He evidently survived being shot and an attempt to explode him with dynamite. During this latter effort, he allegedly ate half his attacker's granddaughter, a sad ending to the story.

There is also Gustave, the Killer Crocodile of Burundi. This fellow, a Nile crocodile (*Crocodylus niloticus*), is about 20 ft (6m) long, could weigh more than a ton (900 kg), and is rumored to have killed as many as 300 people.[5] We suspect there were and are more of these legendary crocs in existence around the world because both they and people frequent rivers in Asia, Africa, and Australia and sometimes run into each other. Nile Crocodiles in Africa, the most common of the crocodylians, are estimated to kill over 300 people annually. Saltwater Crocodiles in both Australia and Malaysia round out the list of deadliest crocodylians.

Most famous lizards are actually dinosaurs, which are not lizards at all (although they are reptiles). Godzilla, for example. Not counting those, most lizards in culture are friendly pets (Ms. Frizzle, driver of *The Magic School Bus*, had one) or spokes-lizards (the GEICO insurance company gecko), and they are, well, not all that famous. Cheyenne people consider it bad luck to kill a lizard, and the Gila Monster is a powerful hero to the Navajo.[6]

The ancient Greeks viewed snakes as sacred, with the ability to shed skin a symbol of rebirth and renewal. You can be born in the Chinese Year of the Snake, which is okay, but most snakes in culture are bad guys. First, according to Judeo-Christian lore, there was that demon-snake in the Garden of Eden that ruined the whole thing by enticing Eve. Rudyard Kipling had villain Kaa in *The Jungle Book*, and Nag and Nagaina in *Rikki-Tikki-Tavi* (if you are familiar with this latter story, it is absolutely true that a mongoose is capable of killing a snake). Harry Potter's nemesis Voldemort had the giant python-viper hybrid Nagini (clearly J. K. Rowling is familiar with *Rikki-Tikki-Tavi*). Even in most Native American cultures, where animals are generally revered, the snake is associated mostly with violence, revenge, and bad luck.

Mara, a being of pure hatred in the television show *Doctor Who*, can manifest as a snake that requires fear from people to survive. This fictitious requirement succeeds in the program because many people are deathly afraid of snakes. You may recall the famous scene from the movie *Raiders of the Lost Ark*, in which our not quite fearless hero, Indiana Jones, drops his torch into a crypt, revealing a mass of squiggling serpents, and says in disgust, "Snakes. Why did it have to be snakes?" His sidekick then deadpans, "Asps. Very dangerous. You first." (The impressive piles of snakes in the pit were mostly legless lizards, which are harmless. Take a scientist with you to the movies!)

Because it is so common, fear of snakes, *ophidio-* or *ophiophobia*, has led to considerable research on whether this fear is innate or learned. The consensus now seems to be that it is both. For primates, the larger group of over 700 species and subspecies to which humans as well as monkeys and apes belong,[7] this fear is probably less innate and more learned. In evolution, primates became predisposed to notice snakes, so learning to fear them comes very easily for humans. A primate, including a human toddler, when presented with a picture of flowers with snakes hidden throughout, will immediately see the snakes.[8] This makes sense evolutionarily. If experience or a parent lets the little primate know that snakes can be trouble, and it notices them easily, then it is more likely to last long enough to reproduce compared to the primate who does not. And once you are afraid of snakes, as with all phobias it is a devilishly difficult fear to unlearn.

As for turtles, in many native creation stories it is a turtle that holds up the Earth. This is a common theme in author Terry Pratchett's *Discworld* fantasy novels, in which a disk-shaped Earth floats through space, supported on the back of four massive elephants, which in turn are supported by an enormous turtle, the Great A'Tuin. After a lecture on cosmology, nineteenth-century philosopher and psychologist William James was accosted by an elderly lady claiming she had a better theory: we live on a crust of earth that is on the back of a giant turtle. James asked what the turtle stood on. "It's turtles all the way down!"[9] Turtles are associated with long life and protection. In the Santeria religion, it is good to live

with a turtle because it absorbs negative energy. If you are a Buddhist, releasing a turtle to the wild confers good karma.[10]

These nonavian reptiles are common in our culture, have been around a long time, and are diverse. We have discussed the general adaptations they have to allow for life on land. Beyond that, the crocodylians, lizards, snakes, and turtles are quite different from one another. Now, let us consider each group separately and look at a few of the impressive predators among them.

WHAT IS A CROCODYLIAN?

Recognizing a crocodylian does not require an academic degree. They all have elongated bodies extending into a thick, laterally compressed tail, and four short but robust limbs. There is a large,

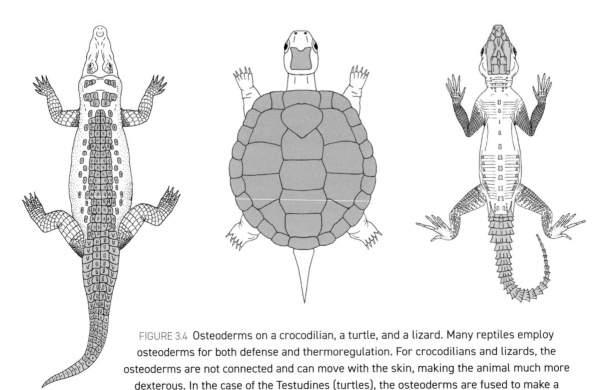

FIGURE 3.4 Osteoderms on a crocodilian, a turtle, and a lizard. Many reptiles employ osteoderms for both defense and thermoregulation. For crocodilians and lizards, the osteoderms are not connected and can move with the skin, making the animal much more dexterous. In the case of the Testudines (turtles), the osteoderms are fused to make a solid outer shell, trading mobility for being able to take your house with you.

heavy skull with a long snout and a jaw filled with teeth. They have scales, but underneath are thick, bony plates called *osteoderms* (bony skin) (fig. 3.4), armoring the body. Not much can take a bite out of an adult crocodylian (except Jaguars; see chapter 5).

There are about 27 species of crocodylians, divided among three groups: the Crocodylidae, Alligatoridae, and Gavialidae (fig. 3.5).[11] Most live in tropical climates, but the American and Chinese Alligators (*Alligator mississippiensis* and *sinensis*), American Crocodile (*Crocodylus acutus*), and Yacare Caiman (*Caiman yacare*) have ranges extending into temperate zones. As we noted above, there are some large ones, but forest-dwelling species less than 6 ft (2 m) long

inhabit Africa and South America. The alligators and caimans are all associated with fresh water and are limited to North and South America, except for the Chinese Alligator. Crocodile species inhabit both fresh and salt water, so they are more widely dispersed than alligators. Saltwater Crocodiles are found in the Indo-Pacific and from the Indo-Australian archipelago into Australia. The Nile Crocodile inhabits the Nile River basin and freshwater marshes and estuaries throughout sub-Saharan Africa and Madagascar. The Mugger Crocodile (*Crocodylus palustris*; the Hindi word for "crocodile" is *maggar mach*) prefers fresh water, including irrigation ditches and backyard ponds, throughout India and Pakistan. The American

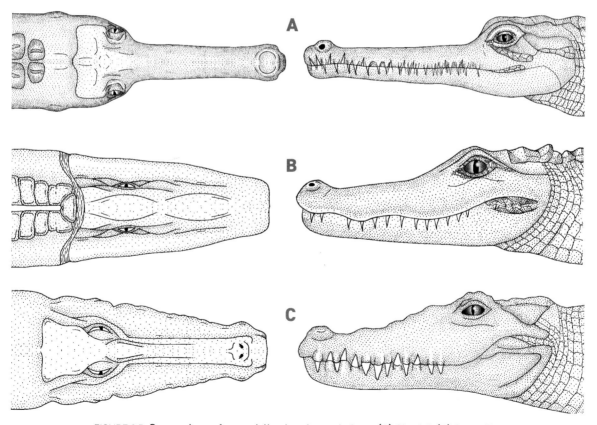

FIGURE 3.5 Comparison of crocodylian head morphology. (A) Gharial, (B) Crocodile, (C) Alligator. Note the difference in skull structures. The thin snout of the Gharial is more specialized for fish, which makes it more dependent on a single food source.

Crocodile is found along coasts from southern Florida through the Caribbean, and into northern South America.

The gavials include just two species: the Gharial (*Gavialis gangeticus*) and the False Gharial (*Tomistoma schlegelii*). The former is restricted to a small area of the Ganges River. The latter is found in the Malay Peninsula, Sumatra, Borneo, and Java. Both are characterized by their very slender snout in comparison with the other two groups (fig. 3.6).

BIOLOGY AND ECOLOGY

Crocodylians all spend most of their time in the water but, as ectotherms, come ashore to bask in the sun and heat up on cooler days. Alligators are the most tolerant of cold temperatures, and American Alligators can sometimes be seen in the winter hunkered down in the mud with their nose poking up through the ice. Chinese Alligators hibernate during winter. The most aquatic of

the crocodylians are the gavials, which are characterized by their front teeth, both top and bottom, protruding outside their closed mouth in their narrow, fragile-looking snout (fig. 3.6).

As we noted, gavials are easily identified by their weird snouts. You can also distinguish alligators and caimans from crocodiles by their snouts (fig. 3.5). Alligators and caimans have a rounded snout, whereas crocodiles all have a narrower snout with a big fourth tooth on the lower jaw visible on both sides when the mouth is closed. Distinguishing that visible fourth tooth requires a rather close look, though. Fortunately, at least for identification purposes, there are not many places in the world where you still find both varieties in nature.

Before you go looking too closely at a crocodile's teeth arrangement, consider that although they are proficient swimmers, with a smooth, lateral motion of the tail propelling them forward while the back feet steer, they are also quite good at moving on land. They can do a *belly crawl* and *belly run* where the legs remain splayed out to the side such that

FIGURE 3.6 False Gharials (*Tomistoma schlegelii*) in Thailand. The Tomistoma (or False Gharial) is a specialized fish hunter, and its snout is adapted for that. Note the protruding teeth of both jaws, characteristic of the gavials.

the belly skims along the ground. This looks a bit weird, but they can move remarkably fast over short distances in either belly mode, 7–9 mph (12–14 kph). For longer-distance jaunts, they deploy the *high walk*, where the legs straighten under the body and move front to back rather than extending sideways. A slower way of advancing, yes, but they can walk for miles this way. In all these gaits, the paired limbs move alternately, with right front and left back going forward together, which is standard tetrapod motion (walk across the room and see what your own limbs do). And that is the entire movement repertoire for alligators, caimans, and gavials. Crocodiles, however, can also *gallop*! The legs move beneath the body as in the walk, but rather than the legs moving alternately, the forelimbs and hind limbs move forward as a pair and the back flexes upward to increase stride length. They can achieve breakneck speeds, at least by the crocodylian standard, of 10 mph (17 kph) or more for a short distance.[12] Luckily for you, they sprint mostly to get back to the safety of the water, and not to chase down prey. (FYI, a fit human can run at between 10 and 15 mph, or 16–24 kph. Time to hit the gym?)

Crocodiles possess a trait unique to them among the crocodylians: salt glands. Other reptiles have salt glands, but not any other crocodylians. Since they live in salt water, a notoriously dehydrating environment, Saltwater Crocodiles or Salties, in particular, run the risk of desiccating or raising their internal salt levels beyond what is physiologically safe. They avoid this potentially fatal condition by having salt glands on their tongues that excrete excess salt with minimal loss of water. Sea turtles have salt glands near their eyes, and they appear to be crying when salt is secreted. You might be tempted to conclude that *crocodile tears*, which also allude to the manufactured tears of someone who isn't really sad, similarly result from ocular salt glands.

However, you'd be wrong! Crocodylians apparently cry when they eat, for reasons that are not understood, since they obviously are not sad to be eating, but nor are they excreting salt.

When you think of crocodylians, the word *frisky* does not come to mind, but they are known to play. An activity can be considered *play* if it requires energy but does not offer any obvious gains for the individual playing. Mammals play, of course, but the phenomenon is not commonly known in crocodylians except among their farmers and zookeepers. A recent summary of these observations categorizes crocodylian play into three types: locomotor play, object play, and social play. As ectotherms, these animals do not have energy to waste, so sustained motion is uncommon, but young American Alligators have been seen to repeatedly slide down mud banks into the water, and an Australian Estuarine Crocodile was observed to repeatedly surf ocean waves. Alligators and caimans snap at drips or streams of water over and over, as well as standing or resting under showers. Captive crocodylians of many species play with balls or other floating objects, seemingly preferring the pink ones. Wild crocodilians similarly push or carry around floating sticks, flowers, leaves, and feathers. They also sometimes roll and toss prey items around and do not always eat them. Social behavior is mostly related to courting in this group, but juveniles are sometimes observed chasing each other around and sometimes giving each other rides on their backs.[13]

During courting, social behavior consists of vocal displays, including hissing, coughing, bellowing, and purring, in addition to attention-grabbing tail and head slaps on the water. Male alligators bellow loudly enough to be heard by other alligators (both rivals and potential mates) over 500 ft (160 m) away. Females can bellow, too, but only males produce the very low-frequency vibrations (~10 Hz, below what a human can hear)

FIGURE 3.7 American Alligator (*Alligator mississippiensis*) bellowing. The territorial bellowing of American Alligators may include an important visual display for potential mates. An alligator will often position its back just below the surface of the water. The low-frequency rumbling produces rapid vibrations through the skin and osteoderms, resulting in water droplets springing upward (sometimes called *dancing water*). Carolyn Hutchins

that travel 0.5 mi (1 km) or more underwater and cause the water to dance at the surface (fig. 3.7). Most sounds are produced by the vocal cords, but these low frequencies are the result of contraction of body wall musculature underwater. These sounds convey information about the size of the male making them, allowing potential challengers to assess their chances before getting involved.[14] Author Rob regularly witnessed the territorial bellowing of a large alligator named George coinciding with the weekly approach of a garbage truck. It seems likely that the low-frequency rumbling of the large diesel engine was perceived as a rival alligator (or potential mate). The truck driver and the alligator would probably both be a little shocked if they ever met face to face.

Male Gharials, also sometimes called Fish-Eating Crocodiles, in southern Asia make sounds a bit differently. Gharials are characterized by a bulbous growth on the end of their snout called the *boss*. This protuberance resembles an earthenware pot called a *ghara* in Hindi, hence the name Gharial. It grows as the animal grows until it overlaps the nostrils, creating a hissing or buzzing sound with each breath. The boss gets used in territorial defense, and males with bigger bosses have the advantage in courting.

Courting crocodylians display a variety of behaviors. They sidle up to one another, raise their snouts, rub their chins on each other, release musky odors, swim along with each other, sometimes ride one another, and sometimes produce those subsonic vibrations. They are polygamous, meaning a female may mate with multiple males, and a male with multiple females. Within a species, the bigger animals are typically most dominant and do the most mating. A young

BOX 3.1 HOW DO YOU LIKE YOUR EGGS?

Both avian (bird) and most nonavian reptiles lay eggs. Nonavian reptile eggs resemble bird eggs, and some are rigid like those while others are leathery. They vary in size according to the size of the species laying them. In the crocodylians, many turtles, and a few squamates (snakes and lizards), the temperature of incubation determines the sex of the hatchlings, a phenomenon called *temperature-dependent sex determination.* For crocodylians, females are produced at low and high temperatures and males at intermediate temperatures. Crocodylian nests are large enough that there is probably sufficient temperature variation within a nest to produce both males and females.[15] Sea turtle sex is also determined by incubation temperature, but in that case, females are produced at warmer temperatures and males at cooler temperatures. Again, temperatures in one nest may vary enough to produce both.

One of the innumerable consequences of global climate change is a skewing of hatchling sex ratios. This has been observed in lizards and sea turtles and will also affect crocodylians. Given that sex ratios are the products of natural selection over long periods, climate change–induced alterations may have profound negative consequences.

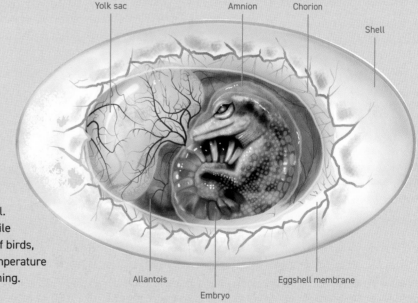

A crocodile embryo inside its shell. Although the anatomy of a crocodile egg very closely resembles that of birds, sex determination depends on temperature during development prior to hatching.

Yolk sac Amnion Chorion Shell Allantois Eggshell membrane Embryo

male American Alligator will not achieve sufficient size to compete and win a mate until he is 15–20 years old, but he may live more than 50 years, with larger crocodilians making it 70 years or longer.

Like many other reptiles, crocodylians are *oviparous*, meaning they lay eggs. They all build nests, some of which are mounds of soil and vegetation, and others are depressions in vegetation or are dug into stream banks. Nest building is a critical task because the location and structure of the nest determine the temperature of egg incubation, and that determines the sex of the hatchlings (see box 3.1). The female builds or digs the nest and deposits 12 to 50 or more eggs, with bigger females generally laying more eggs, regardless of species. She then guards and defends the nest, although sometimes the male assists. Eggs incubate for about two months. The babies start to vocalize in the eggs (*Let me out of here!* in crocodylian), and synchronization of the calls alerts the parents that it is time to hatch, stimulating one or both to excavate the nest, sometimes delicately breaking the eggs with their teeth to assist the hatchlings. One of the parents

then ferries the offspring in her or his mouth to the adjacent pool (fig. 3.8).

Juvenile crocodylians stay together in a group called a *crèche* for more than two years, and a single crèche may contain the offspring of multiple parents (fig. 3.9). Several parents and other adults may guard the whole group. Gharials are famed for their crèches, which contain up to 1,000 hatchlings, with groups of young males guarding the assemblage. There is constant communication

FIGURE 3.8 Mother crocodile carrying a baby in her mouth. Although typically perceived as killers (which of course they can be), crocodylians are also known to be gentle and protective parents.
Catchlight Lens / Shutterstock

BOX 3.2 ▸ CALLING ALL ALLIGATORS!

When Steve, the dive safety officer at the university where your authors work, was an undergraduate, he had the job of collecting errant golf balls from the ponds along the multitudinous golf courses of Myrtle Beach, South Carolina. The climate of South Carolina is warm enough to support healthy populations of alligators, and one of their favorite habitats is golf course ponds. The makings of a yet another superpredator horror movie? Nah, but if you need to dive into these murky waters to collect golf balls, it is good to know who is home. What to do? Steve took advantage of the communication skills of alligators, using a hidden talent few possess: he is an alligator whisperer. He can do a marvelous, and apparently convincing, imitation of a juvenile alligator calling in distress. It's a quiet, rather whimpering grunt, sort of like a newborn puppy makes, and will bring any adult alligator in the vicinity right over to see what the trouble is. If no bigger alligators respond, none are nearby, and it is safe to dive right in! You first.

FIGURE 3.9. Gharials (*Gavialis gangeticus*) grouped in a *crèche* along India's Chambal River. Safety in numbers is the name of the game for young crocodylians who might become a meal for other predators. kunaljain7/Shutterstock

between the juveniles and adults (see box 3.2), keeping everyone together and informed about threats and food sources.[16] This nesting behavior and parental care, rather like that in birds, served as an early clue that crocodylians and birds were closely related on the reptile family tree.

CROCODYLIANS AS PREDATORS

The crocodylians as a group are primarily generalist aquatic predators, willing to eat whatever is convenient, but the Gharial eats mostly fish. The narrow shape of its jaw facilitates a quick sideways grab for fish, which are easily snagged on the teeth, although the Indonesian species is more of an ambush predator of animals on the stream bank, like the other crocodylians. Most crocodylians will hunt day or night, although they are typically more active at night. African Dwarf Crocodiles (*Osteolaemus tetraspis*) are strictly nocturnal hunters.

One of the easiest ways to locate crocodylians is with a spotlight at night. Their eyes have a tapetum lucidum, like those of sharks, deer, raccoons, and others, enhancing night vision and conveniently glowing when you shine a light on them. But crocodylians do not depend on vision at night to find prey. Instead they may rely on small bulges called *integumentary sensory organs*, which cover the heads of alligators and caimans and the entire bodies of crocodiles and gavials. These are extremely sensitive to pressure. In the dark, an American Alligator can locate the point of impact of a mere drop of water (consider that before you relieve yourself off a stream bank or boat). They must use these to sense fish passing within grabbing distance.

Since they are also active during the day, crocodylians have decent visual acuity and may be able to see color. Studies of the eyes of American Alligators indicate the presence of both rods and cones in the retina, facilitating both night and some level of sharp color vision. Certainly, they can see each other's courtship displays, and swimming alligators are known to stalk animals walking onshore.[17]

Crocodylians are ambush predators, meaning they rely on stealth to get close to prey. How do such big animals hide? In plain sight, but mostly underwater, thanks to a secondary palate that lets them breathe with just their nostrils (and eyes) emergent, with a flap of tissue blocking water from entering their lungs. With a gentle swing of the tail they move forward, leaving hardly a ripple. Prey simply do not see them coming.

And when they are close, look out! With a bite force and tooth pressure correlated to body size, regardless of species, and higher than that of any other living animal (up to 8,983 N),[18] once the prey is gripped, it is pretty much *game over*! A crocodylian will drag the prey underwater to drown it and then bite off chunks. If the prey is large, like a juvenile Cape Buffalo,[19] the crocodile will bite into it and then execute what is commonly called a *death roll*, rotation around its own body axis, tearing loose the piece it has a hold on. It may also stuff the prey under submerged vegetation to hold it in place while eating, or in order give it sufficient time to decay a bit, making it easier to get pieces off.

Caimans and alligators have been observed hunting in groups, although not exactly cooperating. It is more likely that they take advantage of the likelihood that a fish fleeing one mouth will swim right into one adjacent.[20]

American Alligators have been observed using twigs to lure birds. Birds will often nest in trees above crocodylian territory, which seems like a risky proposition. However, not only are crocodylians incapable of climbing trees, but these toothy predators also tend to prevent other predators in the vicinity from doing so. The trade-off for the nesting birds is losing an occasional chick to a crocodile, but overall it seems to be a winning strategy for the birds. At the St. Augustine Alligator Farm Zoological Park in Florida, egrets nest in trees over the crocodylian habitat, and during nesting season they collect sticks to construct their nests. During this time, the cunning alligators will float for lengthy periods with a stick or two balanced across their snouts, moving carefully to avoid dislodging the stick. And when a distracted egret reaches to grab one of these sticks, *chomp*! It doesn't happen often, but it happens, and only during egret nest-building season.[21]

HOW ARE CROCODYLIANS DOING?

The IUCN Crocodile Specialist Group reports that the conservation status of most of the well-known crocodylians is of Least Concern. These include the American Alligator, all the caiman species, the Nile and Saltwater Crocodiles, and the Australian and New Guinea Freshwater Crocodiles (*Crocodylus johnsoni* and *novaeguineae*). Critically Endangered species are the Gharial; the Orinoco, Siamese, Philippine, Cuban (*Crocodylus intermedius, siamensis, mindorensis,* and *rhombifer*), and African Slender-snouted (*Mecistops cataphractus*) Crocodiles; and the Chinese Alligator. The Indonesian Tomistoma, or False Gharial, the American, Mugger, and Dwarf Crocodiles are listed as Vulnerable. Considering the dismal human history of maintaining predator populations, notwithstanding the species on the verge of extinction, as a group crocodylians are mostly not as depleted as they might be. Why?

It is not that humans haven't hunted them. By the early 1970s nearly all the crocodylians listed by the IUCN were in trouble. People were hunting them worldwide, mainly to meet demand for their skin. Science journalist David Quammen, in his book *Monster of God*, notes that three million American Alligators were killed in the state of Louisiana, USA, alone between 1880 and 1933, and beginning in 1950, seven million Caiman skins were shipped out of the state of Amazonas in Brazil in just 15 years. Generally, tanneries and customers were not located where the crocodylians were being harvested, so international trade was required. Ironically, this disconnect between the location of the raw material and the manufacturing and market sectors may have contributed to the stability of wild populations, because in 1975 the Convention on International Trade in Endangered Species of Wild Fauna and Flora (CITES) went into effect, and all that trading became illegal.

Crocodylians turn out to be quite resilient, too. Females of larger species like the big crocodiles, American Alligators, and caimans can produce 50 eggs a year. If sufficient food and habitat are present, populations tend to recover. For the seven critically endangered species, reintroduction plans for captive-raised specimens are under way.[22] Crocodylians can be raised in high densities in captivity (fig. 3.10), euphemistically called *farms.* While many people find these operations disagreeable at best and horrifying at worst, they take the pressure off wild populations by providing reliable sources to manufacturers. One widely used alternative to real skin is some version of plastic, with its own issues. Another alternative is, well, just don't wear pretend crocodylian togs.

Farms or zoological parks also introduce people to crocodylians, and they become tourist attractions. As alligators made a comeback in the southern United States in the late twentieth century, so did alligator hunting. US alligator populations are carefully managed, hunting licenses are limited and pricey ($200 in South Carolina for a nonresident), and the animals have become worth having around for hunting, as well as for ecotourism. As you will learn in

FIGURE 3.10 A pen in a crocodile farm. We currently use captive-bred crocodiles for everything from skins and meat to tourism and education. In some cases (as with the endangered Tomistomas in fig. 3.6), captive-bred specimens are being reintroduced into the wild.

subsequent chapters, this argument works less well with predatory mammals that are cuddlier than crocodiles and reproduce more slowly, but we almost killed off our American Alligators and now they are doing fine. We do not believe an animal's worth (or a plant's, for that matter) should be tied to its economic value, but sometimes that can work in its favor. For most crocodylians we have, to a large extent, undone the damage we did to them. Even though some of them still can and do eat us!

WHAT IS A LIZARD?

Lizards, snakes, and the single species Tuatara (living only in New Zealand) make up the *lepidosaurs* (*lepi* = scaled, *saur* = lizard), the most diverse group of nonavian reptiles. Within this group are the *squamates* (*squama* = scaled), the lizards and snakes. There are nearly 7,150 species of lizards and over 3,950 species of snakes. They are found on all continents except Antarctica. Both have scaly skin and most have four legs, except the snakes, of course, and a few lizards also lack limbs. As you saw in figure 3.1, the lepidosaurs are a sister lineage to the crocodylians and birds.

BIOLOGY AND ECOLOGY

Lizards and snakes exhibit *determinate growth*, meaning their size is genetically programmed such that they stop growing at some point. Crocodylians and turtles have *indeterminate growth* and thus will continue growing throughout their lives, although not at a constant rate. Determinate growth explains why most adult lizards now weigh less than a mouse. Very different from the situation prior to the asteroid hit that did in the dinosaurs 66 million years ago! Smallness was a key to surviving that event because of the dramatic loss of productivity—that is, food—that gave a competitive advantage to smaller organisms. Luckily for us, early mammals were small too. It took a while for bigness to make a comeback. And it never really did with lizards.

Male and female lizards and snakes are typically difficult to distinguish based on external characteristics. Male squamates have *hemipenes*, or a dual penis, which everts only when engorged. Only one side is used at a time, and it has been observed that a male will use his right hemipenis if the ovum ready for fertilization is on the right side of the female (fig. 3.11). How either sex might know the location of the most fertile ovum is a mystery. In temperate climates, mating occurs in the spring so that embryos develop during warm months. Where it is warm year-round, there may be no defined seasonality to reproduction.[23]

Most squamates are oviparous (lay eggs), although some are viviparous (give birth to live young). The sex of the offspring of most lizards and snakes is determined genetically rather than by temperature as in crocodylians. Few lizards or

FIGURE 3.11 Green Anoles (*Anolis carolinensis*) mating. Female lizards have two oviducts. It has been suggested that the male will mount the female from the side (lest her tail get in the way), corresponding with whichever ovary is currently producing eggs.

snakes exhibit parental care. Those that do tend to be species capable of defending the nest (see below).

Many smaller lizard species are insectivores, so they are predators, but hardly *top* predators. The biggest lizard species tend to be herbivores, or plant eaters. Iguanas, with which you are probably familiar, fall into this category. There are some big predatory lizards out there: the Heloderma (including the Gila Monster and Mexican Bearded Lizard), tegus, and monitor lizards. Gila Monsters grow to about 2 ft (60 cm) long, and Mexican Beaded Lizards, 3 ft (1 m). Tegus are generally about 3 ft (1 m) long; however, the Black and White Tegu (*Salvator merianae*) can reach 4 ft (1.3 m). The smallest monitor lizard is somewhat diminutive: the Pygmy Goanna (*Varanus caudolineatus*) is 9 in (2.3 cm). Komodo Dragons, however, reach 7.5 ft (2.3 m) long and can weigh as much as 175 lb (80 kg). We will consider each group individually.

HELODERMA: GILA MONSTER AND MEXICAN BEADED LIZARD

The Gila Monster (fig. 3.12) and Beaded Lizard both feature rows of rounded scales that look like beads (*helo* = head of a nail or stud, *derma* = skin). They are found along the west side of Central America and up into the southwestern United States. The Gila Monster is the state animal of Arizona. They are active during the day, with their behavior pattern dependent on temperature, as it is in all ectotherms. Recall that ectotherms like most sharks and nonavian reptiles control their body temperature more externally than via metabolic heat. But this does not mean they have no control over their body temperature. Like every animal, including you, lizards have an optimal body temperature. For a Gila Monster, that is 85°F (29°C), which seems low for a desert dweller. How

FIGURE 3.12 Gila Monster (*Heloderma suspectum*). This "monster" has the distinction of being named "the most venomous lizard in the world" by the *Guinness Book of World Records*. reptiles4all/Shutterstock

do Gilas maintain such a reasonable temperature in the face of summer daytime temperatures routinely exceeding 104°F (40°C)? The Gila Monsters in one study in the Sonoran Desert were active only from mid-April through September, and they were mostly nocturnal when temperatures were cooler. Otherwise, they were underground in burrows or rocky shelters.[24] Gila Monsters can eat a third of their body weight in one meal, and it has been suggested that at that rate they would need to eat only three or four times a year.

The Gila Monster and Mexican Beaded Lizard both have well-developed venom glands, but they do not have fangs. Instead, large, grooved, razor-sharp teeth in their lower jaws channel the venom into prey as chewing squeezes the venom sac, which is also in their lower jaw. They might hang on to their prey for over 10 minutes. Before you decide to pick one up, note that the venom is about as toxic as that of a Western Diamondback Rattlesnake (*Crotalus atrox*) or cobra, although they do not release as much. Also consider this: there is no antivenin (also called antivenom). As is often the case with venomous animals, these lizards are rather sluggish, so they typically ambush prey, or they dig up eggs. Their diet consists of eggs, small birds and mammals, other lizards and amphibians, and carrion.

The Gila Monster is considered Near Threatened but is decreasing, and it is at risk from climate change and habitat loss. There are no IUCN data for the Mexican Beaded Lizard; it falls into a surprisingly large category called Data Deficient. This does not mean that they are not threatened, but rather that too little is known about them to provide an accurate assessment.

FIGURE 3.13 **Tegu skull diagram showing the heterodont teeth. Tegus are omnivorous and have different specialized teeth for both catching prey and eating fruit.**

TEGUS

There are eight species of teiid lizards, commonly called tegus, native to Central and South America. They inhabit rain forests, savannas, and deserts. Tegus have forked tongues, which they flick in and out for chemoreception, essentially smelling the air for aromatic signs of food or mates. Tegus are generalist feeders, with *heterodont* teeth (fig. 3.13), meaning different teeth specialized for biting, stabbing, and grinding, like yours. This is highly unusual for a reptile, in which the opposite condition, *homodonty*, is the norm. For example, crocodilian teeth are all conical.

Tegus put those teeth to uses other than eating. Studies of the Black and White Tegus show that both males and females are willing to fight. Males fight over territory and females, females fight over nesting sites, and both fight over food and stand their ground against predators. In both sexes, the bigger the animal, the stronger the bite force, the more aggressive they are, and the less likely they are to run away. A big male might have a bite force up to 1,000 N, comparable to a dog or a few of the crocodilians, although it may first try a tail whip. However, these bigger animals are also relatively slow, suggesting a trade-off between fight and flight behaviors.[25] If a tegu is big enough to fight back, it is probably too big to be able to run fast.

The tegus have another mammal-like feature, also highly unusual in reptiles. We mentioned that lizards are ectotherms but can control their

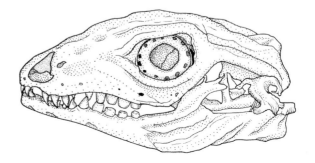

temperature behaviorally. Reptiles bask when they need to heat up and hide when they need to cool down. Tegus take this a step further and can actually lower heat loss and increase metabolism, as described below, to raise their body temperature about 18°F (10°C) above ambient, just like an endotherm, or "warm-blooded" animal. We note again that biology resists neat categorization. This is especially interesting because it offers a possible step along the evolutionary path toward the true endothermy of birds and mammals.

This thermoregulatory behavior has been observed in the Black and White Tegu. In spring and summer in Brazil, late September through March, tegus are active diurnally, retreating to burrows at night. They are dormant in autumn and winter, from early April to early September. In early September, testosterone levels in males rise, their biting muscles increase in mass, and they become more aggressive to defend territories and win mates. Females spend time building a nest and laying eggs from October to November. With a clutch of eggs representing up to 40% of the female's body mass, and subsequent aggressive behaviors related to nest defense, these physiological, anatomical, and behavioral modifications add significant outlays of energy for both sexes. And yet it is during this reproductive season that tegus maintain their body temperature overnight. It turns out that their morning metabolic rates during this season are five times higher than those of a dormant tegu the rest of the year. Heat engines like you find in mammals and birds! And then that heat is maintained by tegus' blood flow being directed away from extremities and toward their core. Your body does the same thing when you are in the cold, making your hands, feet, and nose feel even colder. That, plus the insulative properties of their underground burrows, keeps tegus warm. How is this helpful? For females, since they stay with their eggs until hatching, the additional heat provides the developing embryos extra warmth, speeding their maturation. For males, waking up warm and primed for activity enables them to skip basking, thus getting a jump on the competition. Natural selection chooses the adaptations that improve reproductive outcomes, and this tendency toward endothermy seems to do just that. And that may be the same reason endothermy evolved in birds and mammals.[26]

Finally, at least one species of "tegu" has another unusual trick up its sleeve. (Well, first, these aren't technically tegus because they are members of the Gymnophthalmidae family, a group close to the Teiidae. This is why scientists frown on common names.) As far as scientists can tell, there is no such thing as a male Müller's Tegu (*Leposoma percarinatum*). What? These tegus reproduce via *parthenogenesis*, where females produce eggs with two sets of chromosomes, essentially cloning themselves without needing a male. This type of asexual reproduction is rare but not unknown in vertebrates; it has been documented in a few species of lizards and snakes (as well as a few bony fish, sharks, and birds). From an evolutionary standpoint, it is surprising that more organisms do not go this route more often because, if you look at only the numbers, asexual reproduction is much more efficient than sexual. Each member of the population produces females like herself, so she gets 100% of her genes into the next generation, and all her offspring can produce more females. And there is no need for the energy expenditure involved in finding and mating with a male. It is a risky strategy, though. The big advantage of sexual reproduction, the evolutionary "name of the game," is that new combinations of genes are created every generation. This greatly enhances the adaptability of the species to a changing world, so much so that the vast majority of organisms take the trouble to have sex. But apparently not Müller's Tegu. To each his, er, her own.

All the tegu lizards listed by the IUCN are of Least Concern. As far as is known, their populations are stable. And in fact, in at least one case, there are too many! This stems from the popularity of tegus as pets. Much like the young of the Green Iguanas (*Iguana iguana*) that populate pet shops, the babies are very cute. Tegus even have the reputation of being affectionate with their owners, as a mammal might be. But like those iguanas, they grow, and grow, and grow. Most people lack the desire and ability to house a 3 or 4 ft (1–1.5 m) lizard, no matter how cuddly. And so, as happens with the iguanas, and the pythons we will get to later, people release these animals. Now we have tegus showing up in Florida and Georgia, where it is warm enough for them to survive the winters. In fact, at this point Florida has more established (reproducing) nonnative lizards than it does native lizards.[27] The Black and White Tegus found there can lay as many as 30 eggs per year and eat eggs of other vertebrates themselves, posing a risk to American Crocodiles, sea turtles, Gopher Tortoises, and ground-nesting birds. They also eat the young of all these. A mobile app, IveGot1, has been developed specifically to assist the public in reporting this invasive species and others in Florida, and it's not just there. In 2021, in an effort to avoid this very problem, your authors' home state of South Carolina banned the import and breeding of Black and White Tegus and now requires those here to be registered by their owners.

MONITOR LIZARDS

Finally, there are the largest of the lizards, the monitor lizards (fig. 3.14). There are 74 species of monitor lizards, all in the genus *Varanus*, ranging in size from the 9 in (2.3 cm) Pygmy Goanna to the 7.5 ft (2.3 m) Komodo Dragon. Monitors are found

FIGURE 3.14 African Rock Monitor (*Varanus albigularis*). Monitors have fairly stocky bodies and strong, powerful jaws. With these and a mouth full of large, needlelike teeth, they demonstrate their place in nature as a serious predator.

in Africa, across southern Asia, and through the Indo-Australian archipelago to Australia.

Monitor lizards have a characteristic body shape regardless of size, and most are active predators with strong jaws and sharp, recurved teeth. Only the Panay and Gray's Monitors (*Varanus mabitang* and *olivaceus*) eat fruit. These species have a forked tongue for chemoreception like tegus. The active lifestyle of these large lizards demands higher oxygen uptake than other lizards, so they have larger, more structurally complex lungs and use throat muscles to pump air into the lungs.

Like tegus, monitors are mammal-like, with high metabolic rates and a reputation for curiosity and intelligence. At the Dallas Zoo, Black-Throated Monitors (*Varanus albigularis ionidesi*) were studied to explore their interest in enrichment activities of the sort commonly provided only to mammals in zoos. They were given a tube with prey that they could see and smell but could access only through a hinged door. They were interested in this and quickly figured it out, getting more efficient each time it was presented.[28]

All monitor lizards lay eggs and guard their nests. In the case of the Lace Monitor (*Varanus varius*) in Australia, the large females (5–6 ft, or up to 2 m) dig into termite mounds to lay their eggs, and when the termites repair the mound, those hard walls protect the eggs and help maintain temperature and humidity. Typically, the females then return to help the offspring out of the mound when they hatch. Other than this there is no parental care. Like Müller's Tegu, Komodo Dragons can exhibit parthenogenesis, but this seems to occur only in captivity, as far as is known.

Since the Komodo Dragon is the biggest of the monitors, let us look at predation in the species. They do not have nearly the bite force of a crocodylian of comparable size (39 N versus 252 N, considerably less than tegus), but their skulls are designed to be able to resist prey twisting and pulling away, and they themselves are able to twist and pull. In one study, their technique of grabbing prey was described as comparable to that of some sharks.[29] This "grip and rip" causes nasty wounds that bleed profusely, and the Komodo augments this by introducing venom with anticoagulant qualities. The composition of the venom suggests it also induces hypotension (low blood pressure) and muscle cramping, leading to shock. This minimizes resistance from even large prey animals. Note that a Pleistocene (around two million years ago) ancestor of the Komodo, *Varanus priscus*, was essentially a giant Komodo and the largest venomous animal to have ever lived, as far as we know. It was more than 1,200 lb (575 kg) and 18 ft (5.5 m) long.[30] A dragon indeed!

Komodo Dragons live on five islands in eastern Indonesia, four of which are included in Komodo National Park, a World Heritage Site (fig. 3.15). They are generalist feeders, willing to scavenge or eat whatever they happen upon or can catch. They will go after other lizards, deer, water buffalo, and livestock. It is thought that Komodo Dragons probably preyed on Pygmy Elephants in Indonesia before those were driven extinct by humans.[31] They will even eat each other and are immune to each other's venom. They are quite efficient eaters, gobbling up 88% of a prey item, including bones, hooves, and skin, and they can consume 80% of their body weight in one meal.

HOW ARE LIZARDS DOING?

Monitor lizards are routinely collected, farmed, and traded for food, skin, and as pets, both legally and illegally. The IUCN does not list all monitor lizards, but those listed are mostly of Least Concern, except the Komodo Dragon and the Panay Monitor, which is endemic to just Panay Island in the Philippines; both are considered Endangered (the Komodo Dragon was designated so in 2021).

FIGURE 3.15 Author Rob photographing a wild Komodo Dragon (*Varanus komodoensis*) on a beach on Indonesia's Komodo Island.

The Komodo Dragon is protected mostly by living in a national park isolated by water. The Indonesian government was seeking to have more than 21.5 million tourists come to see the lizards between 2015 and 2019, up from about 170,000 annually. This type of ecotourism, if carefully managed, can provide jobs for locals and protect wild animals. However, supplemental feedings that attract the Dragons for people to see are causing the Dragons to lose their fear of people. In addition, their bodies are changing because they are less obliged to hunt, at least in the subpopulations where they are clustered by tourism. The largest animals are being selected for because those are the ones people want to see, and paradoxically, these animals have developed inferior body condition compared to that of more wild populations. This may be fine for now, certainly better than species loss from hunting or habitat loss, but it is also making certain Dragons dependent on people.[32] Do we want this World Heritage Site to essentially become a zoo? Conservation is always complicated. Tourism means money, which means at least some level of protection for Dragons, but how much is human activity affecting the species? And what are the options? Further study and careful management are warranted.

WHAT IS A SNAKE?

As you saw on the family tree of the nonavian reptiles, snakes are derived from lizards. There are legless lizards, but they are not snakes. Aside from lack of limbs, which we will consider later, snakes have other unique features. First, a snake can have 120 to 400 precloacal vertebrae (the cloaca is the posterior opening for the reproductive, urinary, and

FIGURE 3.16 African Rock Python (*Python sebae*) swallowing a large Impala. With a large meal, some snakes may need to eat only once or twice per year. (See box 4.1 for a related story.) Gulliver20/Shutterstock

digestive tracts), and that does not include the tail, which can have an additional 205! You have only 33 vertebrae total. Tails in snakes do not regenerate if they break, as happens in some lizards. There are also many differences in the skulls of snakes and lizards, most having to do with the enhanced skull kinesis (essentially mobility) in snakes, which endows them with the ability to expand the head to accommodate food of larger diameter than the snake itself. While lizard skulls generally *crush* prey, snake skulls *engulf* it (fig. 3.16). The skull bones of snakes are loosely connected, so much so that the left side of the lower jaw can outright separate from the right side, in some species up to 20 times the resting distance. Together with their flexible skin, this allows for huge mouthfuls. Their recurved (inward-facing) teeth hold on to the prey, reaching an extreme in the jaws of the largest constrictors. The snake swallows its prey headfirst, perhaps to tuck the limbs out of

the way, and with these recurved teeth, alternates sides and sort of walks its head down over the prey. Once the prey reaches the throat, muscles there push it along into the stomach.[33]

There are now about 3,950 species of snakes, ranging from insect eaters 4 in (10 cm) long to constrictors over 30 ft (10 m) long. Most are harmless to humans and are secretive; however, all are predators. The biggest snakes and the larger venomous snakes eat bigger animals and/or largely avoid being eaten themselves, so we will consider these top predators and take a closer look at some.

BIOLOGY AND ECOLOGY

Most obviously, snakes lack legs. The fossil record on snakes is sparse because their bones are so delicate that they do not tend to endure the passage of time. But their ancestors possessed legs. Some of the more primitive extant snake species, boas and pythons included, have what are known as *cloacal spurs*. These are small keratin protrusions on either side of the cloaca. The cloacal spurs are connected to vestigial leg bones (inside the snake's body) and are the only external evidence that snake ancestors once had legs. So, what happened to them? What we know has come mostly from molecular studies. It turns out that changes in a regulatory factor in the famous limb-enhancing gene known as Sonic Hedgehog or Shh are responsible. (Sonic Hedgehog? "Hedgehog" genes, first identified in fruit flies, produce a protein that is spiky looking, like a hedgehog. As multiple versions of the gene were discovered, more identifiers were needed, and the video game Sonic Hedgehog was popular in 1993 when this version was found. Who says scientists are no fun?[34]) In most vertebrates, Shh is highly conserved—that is, it has not changed much over evolutionary time. If you transplant a fish or a human Shh gene into a mouse embryo, the mouse develops normal limbs. (Think about that for a minute. The mouse embryo is completely okay getting instructions from a fish gene to set up its legs. Wow!) If you put the snake version of Shh into the mouse, though, the mouse develops severely reduced limbs.[35] So, it is changes in this gene that result in snakes not having legs. That's the *how*. But *why*, since legs would seem so useful?

It turns out that if you are a burrower, legs are not useful unless you are digging the burrow. If you are pushing your way through soil or even dense grass, it is easier to do so without limbs in the way, and under those constraints, the legless body plan was a winner (see box 3.3). Evolutionarily, snakes originated from a group of burrowing lizards, losing limbs and gaining length and flexibility.[36]

Snakes hunt for signs of prey, as well as mates, using their chemical sense. That flicking forked tongue passes molecules of aromatic chemicals back over their *vomeronasal organ*, as in the monitor lizards, allowing the brain to identify the presence of both prey and other snakes. In some snake species, the tongue is longer and more deeply forked in males than in females, perhaps reflecting its role in finding females. Males can follow the pheromone trails females leave behind, and they can distinguish their own species from other species, which may be the key mechanism to maintain reproductive isolation of one species from another where their home ranges overlap. The male snake in most species rubs his chin over the female to initiate courtship, eventually using his tail to locate her cloaca. Snakes have the same hemipenes as lizards have and fertilization is internal, even in aquatic snakes.

Most snakes, again like most lizards, are egg layers. Snakes generally lay eggs in cavities on the ground or in trees, or buried in debris. Parental care is rare, with some exceptions we will discuss. Sex of the offspring is determined genetically, not by temperature.

Let's look at a few examples of top predators among the snakes.

VIPERS: GABOON VIPER (*Bitis gabonica*) and EASTERN DIAMONDBACK RATTLESNAKE (*Crotalus adamanteus*)

Ground-dwelling vipers like these two species tend to be heavy bodied. They are ambush predators with venom, so they do not need to be fast beyond striking distance. If they find a productive place to wait for prey, they will stay in the same area for weeks or months at a time.

Gaboon Vipers (fig. 3.17) are *true vipers*, the largest members of what may go down in history as the most appropriately named genus among animals: *Bitis*. They are found in rain forests and wet areas of Central, East, and West Africa and can

BOX 3.3 IT'S A SNAKE! GET MOVING!

Snakes can move in one of four different ways, sometimes more, depending on their size and habitat. They all can employ *lateral undulation* (also known as *serpentine locomotion*), where they move in a series of horizontal S curves, with scales on their bellies and uneven ground providing traction. A specialized version of this mode of movement is found in the *sidewinders* (unsurprisingly known as *sidewinding*). This group is found in deserts or other environments with loose, often hot, sandy soil and not much to grab on to. Their undulation is also an S-curve motion, but the curves are nearly perpendicular to the direction of motion, and with only two or three points of the body in contact with the ground at a time, these snakes sort of throw themselves forward. In a narrow passage, *concertina* locomotion works. Here the snake alternately anchors a back portion of itself against the wall of the passage and pushes the front forward, then anchors the front and pulls the back forward. In *rectilinear* locomotion the snake's body forms vertical rather than horizontal S curves. By lifting alternate sections of the body off the ground, these snakes move forward in waves of contractions from front to back, propelling the snake forward. This is typical of big, heavy snakes. It is slow and subtle and so is good for stalking prey.[37]

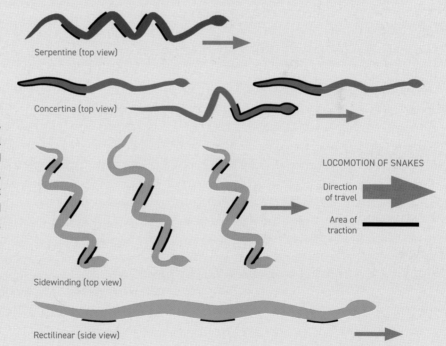

Styles of snake locomotion. Although snakes lack external limbs, they have developed effective means of locomotion, each method often dependent on their morphology and preferred environment.

Serpentine (top view)

Concertina (top view)

LOCOMOTION OF SNAKES

Direction of travel

Area of traction

Sidewinding (top view)

Rectilinear (side view)

get to 22 lb (10 kg) and 6 ft (1.8 m) long. Up to 20% of this weight can be fecal matter maintained in the gut, since they defecate only a few times a year (contemplate how uncomfortable that might be!).

The stored feces are thought to serve as ballast, anchoring the strike motion. The biggest Gaboon Vipers have heads up to 6 in (15 cm) across that look strikingly like a leaf lying on the forest floor.

FIGURE 3.17 Gaboon Viper (*Bitis gabonica*). While potentially deadly, Gaboon Vipers are typically not aggressive. Most venomous snakes inject venom only while hunting or for self-defense, thus saving valuable energy.

A pair of venom glands takes up the space on either side of that enormous head. The nearly 2 in (5 cm) fangs of the Gaboon Viper are the longest known. In all vipers, those fangs attach to the upper jawbones and fold back against the roof of the mouth until they deploy during a strike. Gaboon Vipers release a volume of venom second only to that of King Cobras (*Ophiophagus hannah*). They can control how much venom is released, depending on the size of the prey. Gaboon Vipers eat small to medium-sized mammals and birds. Unlike other snakes, they don't bite and let go. Instead they hang on until the prey is dead, making slow chewing motions while they wait. This injects more venom, which includes enzymes that assist in breaking down the body of the food item.

Venom requires energy to produce, so venomous animals will not waste it. Gaboon Vipers are not aggressive (whew!). They are so well visually camouflaged that you might step right over one while it just sits still and lets you walk on. The Puff Adder (*Bitis arietans*), a close relative, has even been found to be scentless, giving it camouflage

against predators that would find it by smell.[38] If you see one of these African vipers and give it space, it may rise up a bit and warn you with a hiss, or just go on its way. Only if the snake is on the move and you persist in bothering it (what is wrong with you?!) will it strike out. Or if you are food, which a person is not. Again, venom requires energy to produce, so it makes biological sense for the snake to use it only to subdue food or in a life-or-death situation.

Gaboon Vipers live about 20 years and can have 50–60 offspring annually. Perhaps because of this high reproductive capacity, they are of Least Concern according to the IUCN. They are protected by CITES because of interest in their skins, their use in traditional medicine, and, oddly, demand in the pet trade.

Eastern Diamondback Rattlesnakes (fig. 3.18) are *pit vipers*, so named because they have a heat-sensitive pit on the face that helps them identify their typically warm-blooded prey. They are the largest venomous snake native to North America, typically 3–6 ft (0.8–1.8 m) long, but can get up to

8 ft (2.4 m). They are endemic to the southeastern United States, preferring coastal regions and pine forests. Good swimmers, they are often found on barrier islands near the mainland.

Rattlesnakes are easily identified by their rattles (whose sound makes a dandy, if fear-inducing, cell phone ringtone). If any of the 36 species of rattlesnake conclude that you are not what they would prefer to waste their venom on and/ or they would like to scare you away before you try to catch them, they coil their bodies (this is a defensive posture but allows them to strike if necessary) and rapidly shake their tails, creating a rattling, buzzing sound not unlike that of a hive of very angry bees. Most people (and animals) instinctively know that this is a warning and back away. This is such an effective deterrent that many nonvenomous snakes will shake their tails like a rattlesnake even though they lack rattles. Doing this in a forest littered with dry leaves results in an eerily similar warning buzz. A rattlesnake's rattle is made of a section of external scales at the end of the tail, left attached each time the animal sheds its skin. Therefore, a longer rattle indicates an older snake, but they are not annual segments and

they sometimes break off, so you cannot precisely age a snake based on this. And if rattling does not rattle you enough, rattlesnakes also hiss.

If you hear a rattlesnake, back off! These snakes, even juveniles, can inject a very potent dose of venom. The good news is that a coiled snake can strike a distance only two-thirds its body length. This means that a 6 ft (2 m) rattlesnake cannot lunge at you from across a room. Envenomation can kill a human, but since many rattlesnake species are found in well-populated regions, there is usually antivenin available, and thus fewer fatalities than might be expected. The entire strike sequence is over in a flash, about 200 milliseconds from start to finish. Unlike Gaboon Vipers, rattlesnakes release once they bite, allowing the prey to go off and die before retrieving it. This strategy saves energy and minimizes risk to themselves. Mostly they eat small mammals and lizards.

Male snakes compete for females by rearing up, entwining, and trying to push each other over. Females are live bearers and will watch over the 6–21 neonates until they disperse. Eastern Diamondbacks are of Least Concern according to the IUCN.

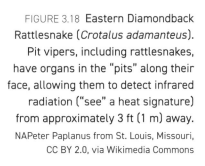

FIGURE 3.18 **Eastern Diamondback Rattlesnake (*Crotalus adamanteus*).** Pit vipers, including rattlesnakes, have organs in the "pits" along their face, allowing them to detect infrared radiation ("see" a heat signature) from approximately 3 ft (1 m) away.

CONSTRICTORS: BOAS AND PYTHONS

Green Anacondas (*Eunectes murinus*), also known as Common Anacondas, are in the Boidae family, commonly known as boas. The Green Anaconda (fig. 3.19) is the heaviest living snake (but not the longest; see below), weighing as much as 200 lb (91 kg) in the wild but up to 1,100 lb (500 kg) in captivity. They can grow to over 17 ft (5 m) long, and a big female (females are larger) might be 1 ft (0.3 m) in diameter. Green Anacondas are found in Trinidad, in tropical South America east of the Andes, and as far south as Paraguay. They are aquatic and wait for prey while submerged, relying on surprise. They can remain underwater for 10 minutes, and their nostrils and eyes are on top of their heads to facilitate seeing and breathing while swimming. They are known to eat fish, amphibians, turtles, caimans, birds, and large mammals like Pacas, Capybaras, deer, sheep, dogs, and sometimes a Jaguar. Adult Green Anacondas are large enough to eat a human, but there are no credible reports of this happening.

Females are live bearers and may produce a clutch of up to 50 offspring that are 2–3 ft (up to 1 m) long. The IUCN has not evaluated these snakes.

The Green Anaconda may be the heavyweight champion of the snake world, but the Reticulated Python (*Malayopython reticulatus*) (fig. 3.20) is its Robert Wadlow (the tallest person in the world)—that is, the longest of the living snakes, reaching nearly 30 ft (9 m). Second place goes to the Burmese Python (*Python bivittatus*) of South Asia and the African Rock Python (*Python sebae*) from Africa, which reaches 26 ft (8 m). This family of snakes was once classified with the boas but

FIGURE 3.19 Green Anaconda (*Eunectes murinus*) in water. As with an iceberg, what you see above the water does not fully reflect the danger that lurks below. Vladimir Wrangel / Shutterstock

FIGURE 3.20 **Various color morphs of captive-raised Reticulated Pythons (*Malayopython reticulatus*). Reticulated Pythons hold the current record for the longest snakes in the world. For size reference, look at animal keeper Jay Prehistoric in the background of this photo.**
Jay Brewer, The Reptile Zoo, Fountain Valley, CA

is now considered to be a separate family. They share the heat-sensitive sensors and pelvic spurs of the boas but are not closely related based on genetic comparisons. They are big enough to eat medium-sized antelope and humans. Between feeding events, their digestive activity and the mass of their digestive organs and heart decrease considerably. Once they eat, there is a 40-fold increase in metabolism and tissue regrowth.

Pythons lay the largest eggs of all snakes, up to 10.5 oz (300 g), and many construct nests. Female pythons coil around the eggs and, in some species, produce heat with muscle movements. Or they may leave the nest to bask in the sun, returning to transfer their own body heat to the eggs. They may also leave the nest to cool off in water and then return to cool the eggs. These behaviors regulate incubation temperature, reduce water loss, and deter predators.

Pythons and boas are constrictors. Contrary to common belief, they do not crush their prey. Precisely put, they constrict their prey until it can no longer breathe, although the Green Anaconda may also drown its prey. In constriction, the snake wraps itself in loops around the prey, and each time the prey exhales, the snake tightens the loops. This restricts oxygen intake, blood flow, and overall cardiac function, resulting in rapid death.

HOW ARE SNAKES DOING?

A farm may represent corn, wheat, or soybeans to a farmer, but to some snakes it represents a rodent buffet. In other words, snakes benefit from farming since it concentrates populations of small prey like rodents. For example, in Florida where vast areas of native habitat have been converted to sugarcane fields, Florida King Snake (*Lampropeltis getula floridana*) populations have increased.

But stories like this are almost anecdotal in scale, since our relationship with snakes is skewed toward considering them mortal enemies to be eliminated rather than revered. Snakes eat chicken eggs and chicks, so KILL THE SNAKE! Snakes scare us by their stealth and the threat we believe they pose, so KILL THEM! I recognize the importance of snakes to a balanced ecosystem, but there is one in my backyard big enough to eat my malti-shihtzu-rottweil-a-poo, so I'm going to KILL IT or hire a snake exterminator to GET RID OF IT! Venomous snakes in the United States kill five people per year on average, far fewer than your household toaster.[39] Around the world, 80% of people bitten never require further treatment. More than half of all snakebites result from people trying to dispose of a snake rather than leaving it alone.

Humans have such a conditioned fear of snakes that even nonvenomous snakes have the potential to be lethal. In 2014 a wild nonvenomous snake fell from a tree and bit an eight-year-old boy at Disney's Animal Kingdom. The boy's grandmother saw the incident, went into cardiac arrest, and died two days later, while the boy who was bitten required nothing more than a small adhesive bandage. Heart attacks and fear of snakes pose a far greater threat to humans than do snakes themselves. As is often the case, we are our own worst enemy.

We also destroy snake habitats. Snakes frequent forests in general and tropical forests in particular, and humans seem hell-bent on deforestation. According to the IUCN, more than half of all tropical forests have been destroyed in just the past 60 years.[40] Roads isolate and fragment habitat so snakes cannot get to each other, reducing mating opportunities and genetic diversity, the sine qua non for survival as a species. And we have seen drivers who deliberately veer to crush snakes that often take advantage of the overnight warmth of asphalt. (Author Rob regularly removes snakes from busy roadways, including a 6 ft [2 m] rattlesnake on southern Florida's Highway 41. Lacking a stick or snake hook, he got out of the car in busy traffic, wearing sandals, and improvised with a rolling suitcase. You should probably not try this, but at least try to avoid running snakes over.) Of the 3,900 species of snakes, the IUCN lists 118 as Critically Endangered, but the reality is we do not know much about most snake populations, so odds are good that more species are in trouble. Although some are causing trouble (see box 3.4).

FINALLY, A LITTLE BIT ABOUT TURTLES

Turtle is a very general term used for members of the order Testudines. The Testudines includes all extant turtles: marine sea turtles with paddle-like flippers; aquatic turtles with webbed feet with claws, like sliders and terrapins that live in brackish and freshwater lakes and rivers; and tortoises with stubby elephant-like feet and a heavily domed shell, which are almost exclusively terrestrial (fig. 3.21). Most people just lump all these together as "turtles," and so shall we.

The biggest turtles are sea turtles, the larger tortoises, and snapping turtles. Tortoises are mostly herbivores. Sea turtles are also herbivores or eat shellfish or jellyfish. Those two groups do not qualify as top predators. Snapping turtles

BOX 3.4 INVASION OF THE ANIMAL SNATCHERS

While some snake species are in peril, others are causing peril. Invasive species of anything, including snakes, can wreak havoc on ecosystems unaccustomed to them. In southern Florida, Burmese Pythons were introduced (likely the result of escaped or released pets) and are now well established. Each female can produce up to 100 eggs per year. This is a massive constrictor, regularly reaching 16 ft (5 m) in length, and is known to eat deer, alligators, and Bobcats in Everglades National Park: an introduced apex predator! These animals have devastated populations of medium- and large-sized mammals in southern Florida. Studies in Everglades National Park now find few Bobcats, Coyotes, White-tailed Deer, Common Raccoons, Opossums, and rabbits. All that is left is small rodents. (We should note here that the snakes may not directly eat many other predators, but they are very much eating up all their prey.) They also eat wading birds, putting grebes, herons, limpkins, rails, and storks at risk.[41]

Unfortunate, but not a threat to humans, right? Guess again. Rodents often serve as reservoirs of zoonotic pathogens (pathogens from animals that have the potential to infect humans). If a mosquito has a variety of mammals from which to suck blood, its odds of sucking on an infected reservoir animal and then sucking on a human are low. But if the only animals it can suck on are prolific infected rodents, those odds go up. One study found that in the Everglades from 1979 to 2016 mosquito meals from Hispid Cotton Rats increased 422% because of precipitous declines in populations of deer, Common Raccoons, and Opossums resulting from python predation.[42] The COVID-19 global pandemic may well have begun with a zoonotic pathogen, so this is something to consider. Humans need intact ecosystems just as much as the animals that inhabit them.

FIGURE 3.21 Turtle, terrapin, and tortoise. Although all are commonly called "turtles," very noticeable differences exist between different members of the Testudines.

warrant a mention, though. Common Snapping Turtles (*Chelydra serpentina*) are typically 10–35 lb (4.5–16 kg) but can reach 75 lb (34 kg) or more in captivity. They are found in fresh water from southern Canada to Texas east of the Mississippi River. They are omnivores, eating whatever they can catch, including small mammals, fish, other reptiles, and birds. They are sometimes referred to as freshwater loggerheads because of their large heads, but they are dwarfed by their close relative the Alligator Snapping Turtle (*Macrochelys temminckii*).

FIGURE 3.22 Alligator Snapping Turtle (*Macrochelys temminckii*). A snapping turtle may rest on the bottom of a body of water for up to an hour without moving, waiting for a fish to swim by.

These larger "snappers" have a narrower range, from western Georgia to eastern Texas and up to Iowa. Alligator Snapping Turtles (fig. 3.22) can weigh more than 200 lb (91 kg) and live for 70 years. They have a huge head and hooked beak and feature prominent dorsal keels on their carapace, making them look quite prehistoric. They have a small pink appendage in their mouth that wiggles enticingly to attract fish to their doom. They are reported to be able to snap a broomstick in half, and while we did not find evidence of that, we did find one recent report of a 15-year-old getting his index finger removed by an Alligator Snapping Turtle's "impressively precise transverse amputation," according to the doctor.[43] Both varieties of snapping turtles are omnivores, eating whatever they can catch.

A Peterson Field Guide describes snappers as "large freshwater turtles with long tails and short tempers."[44] They have a cantankerous reputation and would prefer to avoid you but will bite if they cannot escape. Even as you guide or carry a small one across a road so it won't get hit by a car, the unappreciative little beast will squirm and twist its rather long neck in an effort to bite you. Take care.

Common Snapping Turtles are of Least Concern and Alligator Snapping Turtles are Vulnerable according to the IUCN. They are hunted for food. Author Sharon can vouch for this, as her parents used to enthusiastically eat what looked like rich, thick mud in a bowl with a bit of sherry on top: canned snapper soup from Bookbinder's restaurant in Philadelphia, Pennsylvania. Sharon never tried it.

Unfortunately, all turtles are not doing as well as snapping turtles. Many are still eaten and are illegally traded and sold for food, medicine, or the pet trade, particularly in Asian markets. The rarer they are, the more in demand they are for showy pets or fancy dinners. If you are interested in the status of turtles today, you should read *Dreaming in Turtle* by Peter Lauer.[45]

4 THE RAPTOR CHAPTER

Technically speaking, of course, most birds are birds of prey; that is, they eat animals (worms, insects, and spiders count as animals). But, like the two groups we have already considered, sharks and nonavian reptiles, they are not all *top* predators because most are commonly prey themselves, often of the biggest birds of prey, the raptors (although not all raptors are large). Mostly no other animals routinely eat adult raptors, except bigger raptors, notably owls as they hunt after dark. The arsenal of raptor characteristics that make them such effective predators, such as their large size, their strong, sharp bills and talons, and their impressive flight skills, largely preclude them from being prey. So, many of these impressive beasts qualify as bona fide top predators as much as any in this book. Raptors play key roles in their ecosystems, and they have been unduly persecuted by humans in the same way as most other predators.

As we discussed in chapter 3, birds are reptiles (refer to fig. 3.1). We know that birds date their origins back to the late Jurassic (~150 million years ago [mya]) with a group of dinosaurs called coelurosaurs that included the *tyrannosaurs* (tyrant lizards), *ornithomimids* (bird mimics), which were similar to ostriches, and big-clawed *maniraptorans* (predators that grasp with hands). All were bipedal, fast-running carnivores or omnivores. Like modern birds, these animals had a wishbone, or *furcula*; pneumatic bone (bone with more air space, eventually helpful in lightening the skeleton for flying); a wrist that could fold completely against itself; and bony projections on the ribs that eventually allowed for the flow-

through breathing structure of bird lungs. Birds need highly efficient lungs to take up the oxygen necessary to fuel the high energy demands of flight. And the ancestors of modern birds had *protofeathers*, precursors of modern feathers (fig. 4.1). Fossilized nests containing eggs have been unearthed, indicating parental care similar to that of current crocodylians. Over time, some descendants of this group of animals got smaller, developed specialized feathers, evolved wings from arms, and eventually took flight as what we call the class of modern birds: Aves.

Modern raptors have been categorized into four scientific orders: the Accipitriformes, Falconiformes, Cathartiformes, and Strigiformes

A

Coelophysoids Allosaurids Compsognathids Tyrannosaurids Oviraptorsaurs Dromeosaurids Archaeopteryx Modern Birds

FIGURE 4.1 (A) Evolution of feathers from dinosaurs to modern birds. (B) Fossil of *Archaeopteryx lithographica*. This specimen dates to the early Cretaceous period (140 mya) and displays distinct evidence of early feather development.

Mark Brandon / Shutterstock

(fig. 4.2). The Accipitriformes includes eagles, Old World vultures, kites, and hawks. Hawks include buteos (known as buzzards in England), harriers, goshawks, Secretary Birds (*Sagittarius serpentarius*), and Ospreys (*Pandion haliaetus*). The Falconiformes includes, you guessed it, falcons (sometimes, but not often enough, taxonomists make life easy). But this group also includes caracaras (not to be confused with the dessert oranges of the same name!). These two groups, the falcons and caracaras, are sister taxa, meaning they are adjacent on the bird family tree and share a relatively recent common ancestor. The Cathartiformes contains New World vultures (colloquially known as buzzards in North America),

including condors. The Strigiformes contains owls. If you go back far enough in time, all these orders share a common ancestor, of course (after all, they are all birds). Falcons are fairly removed from other raptors, though, evolutionarily speaking (about 60 million years or so).[1]

There are about 480–505 species of raptors, some diurnal (active during the day) and some nocturnal (up at night; including most owls). Why are we not sure how many species of raptors exist? The answer depends on the definition of *species*. Recall that for animals, a species consists of individuals that will freely interbreed with each other and that will produce viable (healthy and capable of growing) offspring that are fertile (capable of having their own offspring). But with animals as mobile as birds, there is some flexibility with the definition, and taxonomists will argue whether two named species that sometimes encounter each other and reproduce successfully are in fact a single species. Or even three species if you count the hybrid as yet another species. Taxonomists themselves can be classified as *lumpers* or *splitters*, depending on which way they lean on this issue. Really.

Regardless of the exact number of species, raptors are typically large (fig. 4.3). The raptor with the biggest wingspan, at 7 ft (2.1 m), is the Philippine Eagle (*Pithecophaga jefferyi*), which weighs about 18 lb (8 kg). But the California

Condor (*Gymnogyps californianus*), a vulture, beats that wingspan with one reaching 10 ft (3 m). The Andean Condor (*Vultur gryphus*) and Eurasian Black Vulture (*Aegypius monachus*) can both get up to around 30 lb (14 kg). The Harpy Eagle (*Harpia harpyja*) of South and Central America weighs in at about 20 lb (9 kg), and Steller's Sea Eagles (*Haliaeetus pelagicus*) from coastal Asia may weigh this much, too. The Secretary Bird, the only mostly terrestrial raptor, is over 4 ft (1.3 m) tall! Size is the main difference between eagles and hawks.

Not all raptors are big, though (fig. 4.4). The smallest are African Pygmy Falcons (*Polihierax semitorquatus*) and the falconets: the Spot-Winged Falconet (*Spiziapteryx circumcincta*) of South

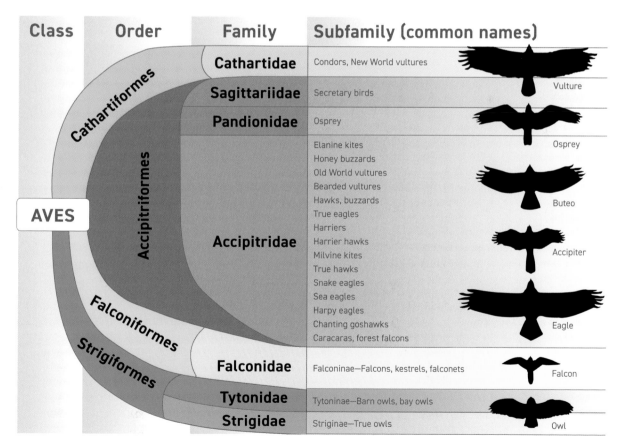

FIGURE 4.2 Phylogeny of modern raptors. Most raptors belong to the Accipitriformes, or eagles, Old World vultures, kites, and hawks.

FIGURE 4.3 (A) Harpy Eagle (*Harpia harpyja*), (B) Philippine Eagle (*Pithecophaga jefferyi*) (C) California Condor (*Gymnogyps californianus*), (D) Steller's Sea Eagle (*Haliaeetus pelagicus*). These are some of the largest raptors and the heaviest flighted birds in the world.   (A), Chepe Nicoli; (B), Ivan Sarenas; (C), Barbara Ash; (D), Ram Kumar T M / Shutterstock

America, and six species from Asia, at 6–7 in (15–17 cm), the smallest weighing under 2 oz (50 g). The Elf Owl (*Micrathene whitneyi*) is even smaller, at about 1.4 oz (40 g) and just 4.9–5.7 in (12.5–14.5 cm).

Raptors are found in most ecosystems, including on numerous small islands, and on every continent except Antarctica. A few species, like the Peregrine Falcon (*Falco peregrinus*), Barn Owl (*Tyto alba*), and Osprey are widespread in numerous ecosystems. They are birds, after all, and birds can cover lots of ground. Some species, however, are restricted to particular continents. The Secretary Bird is found only in sub-Saharan

Africa. Red-tailed Hawks (*Buteo jamaicensis*) live only in North America. The Galapagos Hawk (*Buteo galapagoensis*) is found exclusively in the Galapagos Islands.

Many raptors migrate from one continent to another to breed. Northern Harriers (*Circus cyaneus*), for example, breed across North America, Europe, and Asia, and some stay there year-round, while others opt to winter in Central America or North Africa. The upshot here is that wherever you live, you probably have raptors of some kind as neighbors. They are probably more familiar to people than most of the predators in this book.

FIGURE 4.4 Saw-whet Owl (*Aegolius acadicus*). This diminutive species (7–8.5 in, or 18–21.5 cm, and less than 4 oz, or 100 g) eats small prey, from insects to small rodents. In the North Pacific area of the United States and Canada, this owl is also suspected to eat crustaceans along intertidal zones.
Mlorenz/Shutterstock

RAPTORS AND US

Like other predators, raptors are frequent representatives of strength and power throughout human history. Here are a few examples. If you are a resident of the United States, you are officially represented by the Bald Eagle. Mexico features the Golden Eagle on its flag. Likewise, the German federal crest is known as the *Bundesadler*, or German (federal) Eagle. Raptors are powerful totems to many indigenous clans of the Americas. The US Air Force Osprey is a flying vehicle, sort of a cross between a plane and a helicopter, in homage to the stellar flying and hovering skills of its namesake. If you are a modern American sports fan, you can root for the Philadelphia Eagles, the Atlanta Falcons, and the Seattle Seahawks, and that is just in the National Football League. Eagles, hawks, falcons, and to a lesser extent owls are common college and high school mascots. The

sport of falconry, where people use raptors to hunt, is one of the most ancient human-animal associations, dating back thousands of years. Author Rob is a falconer (box 4.1). Mut, the mother goddess worshiped in ancient Egypt, is depicted in hieroglyphics as a vulture. We cannot think of any other vulture namesakes except "vulture capitalists" and Vladikoff, the heartless vulture in Dr. Seuss's *Horton Hears a Who!* Actual vultures would probably prefer not to be associated with either of these.

Owls often represent wisdom: the wise old owl. The ancient Greek goddess of wisdom, Athena, had a small owl as her symbol. In some traditional cultures the presence or mere sound of an owl foreshadows death or disaster. Owls make kinder appearances in TV programs and books for kids, such as *New Zoo Revue* (Charlie), *H.R. Pufnstuf* (Dr. Blinky), *Mr. Rogers' Neighborhood* (X), *Sesame Street* (Hoots), and Harry Potter's companion Hedwig.

BOX 4.1 HOW THE HECK DO YOU CATCH A HAWK?

Author Rob tells the tale:

When I became an apprentice falconer over a decade ago, capturing your first bird from the wild was mandatory. According to the laws of South Carolina, where I reside, I was permitted to capture either a juvenile, or *passage*, Red-tailed Hawk, or a Kestrel (*Falco sparverius*, a small, ubiquitous falcon). I chose to pursue a Red-tailed Hawk because it is a much more robust species and capable of capturing larger game.

Techniques for trapping a raptor vary by species, habitat, and time of year. With the help of my mentor, I chose to use the ancient *bal-chatri* trap, which I had to construct myself. The bal-chatri, which originated in East India, is a small, wire-mesh, weighted cage with an access door on the bottom. The top of the cage is covered with hand-tied nooses of monofilament fishing line, such that it looks like a bowl of translucent spaghetti. The concept is simple. Put a lure inside the trap, place the trap in the vicinity of a hawk, and monitor it continuously. If the raptor is hungry it will fly to the trap, use its talons to grab at the bait, and get its toes caught in the nooses. At this point the falconer rushes in with a pair of thick gloves to complete the capture.

Next, find a hawk. Trapping with the aid of a vehicle is a bit different from ancient techniques, but the principle is the same. After spotting a potentially suitable bird, I slowly drove to the general vicinity and tossed the trap out the passenger-side door onto the grassy shoulder of the road. It is important to keep the vehicle moving, since these raptors see cars and trucks passing underneath them every day, and a stopped vehicle, especially one with a moving human, is enough to create suspicion. After deploying the trap, I drove about half a mile, turned the vehicle around to face the trap, and waited, binoculars in hand. Most often, the raptor was not interested. An hour of waiting, pulse racing and

leather gloves standing by, often resulted in anticlimactic resignation: the bird was not hungry. Several times, though, a hungry raptor was interested and immediately attempted to catch and consume the lure. In these cases, excitement turned to disappointment when I discovered the bird had a bright red tail, indicating an adult, not a young-of-the-year passage, as was stipulated by law. After a quick assessment of the bird's condition, and often a treatment of topical parasite preventive, I released the bird back to the sky, probably leaving it thinking *What just happened?*

Trapping raptors is allowed only during a two-month season, established to coincide with the winter migration. My first season as an apprentice falconer coincided with an extremely late onset of cold weather along the Eastern Seaboard, and the northern birds had not made their way south by the end of the trapping season. Year one: no bird.

The following year I decided I would be better prepared. Falconry permits allow you to trap only in your home state. If you talk very nicely to the departments of natural resources of other states and pay their requisite fees, they may issue a temporary trapping permit for their state. I frequently travel to South Florida, so I got a permit for Florida for the trapping season and, on four separate 700-mile drives back and forth, scanned the trees and billboards up and down the I-95 corridor for a passage Red-tailed Hawk. This effort yielded two more captured birds, but again, they were adults. As the last few weeks of the second year's trapping season approached, I stepped up my game, spending at least four hours a day patrolling South Carolina back roads.

Three days before the end of trapping season, in desperation, I awoke at 2:00 a.m. to drive four hours across South Carolina to be near the border of Georgia when the sun came up, so I'd have all day to work my

Author Rob introduces Dasan, a young Red-tailed Hawk (*Buteo jamaicensis*), to snow for the first time.

way back home. Hour after hour, scanning the trees and signposts, I spotted a number of adult Red-tailed Hawks and a host of other species, but not the elusive quarry. As the day dragged on and I neared home, I spotted a silhouette in an old dead oak tree across a field positioned on an intersecting road. Hastily pulling my car to the shoulder, I lifted the binoculars and spied the easily identifiable head of a Red-tailed Hawk and his *brown* juvenile tail!

I deployed the trap and drove away. I had not gone 50 yd (46 m) before seeing the reflection in the rearview mirror of wings flailing on the ground. I slammed on the brakes (luckily this was a quiet country road with no traffic), spun the car around, and drove within 20 ft (6 m) of the captive bird. Donning gloves as I ran, I quickly crouched down over the trap and gently but firmly grabbed the legs of the bird. I had *finally* done it!

Lest I lose my grip and the bird, I picked him up, feet still tangled in the bal-chatri, and got back into the safety of the car. As luck would have it, the bird offered no resistance at all as I placed a traditional falconer's hood over his eyes (the dark helps calm them down) and untangled his toes. I also put anklets on his legs and affixed a leash. When I was sure everything was as it should be, I gently put the newly acquired raptor in a darkened kennel for the 90-mile ride home.

I eventually named this hawk Dasan, meaning "Ruler of the Bird Clan" in the Pomo Indian dialect. He adapted to his situation far faster than I expected. Because raptors are naturally afraid of humans, *manning* them (habituating them to humans) can be a very long process. The first night Dasan was in my care, he was standing on my glove without wearing a hood. Within 24 hours, he was eating from my hand. Within one month, Dasan was free-flying and coming back to me. He and I have spent many years tromping through the woods together working as a team. At this point, if Dasan is sitting in the oak tree in the front yard, I can whistle and he will fly in through the front door of the house and land on the back of the sofa, waiting for something to eat.

WHAT IS A RAPTOR?

As birds, raptors possess feathers and wings. Since you do not have either, let's take a closer look at both.

Feathers are composed of a protein called *beta* keratin. This type of keratin makes up scales as well, so it is no coincidence that birds have scaly legs and feet. Mammals produce *alpha* keratin, which is a component of soft, flexible skin, fur, and hair (but also claws). In both cases though, the skin serves to protect the animal; the fur or feathers, accoutrements to the skin, provide insulation. This matters because birds, like mammals, are homeotherms and endotherms,[2] maintaining a fairly constant body temperature regardless of the temperature of their environment. Birds accomplish this with their high metabolic rate, which produces heat as a by-product, and feathers that insulate and thus hold on to some of that metabolic heat. Mammals accomplish the same heat trapping with fur, blubber, or, in the case of humans, a jacket.

In the last chapter we explained how birds, as reptiles, are basically dinosaurs, and numerous dinosaur fossils have now been found associated with imprints of feathers, so it seems feathers have been around a while, even before there was evidence of flight. There is no way, for example, that the dinosaur *Yutyrannus huali*, which translates loosely into "beautifully feathered tyrant," at over 2,200 lb (1,000 kg), was capable of lifting off the ground, so the feathers must have first been for insulation and maybe also for impressing potential mates or competitors. Considering dinosaurs as feathery, even colorful, beasts might seem a bit odd, but modern birds use feathers for both insulation and attraction, so it makes sense that the first beings with feathers likely used them in these ways, too. Current thinking is that smaller feathered dinosaurs began employing their feathered limbs to provide lift as they ran up hills or tree trunks, because the present-day group of heavy, ground-living birds known as Galliformes, including turkeys, chickens, partridges, quail, grouse, and pheasants, employ that method still. As natural selection worked on this process, feathers and wing structures became specialized and therefore better suited for their various jobs, and true flight was achieved.

There are seven general types of feathers on most birds (fig. 4.5), each individually anchored to a follicle in the skin: wing feathers (remiges), tail feathers (retrices), contour feathers, down feathers, semiplumes, filoplumes, and bristles.

FIGURE 4.5 The seven categories of bird feathers and the "hook and groove" system that keeps them smooth. Each feather type serves a specific function.

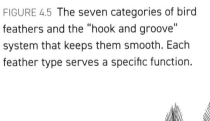

Filoplume   Bristle   Semiplume   Down   Contour   Wing   Tail

Primary flight feathers

Secondary flight feathers

Tertiary flight feathers

Tail feathers

The feathers responsible for lift and propulsion are the wing feathers. There are several categories of these, depending on their position. The primaries and secondaries are large, tapered, stiff, and asymmetrical—all adaptations for flight, which we will discuss. Tertiaries are the innermost wing feathers.

The tail feathers (fig. 4.5) share many of these same characteristics but are more symmetrical in form and are largely responsible for precision flying and steering abilities. When you separate individual strands of wing and tail feathers to create a V-shaped gap, running your thumb and forefinger over this separation will magically stick the feather back together. The individual barbs of these feathers have hooklets and grooves that zip them to each other (fig. 4.5). This interlocking matrix makes these feathers both strong and smooth, able to withstand the forces of wind, turbulence, and debris during high-speed, maneuverable flight. Accipiters have long tails to enhance this maneuverability since they chase woodland birds, while owls have shorter tails since they are not obliged to keep up with flying mice.

Contour feathers (fig. 4.5) cover most of a bird's body and do what their name implies; they smooth the outline of a bird moving through the air. Contour feathers are also responsible for the colors of a bird, and they help with waterproofing. Down feathers (fig. 4.5) are small, fluffy structures that provide good insulation for birds, as well as our own jackets, comforters, and sleeping bags. The colder the home climate of the bird, the more down feathers it will have. A Great Gray Owl (*Strix nebulosa*) of northern latitudes and higher altitudes is effectively a temperate Barred Owl (*Strix varia*) in three down parkas. Semiplumes are intermediate between contour and down feathers and provide fill beneath the contour feathers. If you are ever lucky enough to pet a friendly (emphasis on *friendly*) bird, take your finger and gently

probe into its neck. Your finger will sink in deep before you hit actual neck, passing first through the contour feathers on top, then the semiplumes, and then the down underneath. If you need to be aerodynamic, it would be no good to have a big skull perched on the end of a skinny neck. Flamingos? Okay, it can work, but in most birds that skull-neck connection is filled in to smooth airflow around the bird. About 40% of a typical bird's feathers are on the head and neck. Doing all that with short, fluffy down would be too hot, so the longer semiplumes help as filler.

Bristles (fig. 4.5) are short, stiff feathers around the eyes and on the head and toes, especially on owls. These catch foreign particles and sense somewhat like whiskers, helping some birds catch insects, for example, and repel liquids, handy in a place where preening to enhance waterproofing is not possible. Finally, filoplumes (fig. 4.5) are hairlike feathers with numerous nerve endings in their follicle walls in the skin that sense pressure. These provide information about contour feather position, allowing the bird to optimize feathers for flight, insulation, bathing, or display.

Now, what about wings? Wings are the front limbs of birds (like the pectoral fins of sharks or your arms). If you x-ray a wing—or, for human carnivores, if you pay close attention to your chicken dinner—you will see familiar-looking bones. As shown in figure 4.6, a bird wing includes the *humerus* bone in the upper wing, the *ulna* and *radius* midwing, and then a few small wrist-like bones followed by digits—all very much like your arm, but with added primary (outermost) and secondary flight feathers. So, if you put on feathery sleeves and flapped, off you'd go? You wish! Wing feathers are carefully arranged to adjust airflow over the wings in order to satisfy the laws of physics and provide lift. If you stood in front of an airplane and a bird with outstretched wings, you would see that both wings are shaped similarly.

In fact, the flight feathers themselves are shaped this way as well. They are tapered at the ends and arched slightly, or cambered, downward. This is an airfoil shape and provides lift as air passes over the wing. Next time you ride in a plane and are seated adjacent to the wing, note how various panels lift and slide as the plane takes off and lands, allowing air to pass through the wing as well as steering over it and around, providing lift or brakes. Bird feathers adjust like this too. And that is where the similarities end: planes do not flap (although in the early days of airplane development, some prototypes did indeed flap, albeit without achieving flight).

Flapping is such a complex motion that we hardly understand the physics of it, let alone how to replicate it. Birds flap in a sort of figure-eight motion. It is not just up and down, but somewhat like propellers. In fact, on big birds like eagles and vultures, it is easy to see that the outermost flight feathers on the wing are separated from each other into *fingers* (fig. 4.7). As the bird flaps up and down, the slots between primaries allow the feathers to twist, providing forward thrust and minimizing turbulence at the wing tips. This happens on smaller birds too, although only on the downstroke, but because they flap faster and their feathers are smaller, it is much harder to see.

To flap a wing requires considerable strength, so the sternum on the front of the rib cage of a bird has evolved into a much larger bone than your sternum, relative to body size, in order to anchor the muscles necessary for flight. If you consider a roasted chicken again, the sternum is that large keel running across the top as it sits on the platter upside down. Those breasts on either side are the big flight muscles required to work the wings. Commercial chickens have been bred to have freakishly large breasts since this is the priciest meat, and consequently the poor birds can barely stand up, let alone fly, but these are big muscles

in any bird. Flying takes a huge amount of energy, especially if you are a big bird like many raptors.

Most birds can fly, of course, and flying requires a copious supply of energy, which is obtained from food. At the same time, energy must also be conserved, since locating food and successfully eating enough calories is never guaranteed. How does a raptor conserve energy? One energy-saving adaptation is the ability to glide, which does not involve work. The wings of large raptors are long and wide and thus have a large surface area. Although a large surface area increases aerodynamic drag, one of the major forces that impede the movement of cars, planes, and birds alike, during gliding the wings do not flap, which minimizes this drag. Instead, the wings are extended and held at an angle such that the air is deflected downward, which produces lift. The greater the surface area, the more lift. Alas, the drag will inevitably have its way, and the bird will slow. Watch out for falling birds? No, because once airspeed is too slow for flight, the bird can go into a shallow dive. It will lose altitude, but the dive extends the energy savings of gliding. At some point flapping resumes in order to avoid crashing or to regain the lost altitude.

Some birds actually gain altitude when gliding, riding an invisible escalator in the sky. If you live practically anywhere in the Americas you can watch Turkey Vultures (*Cathartes aura*) soar as they glide, since they are common and easily identified (big, black birds with small heads and a characteristic rocking and tilting back and forth as they glide). These large birds roost in trees or on canyon ledges overnight. They wait until the sun starts to warm up their metabolic machinery and muscles in the morning before they take off. Even if it has been a while since you have taken physics, you probably recall that air rises as it heats, which is the principle behind hot-air balloons. As the sun warms the air, the air rises in what are called updrafts or

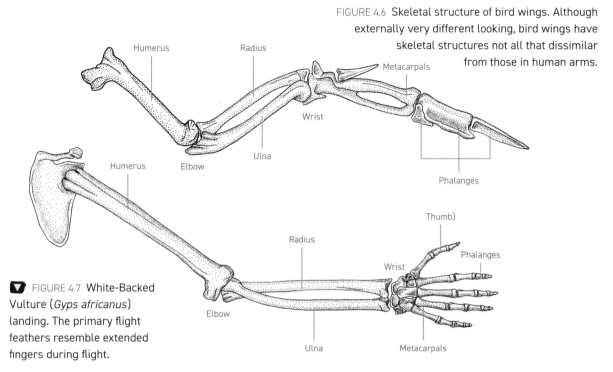

FIGURE 4.6 Skeletal structure of bird wings. Although externally very different looking, bird wings have skeletal structures not all that dissimilar from those in human arms.

FIGURE 4.7 White-Backed Vulture (*Gyps africanus*) landing. The primary flight feathers resemble extended fingers during flight.

thermals. Vultures and other big birds anxious to fly without using too much energy wait for these and then ride them to altitude. Once there they can glide on the breeze looking for lunch with hardly a flap. The larger raptors all can and routinely do ride thermals, except most owls, since there are no thermals at night. Owls have relatively short, wide wings, providing agility through the forest, but they do not glide much. They must therefore flap for extended flight. We will address this when we get to specializations for catching prey.

Bird bones (including those of raptors) have evolved for very specific flight-related purposes, including the aforementioned enlarged sternum, modified wing bones, and so on. These bones also evolved internally, becoming more lightweight and easier to fly with. Contrary to popular belief, bird bones are not hollow, at least not quite. The bones are not a hollow tube, but more like a tube filled with crisscrossing truss-like structures surrounded by air pockets (fig. 4.8). Think of it like corrugated cardboard, strong yet lightweight. By ensuring bone strength while reducing overall weight, birds can use less energy for flight. These semihollow bones are described as *pneumatic*, or containing air pockets. However, not all bird bones are pneumatic, and the number varies by species.

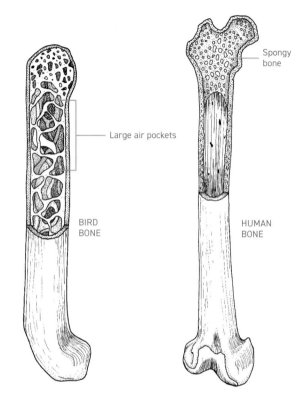

FIGURE 4.8. Bone of a bird (*left*) and human (*right*). Raptor bones have a lower density than human bones because of internal truss-like structures and resulting air pockets. While lightweight and strong, they are more rigid than human bones. As a result, birds are quite agile in the air but walk a bit stiffly on the ground.

RAPTOR BIOLOGY AND ECOLOGY

Male and female raptors have similar features, including coloration, unlike many other birds, except that all raptors exhibit reverse size dimorphism, meaning that females are bigger than males (fig. 4.9). They are about 15%–20% heavier, but in a few species the difference can range up to 60%–75%. Why are females bigger? There are numerous hypotheses. Maybe larger females can focus on larger prey than males do, thereby minimizing competition for food between males and females? Large females can lay larger eggs, resulting in stronger nestlings? Large females can better incubate eggs and protect eggs and nestlings? Large females are less likely to starve if food is diminished? Small males use less energy hunting? Small males are more maneuverable so they can better protect a nest? Large females are somehow better at maintaining pair bonds? We don't really know, but it is the case for raptors, a few shorebirds, and hummingbirds.

Raptor pairs are generally monogamous, at least for the duration of the breeding season (makes

FIGURE 4.9 Male and female European Kestrels (*Falco tinnunculus*). Most raptors display reverse size dimorphism. Note that this female kestrel is significantly larger than her mate. Since coloration does not differ between sexes in most raptor species, scientists and falconers often resort to DNA testing of feathers to accurately determine whether a raptor is male or female. FJAH/Shutterstock

you rethink the definition of "monogamous," no?), and raising nestlings to leave the nest, or *fledge*, requires two parents. The breeding season generally commences in the winter with the regrowth of the birds' gonads. In order to minimize weight, the ovaries and testes of birds regress and are quite small compared to those of other vertebrates when it is not breeding season, but they develop when they are needed. Females normally have only one ovary, and most males lack a phallus. Like most other reptiles, most birds have a single opening, the cloaca, for excretion of wastes and eggs and sperm, and mating consists of what is described as a quick *cloacal kiss.*

In temperate climates, raptors breed in the winter so that the offspring hatch in the spring when food availability is high. So, for example, in December in North America you might see pairs of Red-tailed Hawks soaring over their previous nesting territories prior to breeding. Although they "talk" in what the Cornell Lab of Ornithology website describes as the "Hollywood hawk scream"[3] (whenever any hawk or eagle appears onscreen, the soundtrack is typically the screech of a Red-tailed Hawk), their courtship has more to do with demonstrating flight skills than bad singing. Nor do they sport pretty colors to attract a mate. Instead, they become aerial acrobats. The most dramatic method of this is cartwheeling, where two birds fly up and one flips over so it can grab the other's talons while they cartwheel down briefly, separating in time to fly off before crashing. We suppose if a bird crashes its partner, there will be no mating!

Usually, cartwheeling is an aggressive behavior between males, but it has also been observed as a courtship behavior, perhaps most famously in Bald Eagles (*Haliaeetus leucocephalus*) and Marsh and Montagu's Harriers (*Circus aeruginosus* and *pygargus*), which are known for their sky dancing. In fact, one advantage of keeping the same mate

over time for the harriers is that they get better at using cartwheeling to transfer prey from one to the other. Booted Eagles (*Hieraaetus pennatus*) are also known for elaborate sky dancing, consisting of soaring upward and then diving repeatedly with half-closed wings as if on a roller coaster. Both members of the pair may do this, perhaps both bonding and demonstrating fitness.

Another advantage of keeping the same mate from one year to the next is that often you can keep the same nest. Not all raptors build nests. Some, including all owls, usurp the nests of others or use any flat surface available, such as on a cliff. But the bigger hawks and eagles build large stick nests, which they add to each year. Bald Eagles are well known for this. An average Bald Eagle nest is 5 ft (1.5 m) across and 3 ft (1 m) deep, but one in Ohio that was occupied for 34 years measured almost 9 ft (2.7 m) across and 11 ft (3.3 m) deep, weighing close to 2 tons (1,800 kg)![4] In North America, Great Horned Owls (*Bubo virginianus*) breed very early in the season, often when there is still snow on the ground, and will take over used nests of Red-tailed Hawks.

Where there are no predators, or where trees are in short supply, raptors may nest on the ground (fig. 4.10). Arctic Peregrine Falcons and Snowy Owls (*Bubo scandiacus*) build ground nests, and at high latitudes, Bald Eagles, Kestrels, and Merlins (*Falco columbarius*) will also do so. Likewise, on islands where nest predators have not existed, ground nesting is common. Egyptian Vultures (*Neophron percnopterus*) ground nest on the Canary Islands, and Ospreys do this on islands in the Sea of Cortez in northwestern Mexico. Most New World vultures nest on the ground, regardless of predators, as do all harrier species.[5] Harriers live in shrubby, treeless habitat, and while about half their nests fall victim to predators, harriers have some defenses. They nest in loose colonies, so groups can mob and chase off predators. Their well-camouflaged nests are also extraordinarily

FIGURE 4.10 Montagu's Harrier (*Circus pygargus*) chicks in a ground nest. This is one of a number of species of ground-nesting raptors. Unfortunately, unaware Good Samaritans often try to help ground-nesting birds that they assume are abandoned. More often than not, this does more harm to the raptor than good. WildeMedia/Shutterstock

difficult to find. And that cartwheeling we described above may also come into play. The male calls to his mate as he nears a nest, so rather than giving away the exact location, the female meets him in the sky, collects the prey from him, and slips back to the nest without giving away the location, she hopes. These strategies evidently work because these birds are found on every continent where there are raptors.

Finally, many raptors, especially Ospreys and Peregrine Falcons, take good advantage of human-made structures. We are good for something! Ospreys often nest on nesting platforms built specifically for them, in addition to taking advantage of buoys and channel markers, since

they eat fish and thus nest over or near water. And Peregrines famously nest on ledges on high-rises in New York City, Cape Town, and other cities where the building landscape evidently seems much like the cliffs and canyons they evolved to inhabit. Of course, there is also no shortage of pigeons—a veritable buffet—to be had in a city.

Our raptor has a mate and a nest, so next up is egg laying. Smaller raptors like Barn Owls may lay up to 18(!) eggs, but many raptors, especially the largest ones, are not prolific egg layers, laying perhaps 2–6. Meeting the energy demands of even one raptor chick requires a significant expenditure of time and energy to find and catch prey. Like every growing juvenile, baby birds eat a lot.

Nesting success may correlate with prey population cycles. For example, when it is a good year for voles, small rodents in the Arctic, most pairs of Rough-legged Hawks (*Buteo lagopus*) breed and successfully fledge offspring. In fact, a Snowy Owl couple might build a nest out of dead lemmings if they are in good supply.[6] In a bad year, few nests even get started. Likewise, weather is important, with spring cold snaps or snowstorms reducing nesting success. And while these top predators are not often preyed on as adults, their young are fair game. Great Horned Owls, Snowy Owls, Northern Goshawks (*Accipiter gentilis*), and Cooper's Hawks (*Accipiter cooperii*), among others, are serious threats to the nestlings of other birds of prey.

Eggs are typically incubated for a month or longer. This task normally falls to the female, while the male is responsible for providing her with food. Once the eggs hatch, both parents hunt and deliver food. Raptor hatchlings are helpless at birth, with closed eyes and fluffy down feathers, although that does not prevent them from being immediately demanding! Nestlings remain in the nest eating and growing for four to six weeks, although very big raptors like eagles, condors, and vultures may take more than three months. While the offspring are in the nest, the parents can easily watch over them and defend them if necessary. Once they leave the nest, or fledge, they will typically stay around the nest and collect food from their parents for a few weeks or months while they practice hunting or scavenging. At this stage they are called *hoppers*, as they hop around and make a fuss to get the attention of their parents, even though they can fly. New fledglings are rather klutzy fliers though, and as they roam around focusing on potential prey and not necessarily potential predators, their parents cannot keep as close an eye on them. Much as with human adolescents, this is probably the riskiest time for the youngsters.

RAPTORS AS PREDATORS

The prey of raptors ranges from insects right up to medium-sized mammals, and if we include previously deceased meals, well, the sky's the limit. In general, the larger the raptor, the bigger its prey. There are raptors that eat fish, other birds, monkeys, rodents, snakes, lizards, and pretty much any kind of young animal they can fly off with (fig. 4.11). Does that include people? Generally, raptors attack only biologists getting too close when trying to study them.[7] But do not relax completely. A 2007 study supported the conclusion that a two-million-year-old *Australopithecus africanus* child (the same hominid species as the famous Lucy fossil skeleton) had his skull pierced by raptor talons and was then eaten by a raptor, based on what modern African Crowned Eagles do to monkeys.[8] That incident occurred long ago, before modern humans arose, but more recently there have been folk tales of Haast's Eagle (*Hieraaetus moorei*), an inhabitant of New Zealand before human colonization led to its extinction 500 years ago, carting off Maori children and adults. Given that these birds evidently killed 440 lb (200 kg) Moas (extinct large flightless birds endemic to New Zealand), even if they could not carry humans around, it is not impossible that they posed a threat.[9] Experts caution you to consider stories and videos of extant eagles grabbing little kids with a healthy dose of skepticism. Even the biggest eagle can lift only a few pounds, and we cannot imagine parents would just stand there and watch while the bird labored to fly with that large a quarry, much less film the event! We have seen an extraordinary, almost unbelievable, video clip from a movie (not a documentary) called *Brothers of the Wind* featuring authentic and not re-created footage of a Golden Eagle (*Aquila chrysaetos*) wrestling a Chamois (European antelope) down a rocky hill in the Alps.[10] The eagle seems latched onto the back of the Chamois as the two roll and tumble down the hill. Although the Chamois eventually escapes, it is amazing that the smaller eagle is able to walk away (or in this case, fly away) from an ordeal that one would expect to have crushed its skull, or at least broken its wings. In addition to showcasing the toughness of this raptor, this video clip illustrates the size of prey some raptors are willing to hunt, although this episode is certainly exceptional.

The features that characterize raptors are mostly adaptations to facilitate their methods of catching and eating prey. These include their strong, sharp, large talons as well as their similarly strong, sharp, large beaks. All raptors are strong fliers with excellent eyesight, and most have acute senses of smell and hearing. How do the birds use these features to locate, subdue, and capture their prey?

FIGURE 4.11 Osprey (*Pandion haliaetus*) with fish. Most raptors hunt prey they can fly away with, particularly when they have chicks to feed in a distant nest. This Osprey was photographed near its nest along a freshwater river in South Carolina. It had likely captured its prey, a Spanish Mackerel (a saltwater fish), in the nearby ocean about 4 mi (6.4 km) away.

SENSING PREY

In general, birds rely on vision to find prey, although a few raptors also depend heavily on hearing and olfaction (smelling). But all can see quite well. Their eyes are much like yours and those of other vertebrates, but theirs are much larger, taking up at least 50% of the volume of the skull and displacing the brain backward. For an owl, which needs extra-large eyes to see in the dark, the eyes are tube shaped and bulge out of the head in order to fit in the skull without taking over essential brain real estate. If a human had eyes like an owl, they would be the size of grapefruits. A Red-tailed Hawk, weighing in at about 3 lb (1.4 kg), has eyes about the same size as a human's. And while the term "eagle-eyed" suggests particularly sharp vision, the visual acuity of eagles and other raptors is only marginally superior to ours. What is better is their ability to "see fast," which is a necessity since they must simultaneously fly through, say, a forest, dodging trees and flying over undergrowth while staying focused on a sprinting, darting squirrel that is using its own own arsenal of predator-avoidance maneuvers. If a raptor cannot resolve the often complicated visual stimuli of its surroundings and its prey and react to them very quickly, it faces two equally bad outcomes: crashing into an object while in pursuit, or starving to death.

As a raptor pursues its prey, it looks ahead and laterally, and each eye has a field of view of about 150°. Owls are the champs at this, being able to rotate their heads 270° to the left and right (a total of 540°), which allows them to see completely around themselves (fig. 4.12). Owls owe their ability to turn their heads backward to the 14 vertebrae in their necks (compared to the 7 that you have). Even diurnal raptors looking straight ahead have only a narrow blind spot, about 20° wide, directly behind their heads (and that disappears when they turn their heads), whereas the human blind spot is a half circle 180° wide! You are so easy to sneak up on that you might not notice a herd of elephants if they were quiet and in your blind spot. You cannot sneak up on a raptor.

Diurnal raptors (those active in the daytime) have tetrachromatic (four-color) vision, meaning they can see very deep blue or ultraviolet light waves, in addition to the red, green, and blue wavelengths you can see. It turns out that rodent urine, some songbird feathers, and even some mammals reflect ultraviolet light, and raptors that prey on these animals may be able to detect it. In addition, some raptor feathers may also reflect ultraviolet light that can be used in mating displays, a phenomenon documented in Eurasian Eagle-Owls (*Bubo bubo*).[11]

Structurally, raptor eyes have a few other helpful features. The muscles controlling their pupils and lenses are voluntary, meaning the birds control them, while for you they are automatic. The only action you can do voluntarily to adjust light entering your eye is to squint. Birds can, in effect, squint their pupils, meaning they can adjust the focus and the level of light getting into the eye intentionally and much faster and more precisely than you can. They also often have eyebrows of specialized feathers, or a stripe of dark feathering running down the corner of their eyes like a Cheetah or baseball player. Eagles have built-in glare-reducing *visors* above each eye, contributing to their fierce visage. Raptors have both upper and lower eyelids, plus a third translucent lid or nictitating membrane (like some sharks, nonavian reptiles, Aardvarks, and camels) that keeps the eyes clear of debris and provides antibodies to protect the lens. Finally, raptors are bifoveate, meaning they have two foveae on each retina, the inside surface of the back of the eye. Foveae are concentrated areas of cone cells that allow for color vision and enhance acuity and depth perception. These two areas have a higher density

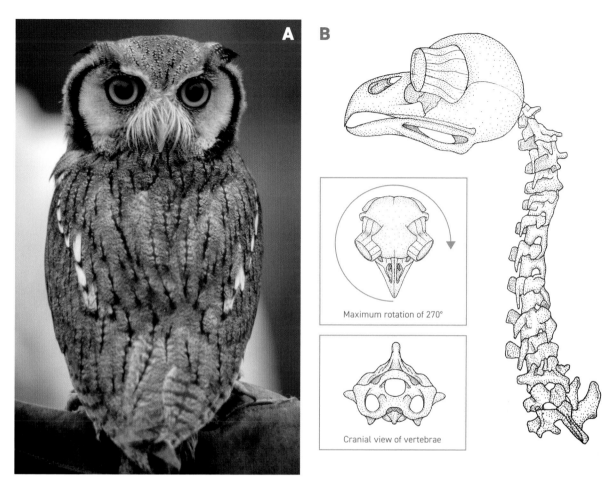

FIGURE 4.12 **(A)** Rearward-facing White-Faced Scops Owl (*Ptilopsis leucotis*). The cervical (neck) dexterity of the owl is largely the result of an increased number of vertebrae in its neck. Thanks to this ability and its wide angle of vision, not much can sneak up on (or away from) an owl. **(B)** An owl's upper skeletal structure, illustrating the increased vertebral count. A, J R Price / Shutterstock

of cone cells than is found in human eyes.[12] Diurnal birds of prey would not be expected to have high concentrations of rod cells on their retinas because these enable vision at low light levels, something most diurnal raptors do not need because that is not when they hunt. Humans are better at night vision than most raptors, except for most owls. Nearly all owls, of course, are masters of night hunting, and part of being a successful hunter at night is being able to see well in the dark. Their retinas have high densities of the rod cells that

facilitate vision at low light levels, and they lack the high-resolution color vision of diurnal raptors.[13]

Raptors also listen for prey. Consider owls again. What they lack in vision they compensate for in hearing, a more useful sense in the dark. Their most obvious adaptation for hearing is a face like a satellite dish. No accident, that. Like a satellite dish, that parabolic facial disk captures sound and directs it to their ears, which are not the tufts on top of the heads of some owls, as many mistakenly conclude. Rather, the ears are large vertical openings on either side of the head, with exceptionally large inner ears, all protected by feathers. In fact, all bird ears are covered by feathers. It would be impossible for a bird to hear if the wind were whistling past exposed ear holes as it flew. You have probably seen the fuzzy-looking covered microphones reporters sometimes thrust toward their interviewees when outside. Same idea. Additionally, owl ears are arranged asymmetrically on their heads (fig. 4.13). Why? Well, when you hear a sound, you can tell the direction the sound is coming from thanks to the sound waves hitting your ears at slightly different times because your ears are separated. The more separate your ears, the better your ability to detect the source of the sound. So, to maximize ear separation, one owl ear is higher on the head than the other. This unusual offset anatomical position allows an owl to pinpoint sounds in both a horizontal and vertical plane, providing three-dimensional hearing.

Contrary to a popular misconception, owls do not use echolocation at all, which would involve sending out a signal to reflect back, like a bat or dolphin; instead they listen for sounds coming from prey. Experiments have shown that with hearing at least 10 times as sensitive as yours at the frequencies of scuttling mice feet, Barn Owls can hear mice and snag them in pitch darkness, following the sound alone.[14]

Diurnal raptors, especially harriers, can hear too, probably better than you. Harriers have facial disks and large ears, although not quite as pronounced as in owls. Studies have demonstrated that Northern Harriers in the American Southwest can catch voles just by sound, like Barn Owls.[15] Northern Harriers living in farm territories hunt by gliding low over open fields, listening for potential prey in the grass. Diurnal raptors listen to each other as well, most often conveying threats, communicating between mated pairs, or maintaining a migratory group. The most vociferous is probably the African Fish Eagle, whose scientific name, *Haliaeetus vocifer*, means "noisy sea-eagle." In the wetlands of sub-Saharan Africa, they talk all day long to mates, neighbors, and rivals.[16] This constant calling leads many people to refer to the African Fish Eagle as the "sound of Africa."

With excellent sight and hearing, most raptors, as far as we know, do not depend heavily on their sense of smell, with one possible exception: New World vultures. Vultures eat carrion—dead things—and it probably does not surprise you that smell

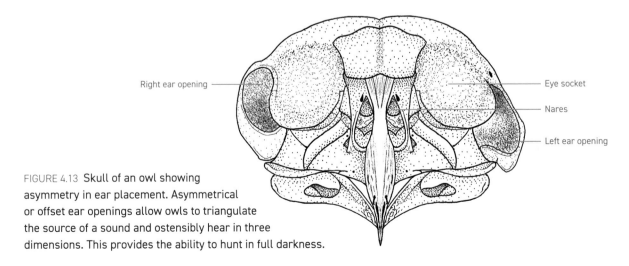

Right ear opening

Eye socket

Nares

Left ear opening

FIGURE 4.13 Skull of an owl showing asymmetry in ear placement. Asymmetrical or offset ear openings allow owls to triangulate the source of a sound and ostensibly hear in three dimensions. This provides the ability to hunt in full darkness.

is an excellent way to locate a dead animal. In particular, the nearly ubiquitous New World Turkey Vultures have been experimentally shown to have a highly tuned sense of smell, with an enhanced ability to identify *ethyl mercaptan*, the volatile and vile-smelling by-product of decomposition. Yum! The scientist[17] who determined this also noted that since the 1930s, the Union Oil Company in California had been introducing this chemical into its gas lines so people could identify gas leaks, and Turkey Vultures would also identify leaks in pipelines for the company, just by showing up and circling overhead in search of a snack. Turkey Vultures have the largest nostrils and largest olfactory bulbs (the part of the brain responsible for the sense of smell) of any New World vultures, including the much bigger condors. Of course, science is not so neat, and later studies have questioned whether the concentrations of odors deployed in the original studies were realistic. Turkey Vultures do not seem to be oversensitive to these decomposition chemicals at the levels they might actually encounter up in the sky.[18] So, the jury is perhaps still out on this issue, but the fact that Turkey Vultures are the most abundant and widely dispersed of the vultures could well be because they can sniff out a carcass anywhere.

Our diurnal raptors and owls, then, have a suite of adaptations to locate prey. It is not hard to catch dead prey, but what about living critters? How do raptors catch what they find?

CATCHING AND KILLING PREY

Raptors hunt in several different ways. Hawks, eagles, and owls living in forests or other dense habitats *still-hunt*, or perch at an elevated vantage point and look and listen for rodents, lizards, snakes, fish, smaller birds, monkeys, or sloths. When they lock in on some likely animal, they will drop quickly and glide or flap in an attempt to orchestrate a surprise attack. One morning author Sharon arrived on our college campus and from her second-floor office adjacent to an oak tree saw a squirrel sitting on a branch, chattering away as squirrels do. Suddenly there was a shape coming from above, and one of our resident Red-tailed Hawks swooped onto the squirrel and continued directly down to the ground with it. For a few minutes it held its ground there, wings outstretched, hiding the prey with a behavior called *mantling*, feet firmly attached to breakfast, with startled students giving it a wide berth. Then it flew off, squirrel dispatched and dangling. That is an example of a successful still-hunt. And an interesting start to a day for Sharon and the hawk, if not the squirrel.

What exactly happened to the squirrel? Well, all raptors except vultures have impressive talons on their toes. A big eagle might have talons 3 in (7 cm) long, rivaling those of a large cat or Grizzly Bear. The rear talon on a Harpy Eagle can be 5 in (12.5 cm) long! In fact, in drawings, the size of talons is usually underplayed because if drawn correctly, they look out of proportion.[19] There are usually four toes, three in front and the *hallux*, or killing toe, in back, which is good for grasping and stabbing. Many raptors have a sharp ridge along one side of the front center talon. This has been nicknamed the *blood groove* by some falconers, in reference to the purported purpose of the fuller groove on many knives and swords. Although not completely understood, this sharp edge may allow blood to flow more freely from a wound with a talon in it, or allow the talon to be more easily removed from the flesh. This groove on a blade also lightens its weight with minimal loss of blade strength and integrity, presumably doing the same for a long talon.

To aid in grasping, raptors have a tendon locking mechanism (TLM). Many birds have this adaptation. If you can imagine having to sleep while hanging on to a branch, you can understand the advantage of having feet that lock on the

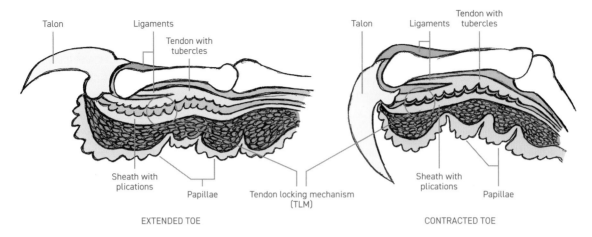

Talon Ligaments Tendon with tubercles

Sheath with plications Papillae Tendon locking mechanism (TLM)

EXTENDED TOE

Talon Ligaments Tendon with tubercles

Sheath with plications Papillae

CONTRACTED TOE

FIGURE 4.14 The tendon locking mechanism (TLM) of raptors allows them to "clamp" their feet in place without expending the extra energy needed to make a fist.

branch firmly without using much energy or requiring any thought. Having feet that lock tightly in place is also useful when you have to grab and retain a moving target like a squirrel, but still control your own motion. The difference between the TLM in most birds versus specifically in raptors is that in the former the TLM involves the legs and feet, while in the latter the TLM is mainly in the toes, freeing up the legs to move while the toes are locked. Small projections on the tendons that flex the toes essentially lock them in place (think of locking pliers; fig. 4.14) without muscles having to work to hold them, and in the raptors investigated, the bigger they are, the more robust the hold.[20] Those long talons impaled the unfortunate squirrel above as he got grabbed, ideally in the head (enabling a quick death), eliminating any struggle.

How do birds hunt at night? Well, nocturnal owls perch and listen more than they look. Remember that nocturnal birds cannot soar on thermals, so they must flap from place to place or down onto prey. Flapping wings should make noise, but owls have fringed feathers, and lots of them. Believe it or not, someone (probably a hapless graduate student) has counted the feathers on a few raptors. At slightly under 2 lb (800 g), a

Barn Owl has about 9,200 feathers, while a Bald Eagle at 10 lb (4,500 g) has about 7,100. This high density of fringed feathers both muffles the sound of the flapping wings and provides a little extra insulation, which is helpful at night when it is cooler. In addition, rather than being sharp edged and stiff, owl flight feathers have fine filaments and a downy surface layer that disrupt or break up air turbulence, the sound you hear when a bird flaps its wings. And owls often have feathery "boots," or feathers extending down along their legs, some even on their toes.[21] These protect them from bites of desperate prey as well as mosquitoes, and from the cold. Rough-Legged Hawks, Booted Eagles, and Golden Eagles also have feathery legs, presumably for similar reasons.

Owls also have a different arrangement of toes, with two facing front and two facing back (*zygodactyl*), and one of the back toes can swivel to the side or front when that is helpful in a grab. And what a grab! (See box 4.2.) The force a Great Horned Owl foot can generate is about eight times stronger than the grasp of a human hand, approaching that of a Rottweiler jaw![22]

As you might imagine, hunting techniques vary by species and by what kind of prey they are pursuing. Owls tend to hunt at night when visibility is minimal, so their preferred method is to bide their time perched on a branch waiting for the food to come to them. Falcons, on the other hand, tend to hunt other birds, so their method of hunting involves high-speed aerial acrobatics. The Northern Harrier flies low and listens for prey. Some raptors excel in wide-open spaces, while others are incredibly adept at flying through dense forests. Harris's Hawks (*Parabuteo unicinctus*) are highly social raptors. They live in extended family groups, share babysitting duties, and often hunt together. In the Desert Southwest of the United States, Harris's Hawks can be seen standing on top of each other's backs (*toteming*) while perched atop a cactus (fig. 4.15). This allows them to monitor multiple directions at once. When they spot potential prey, often a jackrabbit sitting under some scrub brush, one member of the group will walk along the ground and flush out the prey, giving the rest of the family the opportunity to pursue it in open territory. Afterward, they all share the meal. This hunting practice has earned them the nickname "wolves of the air."

Those zygodactyl feet (two toes front, two back) are also found on Ospreys. Why? Ospreys eat fish, and fish are slippery and hard to hold on to, especially during flight. Numerous eagles can eat

FIGURE 4.15 Two Harris's Hawks (*Parabuteo unicinctus*). This is the only known species of raptor to regularly interact in extended social groups beyond a mated pair and their offspring. They are also known for cooperative hunting, another unique attribute among birds of prey. Pierstorff/Shutterstock

BOX 4.2 **A BIRD IN THE HAND, LITERALLY**

Author Rob experienced the force of a raptor's talons while working with a Eurasian Eagle-Owl (the larger cousin of the Great Horned Owl) named Archimedes. During a training session, the owl flew to Rob's glove to receive a piece of food. When he bent his legs upon landing, Archimedes's tendons locked into place and one of the talons found its way through a seam in the glove and directly into Rob's hand like a knife through warm butter. Archimedes was so intent on protecting the food, he neglected to straighten up to unlock those tendons and talons. He kept trying to fly away with his food, but he was firmly stuck to the glove and your author's bleeding hand. Archimedes alternated looks between his feet and straight ahead and took several nips at Rob, thinking Rob holding his feet was why he made no progress. After about 20 minutes (*ouch!!!*) with a talon in his hand, Rob was able to reach behind the owl and push his legs into an upright standing position. The maneuver unlocked the talons and Archimedes promptly flew to a nearby perch. This extreme use of the TLM is not uncommon in owls. When they lock their talons into prey (or their falconer's flesh), it is aptly called *binding*. Sometimes falconers must return owls to the aviary while they are still binding and attached to the glove because the owl refuses to let go. It is for this reason that Rob says he is more wary of owl talons than those of some of the eagles.[23]

FIGURE 4.16 (A) Peregrine Falcon (*Falco peregrinus*) in a stoop. Although small, Peregrines are the fastest animal on Earth, earning them the nickname "jet fighters of the bird world." (B) Open mouth of a panting Peregrine Falcon, showing the tomial tooth. (A), Smiler99; (B), Adventuring Dave / Shutterstock

fish too, but Ospreys are most specialized for it. First, they are very good at the *hover flight* hunting strategy, where they seem to hang in one place in the air, flapping and kiting (using the wind to stay in place). A study of Ospreys in Finland determined that they can hit the water at 44 mph (70 kph) and snag a fish with a single talon.[24] A tactile reflex in their feet enables the claws to snap shut in two one-hundredths of a second. And the feet are covered in short, spiny toe pads to improve grip. If it catches a large fish, an Osprey will turn it so the head faces forward for maximum aerodynamic efficiency, taking advantage of the fish's streamlining. The bird manipulates the fish in the water or in flight, and when the Osprey emerges from the water's surface, it will give its feathers a good shake, like a dog coming out of a bath.

The Osprey's dive is fast and abrupt, but it is the falcons that are the experts at diving (fig. 4.16). Falcons, including Goshawks, eat mainly other birds, which of course have evolved to evade capture. These raptors need to be highly agile fliers, and fast as well. Peregrines have been clocked in level flight at 42–51 mph (65–82 kph). Even migrating, they can go an average of 38 mph (60 kph)—*over a 24-hour period*! But they are most famous for *stooping*, or tucking in their wings and diving after prey, during which they have been recorded reaching speeds of up to 242 mph (389.46 kph)![25] Numerous adaptations allow stooping. Peregrines have strengthened flight feathers, external nasal deflectors to enhance air intake, back feathers that can be erected to act as spoilers to minimize airflow turbulence and drag, and an overall streamlined and aerodynamic shape. These and other adaptations make Peregrines the fastest of all birds. In fact, they are the fastest animal on Earth![26] And to make life more difficult for pigeons, city-dwelling Peregrines, especially, take advantage of city lights and will hunt after dark.[27]

Falcons and some other hawks have a *tomial tooth* at the center of their upper bill (fig. 4.16). Not a tooth at all, it is rather a projection of the upper bill that fits into a groove on the lower bill such that a bite will sever a prey's spinal cord. Peregrines are not very big, so even if they hit hard, a follow-up bite may be necessary to dispatch the prey.

Other raptors follow capture with bites as well, but the champion biters are the vultures. These birds do not require strong claws to hold on to their food. Dead food just sits there, right? But they must be able to rip and tear into a carcass, and animals have robust skins, so vultures have strong beaks and large muscles at the back of their heads and necks. Even so, they often must wait for a scavenger with teeth to arrive first and start the process. There tends to be competition with these scavengers and other vultures, so eating efficiently matters, as well as being able to fight off competitors when necessary. A vulture can gulp down as much as 2 lb (1 kg) of food in 20 minutes! As you might guess, vultures have extra-potent stomach acid to kill bacteria, although they are a bit fussy and will avoid very rotten food.

Alas, vultures are not known for their beauty. But those goofy-looking featherless heads work. If eating involves thrusting your head into a carcass to get at the good bits (fig. 4.17), no use having a fancy headdress, especially if you have no hands to clean it with. They do have facial bristles that repel blood and other liquids. In addition, they are clever. Egyptian Vultures will pick up rocks and hurl them at Ostrich eggs to break them.[28] We are lucky we have this ornithological janitorial staff at work or we would be up to our elbows in carcasses. Humans should show them a little more respect.

Throwing a rock is clever, and it should be noted that all raptors are intelligent. They must be to be able to catch their prey, but they also have to learn and practice the skills required. Young raptors are much less successful hunters than their parents, which is why they hang around the nest a while after they fledge.

FIGURE 4.17 Vultures feeding on a deceased Cape Buffalo. What vultures lack in beauty, they make up for in their janitorial abilities. Vultures are resistant to a number of diseases (including tuberculosis and brucellosis) and harmful bacteria (including fusobacteria and clostridia). By consuming rotting carcasses, vultures play an important role in removing harmful pathogens from the environment.

HOW ARE THE RAPTORS DOING?

In modern times, you might be forgiven for wondering why anybody would have it in for a raptor. We have established that they do not eat people, big or small, but instead prefer mice and snakes and other critters many folks would just as soon not see. And yet, like other predators over the years, they have had bounties on their heads. Why? Well, their nicknames tell the tale, in part. Red-tailed (fig. 4.18) and Red-shouldered Hawks (*Buteo lineatus*), for example, are colloquially called "chicken hawks." These hawks can and do eat chickens, although it is probably usually the small, faster Cooper's Hawks that are responsible for most chicken-pickin', but these aren't as obvious as a big Red-tailed Hawk. Either way, farmers do not take kindly to livestock eaters. Or to competitors for game birds, so Peregrine Falcons, called "duck

hawks" before 1960, as well as Goshawks and Cooper's Hawks were sometimes considered bad guys. Birding as a pastime was coming on strong in the early 1900s, so anybody that ate those pretty songbirds so valued by birders was also frowned upon. And vultures are both creepy and have parts valuable in traditional folk medicine or for casting spells. Same with owls. Maybe the most surprising source of raptor killers, historically, were museum collectors and scientists. If the specimen was, say, a Galapagos Hawk, endemic to just the Galapagos and with a small population, it would not take much to extirpate the species, one of the perils of living atop the trophic pyramid where biomass is scarce. Lesson repeated throughout this book: *It is never easy to be a predator.*

The Migratory Bird Treaty Act of 1918 first protected birds in the United States, and raptors were finally included in 1972. The Bald and Golden Eagle Protection Act of 1940 did what it says, with

FIGURE 4.18 Red-tailed Hawk (*Buteo jamaicensis*) sitting precariously near live electrical lines. This is one of the most common raptor species in North America. It is threatened by farmers, who view it as a danger to small livestock, and by myriad challenges caused by human development, including electrocution. Sundry Photography / Shutterstock

exceptions to allow use of eagle parts for spiritual purposes by federally recognized Native American tribes. The US Fish and Wildlife Service issues permits to use raptors for educational or other purposes, as well as for falconers to possess raptors. Endangered raptors are protected by the 1973 Endangered Species Act. Internationally, the Convention on International Trade in Endangered Species of Wild Flora and Fauna (CITES), with which most countries are compliant (at least on paper), covers endangered raptors internationally. There are also the Ramsar Convention on Wetlands

of International Importance, the Convention on Biological Diversity, and the Convention on the Conservation of Migratory Species of Wild Animals, which include some birds of prey.

So, there are laws in place protecting raptors, and most people now, we would like to think, recognize that a hawk is more valuable to a chicken farmer alive than dead. Nobody wants an exploding rodent population, but a cursory investigation reveals that a common question on Google is something along the lines of "Is it okay to kill the hawk eating my chickens?" (And any largesse virtually never extends to snakes, we have noticed locally, where "chicken snake" is the common name for a Yellow Rat Snake, or, well, any snake in a barn or coop, or your yard. With the same outcome as befell the chicken hawks.) Birders now value a raptor sighting the same way they do any other bird on their life list. Yet these birds are still in trouble.

As of 2017, 52% of raptors had declining populations, although just 21% were classified as Near Threatened or Threatened on the IUCN Red List of Threatened Species. The majority of those live in South and Southeast Asia and sub-Saharan Africa, places with the most raptor species overall. In particular trouble are vultures, especially Old World vultures, in these locations. Over 70% are Endangered or Critically Endangered.[29]

What is the trouble? It is the usual suspects, mostly. A burgeoning human population is taking up their space, as well as the space for their prey, and we indiscriminately poison or kill them in various ways, both unintentionally and purposefully. Again, from the IUCN Red List, agriculture and aquaculture as well as logging and wood harvesting are the main causes of population declines among raptors in general. Electrocution on power lines (fig. 4.18) and, lately, flying into wind turbines cause other mortalities. Hunting and trapping and pollution replace logging as issues for vultures. A more recent category of trouble for birds is climate change.

No doubt you can understand habitat loss as a problem for raptors since habitat loss is a problem for most animals. Every conservation section in this book lists it as a problem. The human population keeps expanding and we demand more and more resources, which often come at the expense of habitat on which other species depend. Birds have the advantage of being able to move to a new place, but only if there is a place left to move to. Now the climate crisis is shrinking those options. But some raptors may be able to relocate

to a higher latitude if it gets too hot where they are. There is evidence from 40 years of Christmas Bird Counts conducted and analyzed by the Audubon Society that 60% of common winter backyard bird species in the United States have shifted their ranges, on average, 35 mi (56 km) northward.[30]

Some birds may be able to adjust to some extent, but the higher latitudes are where warming is occurring fastest, although temperature per se is not necessarily the main problem. Consider the Rough-legged Hawk and Snowy Owl, which normally migrate farther up into the Arctic as the atmosphere warms in spring because that is where lemmings and voles breed, so there is enough prey to support offspring. In a less snowy winter, small rodents are less well protected, so population crashes of these little animals most often occur in warmer winters, and warmer winters are becoming more common. Our unlucky birds thus arrive on the tundra only to find too few baby rodents to support their own hatchlings, and they are forced to forgo reproduction. Arctic rodent populations are cyclical, so mistiming has happened from time to time in the past, but repeated warm winters spell trouble for these predators.[31] What about the harriers that nest low in brackish marshes? What happens to their nesting sites as sea level rises? Yes, in theory the marsh could migrate inland and the birds would do likewise, as they have done with rising and falling sea levels over the millennia, but odds are fairly good that in this particular millennium, there is a beach community in the way. Climate change exacerbates the loss of habitat that humans are already causing.

In addition to our simply taking up space, our infrastructure poses threats to raptors. You have probably heard (or read here) about electrocution on power lines and run-ins with wind turbines killing birds, with raptors among these. It tends to be the bigger raptors that fall victim to electrocution, and thus females more often than males (recall that they are bigger). Golden Eagles top the list in North America, which also includes Red-tailed and Harris Hawks, and the Cape Griffon (*Gyps coprotheres*) and African White-Backed Vulture (*Gyps africanus*) in South Africa. Egyptian Vultures, Spanish Imperial Eagles (*Aquila adalberti*), Bonelli's Eagles (*Aquila fasciata*), Eurasian Eagle-Owls, and Griffons (*Gyps fulvus*) are all common victims in other places. These species are large enough to span electrical components on poles and wires. Bald Eagles are too, but they tend to inhabit forests and shorelines where natural perches are abundant.[32]

Wind farms also present a source of mortality for raptors, in particular vultures and others that rely on soaring and are not necessarily agile fliers. Fortunately the more modern turbines may be slightly less lethal than earlier versions[33] because they are much higher in the sky and rotate more slowly, although the more important factor is how many birds live or migrate near the turbines.[34] Also, artificial intelligence in the form of bird-identifying robots is showing some promise in seeing raptors at a distance and temporarily shutting down rotors.[35] Since both these sources of mortality can be somewhat mitigated by engineering and placement, there is reason for hope as long as companies are willing, or forced, to consider this, but some raptors are going to continue to get zapped by our wires and run into these whirring blades. We hope to get to the point where these interactions are uncommon.

A bigger problem for raptors, and birds in general, is collisions with windows and vehicles. Window collisions are especially common for Sharp-shinned Hawks (*Accipiter striatus*), Merlins, Cooper's Hawks, and Peregrine Falcons. As with wind turbines and power lines, there are engineering solutions to this problem. Making the windows on office buildings nonreflective and visible to birds minimizes strikes. Collisions with vehicles are a tougher problem to

solve, and many raptors fall victim to these, whether they commonly forage along roadways or not. For more than 30% of urban-living raptors, vehicular collisions are the primary source of mortality, and there is not a great difference in this between urban and rural raptors.[36] Curiously, vultures do not often get hit by cars, despite frequently chowing down on roadkill on roadways and not being the quickest of fliers. Evidently, they are watching out or are more easily seen. The problem for other raptors, including owls, is that they are often chasing or swooping down on prey and simply do not notice or do not understand the speed or risk of an approaching car. An unobstructed roadway makes a perfect lane of travel for the low and silent flight of owls hunting at night, but evolution did not prepare them for steel and glass projectiles moving down these lanes at 60 mph (100 kph).

Old World vultures are particularly vulnerable to hunting, trapping, and pollution according to the Red List assessments of the IUCN.[37] These birds are hunted for traditional medicines and by poachers. An increasing problem, particularly in Africa, is the intentional poisoning of raptors by elephant and rhino poachers, known as sentinel poisonings (fig. 4.19). If wildlife rangers see a large group of vultures congregating and flying over a particular area, odds are high that there is something dead beneath them, possibly an illegally poached animal. To cover their tracks, poachers are resorting to poisoning the carcasses of elephants and rhinos (since the real value is the tusks or horns) and letting the vultures feast on the tainted meat. This often kills hundreds of vultures at a time and then allows the poachers to continue their heinous crimes undetected.

Vultures offer a cautionary tale about the ecosystem services to humans a predator provides. In South Asia and Africa in the 1990s, veterinarians started routinely injecting diseased or injured cattle with diclofenac, a cheap, nonsteroidal anti-inflammatory drug (NSAID). If a vulture of the genus *Gyps* dined on the carcass of an animal recently treated with this, the vulture died of kidney failure within days. In South Asia, more than 95% of these Old World vultures, including the White-rumped (*Gyps bengalensis*), Long-billed (*Gyps indicus*), and Slender-billed Vultures (*Gyps tenuirostris*), have died off since then. Think that is inconsequential? In India, the loss of the vultures led to an increase in feral dog populations, increasing the incidence of rabies, and an increase in rats, aiding the transmission of bubonic plague. Without the vultures to scavenge

FIGURE 4.19 A dead Griffon Vulture (*Gyps fulvus*) in a dry riverbed. Poachers commonly poison vultures as a method of masking their illegal activity. FJAH/Shutterstock

infected cattle carcasses, livestock diseases such as brucellosis, tuberculosis, and anthrax increased. Members of the Zoroastrian Parsi community, who normally use "sky burials" for their dead, in which vultures dispose of the bodies, needed to seek alternative methods. Luckily, by 2005 other NSAIDs, safe for vultures, were being identified,[38] and IUCN data suggest that vulture populations had stabilized by 2012, albeit at Critically Endangered levels.[39] We need to be more careful or, failing that, act faster when data start coming in suggesting there is a problem. It remains to be seen whether these vultures can recover.

Vultures perform this cleanup service for us, but what about other raptors? Turns out Barn Owls around agricultural fields in California and oil palm plantations in Malaysia do a good job controlling the rat and mouse populations that would otherwise be gobbling up crops. The New Zealand Falcon (*Falco novaeseelandiae*) has been shown to increase winery output by preying on bird species that eat grapes. Peregrine Falcons in Dubai help control pigeons.[40] Falconers even use raptors at airports to chase other birds away from runways, an environmentally sound method of reducing aircraft collisions with flying birds (known as "bird-strike").

Still, we humans do not seem able to learn from our mistakes. The modern environmental movement was launched largely by the publication of *Silent Spring* by Rachel Carson in 1963 in the United States. This landmark book provided evidence that the rampant use of the pesticide DDT to kill mosquitoes was also preventing the successful reproduction of birds. Poisons like this build up in predators by a process called *bioaccumulation*. If an individual robin, say, eats some DDT-poisoned insects, it may not get enough toxin to harm it, but the toxin gets stored in its body fat. A predator that then eats a bunch of these robins gets a big dose of the toxin. In the case of DDT, it was enough to prevent the successful production of sturdy eggshells. Mother raptors literally crushed these weak eggs while trying to incubate them. Few embryos survived the thinning eggshells. The United States came remarkably close to killing off its own national symbol, the Bald Eagle (fig. 4.20), driving it out of the contiguous states almost completely, along with many other raptors. Rachel Carson very effectively pointed out the problem to the public, and the conservation community rallied to the defense of the birds, eventually leading to the ban on DDT in the United States.

In some cases, however, the damage was already done. Consider the California Condor (*Gymnogyps californianus*). Mortality of these birds increased through the 1960s to the 1980s, possibly because of DDT in combination with lead poisoning from eating carcasses killed with lead ammunition, plus shooting and collisions.[41] By 1982 there were just 22 California Condors left on Earth. Scientists captured every single condor and began raising them in captivity and then releasing them, and by 2010 there were 400 birds, about half of those in the wild, albeit still with potentially dangerous levels of lead in their blood.[42] It was progress, though,

FIGURE 4.20 Bald Eagle (*Haliaeetus leucocephalus*) tending to chicks in a nest. The widespread use of DDT resulted in many raptors producing eggs with reduced shell integrity. Many generations of chicks could not be successfully hatched, and the symbol of the United States was nearly lost. CK_Images/Shutterstock

and as a direct result of the DDT ban, you can now also see Bald Eagles in every state, and they are so plentiful in parts of Alaska that they are disparaged as "dumpster hens." Peregrine Falcons are also again common. We can fix this! That is, we can if we have learned we should not spew concoctions into the environment without considering effects on creatures other than those targeted—in other words, adopting the *precautionary principle*. So far, we have not fully absorbed this lesson, since use, improper disposal, and/or emissions of pesticides and herbicides (e.g., the ubiquitous glyphosate, sold under the innocuous trade name Roundup) in industrial agriculture, as well as persistent and other chemical pollutants, still occur and in many cases are expanding.

Once again we find that predators matter. Not just to their ecosystems, but directly to us. Even the smaller predators without teeth. Let's look more closely at a few of the apex predators (and one vulture) among these impressive birds.

STELLER'S SEA EAGLE
(Haliaeetus pelagicus)

The Steller's Sea Eagle (fig. 4.21) is the largest of the sea eagles, which include the Bald Eagle, the African Fish Eagle, and various other species. As with most other raptors, females are larger than males. A female Steller's typically weighs 13.7–20.9 lb (6.2–9.5 kg). These massive birds are typically a dark sooty color, with a large white tail, white patches on the wings, and a bright yellow beak and feet. As with all sea eagles, the Steller's primary diet is fish, a good strategy as its primary range is along the northeastern coast of Siberia along the Bering Sea.

Relatively little is known about the Steller's Sea Eagle compared to other species because its habitat is so isolated and difficult to access. It can regularly be seen perched on floating sea ice, waiting for the next meal to swim by. The San Diego Zoo, an excellent place to see these birds in person, suggests that the Steller's Sea Eagle is a "glacial remnant," having evolved and survived in the same region over several ice ages.[43]

While the Steller's eats primarily fish, gorging on salmon and trout during spawning season, its palate extends to other delicacies like crabs and puffins. Unlike most of the true eagles that typically eat only live prey, sea eagles will happily scavenge. Its not-too-distant cousin the Bald Eagle regularly picks through human garbage. In fact, a good place to spot a juvenile Bald Eagle who is not very good at hunting yet is your local landfill. As we stated above, large groups of Bald Eagles feasting on garbage in Alaska have earned them the not so majestic nickname "dumpster hens." That is not how most of us want to envision our national symbol.

An estimated 3,600–3,800 Steller's Sea Eagles currently exist, and their numbers are decreasing.[44] One hurdle they face is their limited range. Although they can fly, they have not adapted to different habitats as some other raptors have. Increasing human activity is also taking a heavy toll on their population. Logging is affecting their nesting areas. The construction of hydroelectric dams and the introduction of industrial contaminants into the waters where they fish are affecting them both directly (loss of fish prey) and indirectly (consumption of toxic fish).

ANDEAN CONDOR
(Vultur gryphus)

The Andean Condor (fig. 4.22) is the largest of the New World vultures (Cathartidae) and the only member of the genus *Vultur*. Some consider it to be the largest flighted bird in the world (in this definition, "largest" is a combination of weight *and* wingspan). Unlike in most birds of prey, the male Andean Condor is larger than the female.

FIGURE 4.21 Steller's Sea Eagle (*Haliaeetus pelagicus*). Standing on floating ice offers this species both a chance to take a break from flying and a good vantage point to snag a passing fish.

Devendra Deshmukh / Shutterstock

FIGURE 4.22 Andean Condor (*Vultur gryphus*). A carrion eater and not a hunter, the Andean Condor qualifies as the current largest raptor in the world.

reisegraf.ch/Shutterstock

A large male can weigh up to 33 lb (15 kg), with a wingspan of 10.8 ft (3.3 m).

Like the body of most vultures, that of Andean Condors is covered in mostly black feathers. They are quite distinctive because of a fluffy ruff of white feathers at the base of the neck. In true vulture style, the head is nearly devoid of feathers. Males often have a large comb (*caruncle*) on top of the head. The skin of the head is typically some shade of red, but the color changes with the bird's changing emotional state.[45]

The Andean Condor's range spans most of the Andes Mountains throughout western South America. They can occupy a wide variety of habitats, from open grassy plains to rocky mountainous areas, and occasionally lowland deserts and the Patagonian southern beech forests. These vultures are regularly found along the Andes up to 16,000 ft (5,000 m) in elevation. They have a wide variety of prey, depending on their territory, provided the prey is already dead. Compared to the critically endangered California Condor, with a population of just a few hundred, the Andean Condor population is doing markedly better (approximately 6,700 individuals). However, like the populations of many of the predators we review in this book, that of the Andean Condor is decreasing, and for many of the same reasons, primarily human impact. This includes intentional poisoning by farmers who consider all raptors a threat to livestock. Combine this with a long-lived bird (up to 70 years) that may lay only a single egg every two years, and we may see a continued decline in their population.

GYRFALCON
(Falco rusticolus)

The Gyrfalcon (fig. 4.23) is the largest true falcon. As is true with many animals, those living in colder climates tend to be larger because it is easier to retain heat by being large. In the case of the Gyrfalcon, colder climates include the northern half of North America, and "largest falcon" means up to 4.6 lb (2.1 kg).

Color variations of Gyrfalcons range from snow white to near black. Like most falcons, Gyrfalcons prey mostly on other birds, which is why they must be such excellent flyers. Included in their regular tundra prey are Ptarmigan (a member of the grouse family) and colonial seabirds. Because of their relatively large size, Gyrfalcons hunt mammals more readily than do other falcons. They are also known to regularly fly offshore during winter months to hunt pelagic seabirds. While they are diminutive in

FIGURE 4.23 Gyrfalcon (*Falco rusticolus*). Although not as fast as its smaller cousin the Peregrine Falcon, the Gyrfalcon can reach a very respectable speed of 156 mph (251 kph). Rob Palmer Photography / Shutterstock

size compared to many predators, Gyrfalcons exist in an environment where they can hunt many prey species, but almost no one preys on them.[46]

Gyrfalcons nest in the Arctic and are known to lay eggs even during freezing temperatures. They typically produce two to four eggs in a clutch, and after hatching, the offspring are independent of their parents in three to four months. Things happen much faster in an environment where the summer is short.

The current population estimate is 2,200–3,900 individuals. While this may not seem like many, especially compared to other species, these birds live in an environment that cannot support large population densities. Currently the Gyrfalcon conservation status is considered of Least Concern. While this is certainly encouraging news, the effects of climate change on the Arctic environment they rely on remain to be seen.

EURASIAN EAGLE-OWL
(Bubo bubo)

The Eurasian Eagle-Owl (fig. 4.24) is one of the largest owl species (up to 9 lb, or 4 kg), perhaps surpassed only by Blakiston's Fish Owl (*Bubo blakistoni*). It has a wingspan of up to 6.5 ft (2 m). Its coloration is typically brown to light gray, with interspersed vertical black feathers to give it a mottled look. Like its very close cousin the Great Horned Owl of North America, the Eurasian Eagle-Owl has two telltale groups of pointed feathers on top of its head that look somewhat like horns. The easiest way to distinguish these two species, because they are almost identical at first glance except for their size, is eye coloration. The Eagle-Owl's eyes are bright orange, while the Horned Owl's eyes are a piercing yellow.

The Eurasian Eagle-Owl is one of the most widely distributed raptor species. As its name implies, the full range of this species spans at least 60 different countries throughout most of northern and central Asia, all of mainland Europe, and even (rarely) several countries in northern Africa. With such an enormous range, they do well in various types of habitats, including ocean-side cliffs, mixed forests, marshes, rocky outcrops, mountains, and more. Their prey is equally varied. They will eat insects, amphibians, rodents, lagomorphs (rabbits and related small mammals), birds (including woodpeckers and herons), lizards, snakes, fish, crabs, and even sometimes other predators, including martens, foxes, and other birds of prey, including other owls.

The Eurasian Eagle-Owl has perhaps the widest prey base of any single predator species. While they typically hunt smaller prey (something easy to subdue), there have been reports of Eagle-Owls killing prey up to the size of young sheep and the European Roe Deer. Anecdotal evidence suggests that they have even been seen preying on small Tiger cubs in Russia! While we have not found any verifiable evidence to corroborate this story, it is not hard to imagine a sharp-eyed owl sitting silently in a tree, waiting for a mother Tiger to leave her newborn cubs for the evening to hunt. A 3 lb (1.4 kg) cub would be no match for this feathered assassin. That would be the ultimate in predator eating predator, an example of why the term *apex predator* is hard to define!

The current population of Eurasian Eagle-Owls is unknown, but estimates of up to half a million adult individuals have been suggested.[47] Their conservation status is considered of Least Concern by the IUCN, but their numbers are declining.

In addition to the multitude of wild habitats Eurasian Eagle-Owls occupy, they also do fairly well in the presence of humans, possibly because of their nocturnal predilections. A family of owls could be living in your backyard and you would never know it. These owls are not always shy and reclusive, though. During a 2007 soccer (football,

FIGURE 4.24 Eurasian Eagle-Owl (*Bubo bubo*). This owl sometimes eats other predators, like the Pine Marten (a mammal related to weasels and badgers) shown here.

Ondrej Prosicky / Shutterstock

if you're not American) qualifying match between Finland and Belgium, a Eurasian Eagle-Owl flew into the crowded Helsinki Olympic Stadium and landed in the middle of the pitch while the game was still going on! It looked around for a moment and then flew and perched itself on top of the goal. The game was postponed for eight minutes until the plucky owl took flight again, did several loops around the stadium, swooping toward people along the way, and eventually landed on a barrier railing off the field, after which Finland defeated Belgium 2–0.[48] The video is certainly worth a look.

5

CATS

You might already know something about the cat family, Felidae, because you might very well live with one (or more) of its members. As you will see, that little purring felid who curls up on your bed, or hides under it, is not all that different from its wild predator relatives. In fact, in his book considering the fate of the Earth if humans abruptly vanished, *The World without Us*, Alan Weisman speculates that while domestic dogs would not last long in our absence, our leftover cats would do just fine. It might seem like your cat knows this.

Modern cats, or *neofelids*, have existed since the Oligocene epoch, about 30 million years ago (mya). There have always been fewer predators than prey species; thus fossils of the former are rarer and harder to find. However, scientists have been able to assemble some history of the cats from fossils. Cats as a group arose from a common ancestor of saber-tooths and modern cats. The genus *Pseudaelurus* included the first feline-like species known to inhabit the Northern Hemisphere, including the well-known Machairodontinae (saber-tooths) (fig. 5.1). Here you see why Linnaeus's scientific naming strategy is a good thing. The common name for members of the Machairodontinae (which translates to "dagger tooth") is Saber-toothed Tiger or Saber-toothed Cat. But it is not a Tiger or a cat! A modern cat is in the subfamily Felinae. Our saber-tooths are in a different subfamily altogether, so scientists avoid confusion by agreeing to call them by their

genus, *Smilodon* (or "knife tooth"). If you have been to La Brea Tar Pits, located incongruously right off Wilshire Boulevard in downtown Los Angeles, you've seen *Smilodon* fossils, lots of them. Over 2,000 saber-tooths evidently met their demise trying to snag dinner at these pits. The tar conveniently trapped the prey, but it also trapped the predators. If you have not visited La Brea, put it on your list. And while you are there, consider that had you been born 15,000 years ago, you would have been walking among those (sort of) kitties.

So, when and where did modern cats arise? Unfortunately, for us anyway, we have not found a tar pit chock-full of nicely preserved fossils of Felinae, but scientists have figured out some of their lineage. As we saw with sharks, one feature of a vertebrate skeleton that fossilizes especially well is teeth, because the outer coating of enamel is so hard. Thus, teeth are often used to characterize

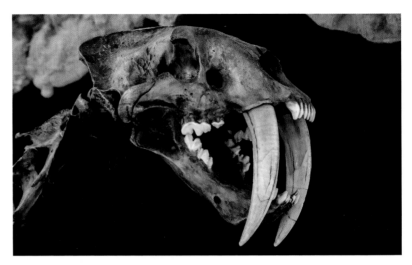

FIGURE 5.1 Skull of a saber-tooth (*Smilodon*). Saber-tooths existed for nearly 40 million years and became extinct about 10,000 years ago. The saber-tooth-free world we inhabit is unusual in recent biological history. Mardoz/Shutterstock

BOX 5.1   BIG CAT CANOODLING

By following the genetic patterns in both DNA and structures from fossils of ancient cats, evolutionary biologists have developed the cat family tree. The oldest known *Panthera* species fossil (known as *Panthera blytheae*) was found in Tibet, suggesting that both *Panthera* and *Neofelis* may have arisen in Asia,[1] although there is some evidence suggesting that Africa may have been the site of origin. It then appears from genetic analysis that the pantherines radiated rapidly during the Pleistocene epoch, between 2.5 and 11.7 mya, about the time of the last ice ages. This is significant because the lower sea levels at that time exposed connections between continents that were previously underwater. (We are seeing the reverse of this now: as the atmosphere heats, ice melts, sea levels rise, and land slips underwater, albeit at a much faster pace than that caused by natural phenomena.) This exposed land allowed migration to new places for some animals (including the American Cave Lion, *Panthera atrox*) and, along with the movements of the continents, probably accounts for this radiation. Speciation of the *Panthera* branch into the current five big cats (if we include the Snow Leopard) occurred after this. So how did we get these distinct species?

For speciation to occur, the population of the stem *Panthera* had to be split in some way such that the separated populations did not interbreed. Let's say a population of the original ancient Asiatic cats took advantage of the low sea level and headed across the newly exposed Bering Land Bridge from Asia to North America. They brought with them just a subset of their Asiatic ancestral cat genes; in other words, they were different in some ways from other members of their species, just as you are different from other people. With the combination of the underlying genetic basis of these differences, genetic drift (each succeeding generation is somewhat different from the previous generation), and different natural selection pressures in North America and Asia, these North American cats evolved differently from those in Asia. The key is that as long as the two remained reproductively isolated from one another—that is, the two populations could not interbreed—they continued to evolve separately. When climate change led to thawing ice and rising seas, Asia and North America were again cut off from one another and these cats became reproductively isolated from one another for the long term, and voilà!—two species from one. Jaguars likely arose this way and were found throughout the Americas until the 1900s. Thomas Jefferson recorded Jaguars as an American species in 1799. The now-extinct American Cave Lion also likely migrated and diverged into different species through the same process.

groupings of fossils, and cats are no exception. Members of the subfamily Felinae all have conical teeth, similar to those of most modern cats. Based on fossils, the ancestor of big cats—meaning those in the genus *Panthera*, including Lions (*P. leo*), Tigers (*P. tigris*), Jaguars (*P. onca*), Leopards (*P. pardus*), and Snow Leopards (*P. uncia*), along with the two species of Clouded Leopards (*Neofelis nebulosa* and *diardi*)—are thought to have diverged from the Felinae around 10 mya. In addition to their conical teeth, *Panthera* cats are identified by incompletely ossified hyoid bones at the base of the lower jaw. This is thought to be what allows them to roar. Genetic studies have further supported and clarified this evolutionary picture (box 5.1).

There are 38 species of wild cats currently recognized in the cat family Felidae by the Cat Specialist Group of the IUCN (International Union for the Conservation of Nature) (table 5.1). Cats are native to every continent except Australia and Antarctica (domestic cats have been introduced to Australia). They live in every sort of habitat, from

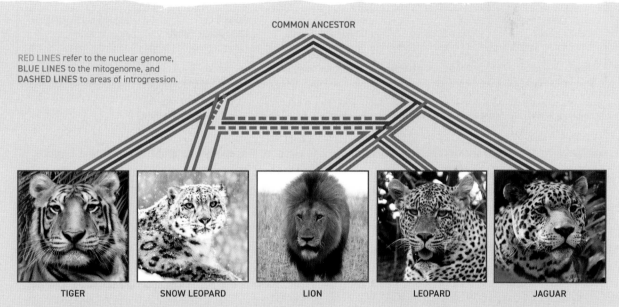

COMMON ANCESTOR

RED LINES refer to the nuclear genome, BLUE LINES to the mitogenome, and DASHED LINES to areas of introgression.

TIGER SNOW LEOPARD LION LEOPARD JAGUAR

Hybridization among big cats. By comparing nuclear and mitochondrial DNA between big cat species, geneticists discovered not only a shared heritage, but areas of genetic admixtures within the genomes.
Diagram re-created with permission of Dr. William Murphy, Texas A&M University. Abeselom Zerit / Shutterstock

But biology is not quite so neat, and recent genetic analysis has revealed a secret: the big cats have crossbred with each other since speciating.[2] How do we know? They share a lot of relatively recent genes. Remember that, unlike they do now, these distinct species once had overlapping ranges. There were Lions in North America that could have bred with the Jaguars there, and probably did. Some regions of Lion and Snow Leopard genomes are far more similar than others, indicating that the two species most likely hybridized after becoming separate species. It appears that with big cat hybridization (as with most mammal interbreeding), the most significant factor is not that the hybrid offspring will ultimately create a new species, but rather that by hybridizing back to parent species, they can pass adaptive traits from one species to another. Big cats are genetically similar enough that they can still interbreed, complicating conservation.

TABLE 5.1 Thirty-eight species of wild felines currently on the IUCN Red List of Threatened Species

Tiger (*Panthera tigris*)	Eurasian Lynx (*Lynx lynx*)	Northern Tiger Cat (*Leopardus tigrinus*)	Wild Cat (*Felis silvestris*)	Fishing Cat (*Prionailurus viverrinus*)
Lion (*Panthera leo*)	Bobcat (*Lynx rufus*)	Southern Tiger Cat (*Leopardus guttulus*)	Black-footed Cat (*Felis nigripes*)	Rusty-spotted Cat (*Prionailurus rubiginosus*)
Leopard (*Panthera pardus*)	Iberian Lynx (*Lynx pardinus*)	Margay (*Leopardus wiedii*)	Chinese Mountain Cat (*Felis bieti*)	Leopard Cat (*Prionailurus bengalensis*)
Jaguar (*Panthera onca*)	Canada Lynx (*Lynx canadensis*)	Andean Cat (*Leopardus jacobita*)	Sand Cat (*Felis margarita*)	Flat-headed Cat (*Prionailurus planiceps*)
Snow Leopard (*Panthera uncia*)	Caracal (*Caracal caracal*)	Pampas Cat (*Leopardus colocolo*)	Jungle Cat (*Felis chaus*)	Asian Golden Cat (*Catopuma temminckii*)
Cheetah (*Acinonyx jubatus*)	African Golden Cat (*Caracal aurata*)	Ocelot (*Leopardus pardalis*)	Serval (*Leptailurus serval*)	Borneo Bay Cat (*Catopuma badia*)
Sunda Clouded Leopard (*Neofelis diardi*)	Puma (*Puma concolor*)	Guiña (*Leopardus guigna*)	Pallas's Cat (*Otocolobus manul*)	Jaguarundi (*Herpailurus yagouaroundi*)
Clouded Leopard (*Neofelis nebulosa*)	Marbled Cat (*Pardofelis marmorata*)	Geoffroy's Cat (*Leopardus geoffroyi*)		

NOTE: Although the IUCN Cat Specialist Group website indicates there are currently 40 species of wild cats, the discrepancy is likely the result of constant refining of species definitions based on newly available scientific data.

wetlands to deserts, from the tropics to some of the coldest places on Earth. They range in size from the tiny Black-footed (*Felis nigripes*) and Rusty-spotted (*Prionailurus rubiginosus*) Cats at under 3 lb (1.4 kg), to the biggest Tigers, weighing in at up to 750 lb (340 kg). Despite this staggering variation in size, all cats are anatomically and behaviorally surprisingly similar, such that you would recognize them all as cats. The bigger species tend to be top predators, but a few of the smaller ones also step into that role or have some interesting feature that we just must share with you. Let's explore these felines.

CATS AND US

In addition to serving as common, if a bit persnickety, animal companions, cats are everywhere in society and have been for a long time. Leopards show up in artwork from ancient Greece, Mesopotamia, and the Byzantine Empire. Jaguar Warriors were members of the Aztec military elite. The Egyptian goddess Sekhmet is usually depicted with the face of a Lion, and mummified Lions have been found in the ancient necropolis of Saqqara. The Sphinx, Griffin, Manticore, Sea-Lion, and Lamassu are all mythical beings who are half Lion and half something else. Lions also show up in religion: the Rastafarian Lion, the Lion of Judah, Daniel and the Lion's den, the Ashokan Pillar of Buddhism, and more. Depictions of big cats have been used by societies around the globe for nearly all of recorded history and still are.

Fur from animals is not the popular status symbol it once was, but cats historically and even currently are turned into coats and rugs. Cat names or images are associated with products way too numerous to mention. Movies and television shows feature real or animated cats. Humans admire these animals and use them to attract, convey strength, and entertain.

FIGURE 5.2 The Lions of Tsavo mounted in the Field Museum of Natural History in Chicago. The pelts from these two Lions were originally turned into rugs. It is suspected that their manes fell out as a result of wear and insufficient preservation before being restored and mounted by the museum.

Of course, cats are also deadly. For most of human history, these massive predators posed a real threat. Between 1898 and 1899, two "man-eating" Lions were famously said to have killed more than 130 railroad workers in the Tsavo region of eastern Kenya, although DNA from *only* 35 different people has been identified from their preserved stomachs. They were the story behind the 1996 movie *The Ghost and the Darkness*. Lieutenant Colonel John Patterson, a hunter and conservationist, eventually found and killed the Lions (fig. 5.2).

There was also the Man-Eater of Mfuwe in Zambia in 1991, which killed at least six people. Because he was seen parading through town with a laundry bag he had apparently taken from a victim's home, locals were convinced he was a demon, but the Zambian government had outlawed hunting except as part of a safari, so they were not able to legally kill the Lion. He was eventually shot by a man on safari and is now also on display in Chicago's Field Museum.[3]

Cats other than Lions sometimes turn deadly. The Man-Eater of Calcutta, a Tiger, was said to have killed more than 200 people before he was captured in 1903. Jim Corbett, a conservationist and hunter and also the founder of India's first national park, is known to have killed, among others, the Panar Leopard, which killed 400 people; the Leopard of Rudraprayag, which killed 126 people; the Champawat Tiger, which killed 436 people; and the Chow Char Tiger, which killed 64 people. There is no way to verify these kills, and the numbers may have been exaggerated, but it is generally agreed that the human predation rate of these particular animals was unusually high. It has been estimated that at least 375,000 people died from Tiger attacks between 1800 and 2009.[4] Currently, it is estimated that about 85 people are killed or injured by wild Tigers annually.[5]

It is not clear that all these killers were also eaters. Cats do sometimes resort to human prey when the cats are impaired by an injury—for example, broken teeth (a problem that sharks, you will recall, do not experience, at least not for very long). That was the case with the Mfuwe Lion. Sometimes attacks may be caused by human encroachment into a cat's territory or depletion of its prey species. The former was likely the case when in 1991, a Puma[6] in Boulder, Colorado, killed a student out jogging, a story related in *The Beast in the Garden* by David Baron. Sometimes there is no clear explanation, and it seems that an aberrant cat just figures out that humans are relatively easy prey.

With the advent of guns, humans have killed many more cats than vice versa. Hunting big cats has been very popular (some of you may recall the recent controversial story of Cecil the Lion; we will talk about that below). Hunting the hunters is nothing new to humans. Tiger hunting was considered the sport of kings. Three Indian kings documented killing an estimated 2,150 Tigers between them during the 1800s. Anecdotal reports suggest that one maharaja of India personally killed more Tigers than exist in the wild today. King George V of England participated in a hunt that reportedly killed 39 Tigers over 10 days, more than 1% of the total number of wild Tigers today. Albert, the Prince of Wales, killed his first Tiger in India in the late 1800s (fig. 5.3).

One big reason that humans continue to kill big cats is the perceived medicinal value of various cat parts (box 5.2), despite a complete lack of evidence suggesting these "treatments" have any efficacy.

FIGURE 5.3 The Prince of Wales and hunting party with the prince's first Tiger (*Panthera tigris*). Library of Congress / LOT 8351, no.116

BOX 5.2 TIGER CLAW FOR YOUR HEMORRHOIDS: PRESUMED MEDICINAL VALUE OF BIG CAT PARTS

Because of its strength and power, the Tiger has played a major role in traditional Chinese folk medicine, which is currently the biggest outlet for poached Tigers. Although no scientific evidence supports *any* effectiveness of consuming Tiger parts for medicinal purposes, the idea is so ingrained in the culture that millions of people still believe in their curative qualities. Changing hundreds, if not thousands, of years of culture is a major challenge in Tiger conservation. We are not sanguine that this can be accomplished before the wild Tiger becomes extinct.

Some of the purported uses include bone for arthritis and rheumatism; bone wine (crushed bones steeped in wine) for strength; blood for a strong constitution; claws as a sedative; fat for dog bites, vomiting, and hemorrhoids; feces for alcoholism; feet to ward off evil spirits; nose leather to treat bites and superficial wounds; skin to cure fever; eyeballs for malaria and epilepsy; brain for laziness; bile for convulsions in children; penis for an aphrodisiac; teeth for rabies; and whiskers to treat toothaches.

TIGER PART USE IN CHINESE FOLK MEDICINE

BRAIN
laziness

EYES
epilesy

BILE
convulsions

FAT
rheumatism

NOSE
snakebite

WHISKERS
toothache

TAIL
skin diseases

TEETH
fever

SKIN
mental illness

BLOOD
willpower

CLAWS
insomnia

BONES
arthritis, headache, dysentery

Description of Tiger parts used as Chinese folk medicine. Various Tiger parts are purported to have healing effects. Although these beliefs are not supported by scientific evidence, adherence to tradition and the resulting demand for Tiger parts is a primary cause of Tiger poaching.

This is especially a problem for Leopards in India and Tigers everywhere. Clearly these animals embody vigor and strength, but imbibing their parts does not confer these qualities to a person. Nevertheless, ancient Chinese medicine, still practiced today, deploys nearly every part of a Tiger. The Convention on International Trade in Endangered Species (CITES) banned cross-border trade in Tiger parts in 1987, and in 1993 China banned domestic trade, but the demand is still significant and the market endures underground. This continues to drive much of the poaching of Tigers worldwide, and the carcasses are frequently smuggled into China. Despite the bans, the Wildlife Protection Society of India reported 846 Tigers and 3,140 Leopards poached between 1994 and 2008 in India[7] alone.[8]

CATS IN CAPTIVITY

Members of the genus *Panthera* and other large cats are big, strong, predatory animals fully capable of killing and eating people. And yet people keep them in captivity. You sometimes hear of owners or zookeepers being attacked by their captive animals. Famously, in 2003, Roy Horn of the Siegfried and Roy Las Vegas magicians, was attacked by his Tiger. It was not clear what precipitated the attack. It rarely is. While the perceived danger is relatively high, on average less than one person dies per year in the United States as the result of contact with one of the 2,300 or so Tigers in captivity.[9] Every person who has been killed by a captive wild cat in the United States was intentionally in the cat's environment, meaning that no Tiger has ever escaped and killed an innocent bystander. Comparatively, the Centers for Disease Control and Prevention reports that an estimated 4.5 million dog attacks happen per year, resulting in approximately 30 human deaths. Of course, there are more than 89 million dogs in

the United States, so your odds of running into a cranky dog are much higher than your odds of running into a Tiger.

While serious injury is far more likely from our house pets than from a captive wild cat, we still carry a greater inherent fear of top predators, a theme repeated throughout this book. One way to avoid the risk entirely is to avoid having wild cats in captivity. Most states in the United States have laws prohibiting private ownership of large predatory animals. So, these animals are captive mostly in zoological parks and sanctuaries. Some people are not fans of zoos, but the role they play in public education and in maintaining genetic diversity through captive breeding is likely to prove necessary if cats in the wild are to be saved.

Some zoos try to avoid direct contact between Tigers and their human keepers, while others go the other way and encourage contact from the time the Tigers are born, with zookeepers essentially raising the cats. Is one way better than the other for the cats? Unfortunately, we cannot just ask them, but their stress levels, which we can measure, should be one indicator. For example, somewhere, unfortunately, you've probably seen a cat pacing back and forth in a cage. Pacing is recognized as the most prominent behavioral form of stress in wide-ranging terrestrial predators, including cats. Counting time spent pacing is one way researchers can quantify stress. A more direct method is to measure levels of the hormone cortisol, which is higher in animals under stress.

It is important to understand what we mean by *stress*. While stress is typically viewed as something negative that should be minimized, research shows that a moderate amount of stress is a key element of a free-ranging animal's well-being.[10] Too little stress produces health and behavioral problems similar to those resulting from too much stress. Similarly, both pacing and cortisol levels can increase when cats are anticipating something they desire (food and mating, not necessarily in that order).

Clearly, however, *too much* negative stress is something zookeepers try to limit. Today, almost all big cats in zoos were born in captivity and are likely to spend their lives with at least some level of exposure to humans. Some zoos believe that interaction between humans and big cats, under safe and positive conditions, can reduce the amount of negative stress those cats experience while living in a human environment. Research has shown that increased interaction between felines and keepers can result in lower pacing and cortisol levels, as well as increased cognitive functioning, reproductive success, and both physical and mental health.[11] Still, these animals (along with crocodiles, Wolves, bears, and more) are top predators and there is a certain level of inherent danger. Disagreement abounds about which style of husbandry is best.

WHAT IS A CAT?

Most cats, including your companion kitty, have similar traits, as we will discuss; however, a few are unique to the big cats. Recall that the big cats are those in the genus *Panthera*: Lions, Tigers, Jaguars, Leopards, and Snow Leopards. In addition to their large size, as we discussed in box 5.1, they also share a close genetic history.[12] Note that other cats, such as Cheetahs (*Acinonyx jubatus*) and Pumas (*Felis concolor*), may be large but they are not in the genus *Panthera*, and for this and other reasons we discuss below, they are not considered "big cats."

Besides size, the trait that is unique to big cats is perhaps the most recognizable and memorable. They, and only they, roar. The ability to roar is the result of modifications of the larynx (or "voice box") in comparison to other cats. There is a trade-off: cats that roar cannot purr continuously like smaller cats. Lions have the loudest roar of the big cats, and to hear a Lion's roar is to experience audio greatness. It can reach 114 decibels, about 25 times louder than a gas-powered lawn mower. You can hear it five miles away. Be glad your companion cat cannot roar! Big cats roar to communicate with each other as a greeting, to keep a group together, or to warn competitors to steer clear or expect a fight. If you have been fortunate enough to hear the chest-rattling roar of a wild big cat, you surely understand that message.

There are complications, however, with designating a cat as a big cat based on its size and ability to roar. Many researchers consider the Clouded Leopard (genus *Neofelis*) to also be a big cat because genetically it seems to be in the same lineage, and it is also rather big. It apparently does not roar, though. Neither do Cheetahs or Pumas, large cats in the subfamily Felinae (i.e., not Pantherinae).[13] The Snow Leopard was recently added to the *Panthera* pantheon because of genetic similarities and the ability to hybridize between genera. Although now classified as a big cat, the Snow Leopard neither roars nor purrs (because of further reduction in the ossification of the hyoid, a trait derived only in the Snow Leopard lineage). Again, biology resists neat categorization, and in fact, your authors cannot entirely agree on whom to put where. Thus, we are going to dodge the issue by focusing henceforth on cats as a group, big and small.

We said that wild cats are much like your house cat, so what makes a cat a cat? Cats, like dogs, elephants, and rhinos, are all mammals and as such can maintain a constant internal body temperature, have fur or hair, and feed their young milk. But what else?

Big teeth are a characteristic of cats, big and small. They all have long, sharp, strong canine teeth. Cats might fairly wonder why we do not call them "feline teeth." The Clouded Leopard has the

FIGURE 5.4 Clouded Leopard (*Neofelis nebulosa*) showing its large gape. Although the gape dimensions of big cats have not been published, the champion appears to be the Clouded Leopard. It also has the largest canine teeth compared to body size of all cats. dalbert Dragon / Shutterstock

longest canines for its size. They are comparable to those of *Smilodon*, or saber-tooths. The Clouded Leopard also has the biggest gape of the cats, nearly 90° compared to about 65° for a Puma (fig. 5.4). Clouded Leopards branched off within the *Panthera* lineage early on, so it is not surprising that, of the modern cats, they are most similar to that big-toothed common ancestor. Members of the Felinae also share a highly domed skull (fig. 5.5), allowing for both strong jaw muscles and a widely gaping mouth for a powerful initial bite ensuring a good, and often fatal, grip on dinner. Give a pat to a cat and a dog and notice that difference in head shape.

Maybe while you are patting your cat's head, she will lick you and you'll feel that sandpapery tongue (fig. 5.6A), a characteristic of all cats. Those rigid projections, or papillae, on the tongue allow the cat to remove fur, feathers, and skin from prey, providing better access to the meat and internal organs while avoiding ingesting all those indigestible coverings. The papillae also facilitate drinking, which we will get to. In addition, the tongue surface is handy for personal grooming. Cats are fastidious about keeping their fur clean and distributing oils from skin glands to keep it in good shape. Some of the oils are also used to communicate and mark territory, which is what your cat is doing when he seems to affectionately rub his face against you. In fact, he is marking you like a cat would mark a tree in his territory. That is also what he is doing if he's spraying around your

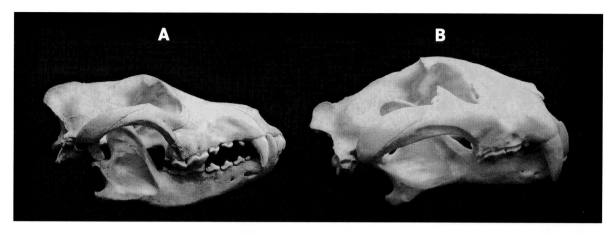

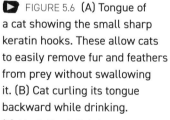 FIGURE 5.5 **Comparison between the skull of a Wolf (*Canis lupus*) (A) and a Tiger (*Panthera tigris*) (B).** Felids have a far more domed skull and large canine teeth, heavily set into the skull. Conversely, canids have a more elongated skull and weaker canine teeth, but they also have more teeth (42 vs. 30 for Wolf and Tiger, respectively), including more carnassial teeth for slicing.

FIGURE 5.6 **(A) Tongue of a cat showing the small sharp keratin hooks.** These allow cats to easily remove fur and feathers from prey without swallowing it. **(B) Cat curling its tongue backward while drinking.**

(A), Myrtle Beach Safari,
(B), Ondrej Prosicky / Shutterstock

house. He just wants it to be clear to everyone that this house is his. You knew that, right?

A cat tongue facilitates drinking, of course, and works the same way for all cats, demonstrating the fine form and balance we tend to attribute to a cat. Rather than merely scooping up water with the tongue, like a dog does, a cat similarly curls its tongue backward (fig. 5.6B), but then only the top actually touches the liquid surface. The cat then lifts its tongue back up, trailing a column of liquid between tongue and surface. The cat closes its mouth, effectively biting off the column, and thus grabbing a drink before gravity collapses the column back into the bowl. A domestic cat repeats this motion about four times per second. Bigger cats go more slowly because they have larger tongues, so the liquid has more surface to adhere to and thus falls back less quickly. Check out this video[14] to see how this works.

A cat's coat provides insulation, identification, and camouflage. Cats can have a solid coat color or are spotted or striped, sometimes changing from patterned to solid as they mature. (See box 5.3 to learn how to identify cats by their spots.)

BOX 5.3 ▸ SEEING SPOTS

Perhaps the most common camouflage technique used by felines is spots, but all spots are not the same. The three largest spotted cats are often confused with each other. Cheetahs have relatively small solid polka dots. Jaguars, on the other hand, have large open rosettes of nonconnected black markings, typically with darker tan fur and a few small solid dots within the rough circle. Leopards seem to be somewhat in between the two, with medium-sized open rosette patterns made up largely of small solid polka dots. All these patterns offer good camouflage, contributing to the cats' ability to be effective and lethal predators. Now you should be able to tell the difference between these species solely by seeing their spots.

Comparison between spots of Jaguars (*Panthera onca*), Leopards (*Panthera pardus*), and Cheetahs (*Acinonyx jubatus*).

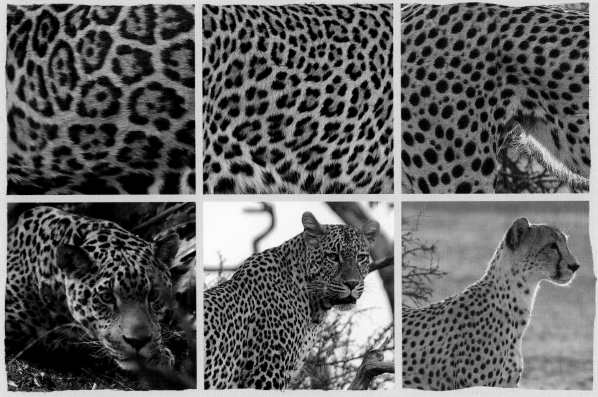

JAGUAR   LEOPARD   CHEETAH

Several cats are black, and black Jaguars and Leopards are sometimes mistakenly called "black panthers" (fig. 5.7). In fact, there is no species named "black panther"! But there are black Jaguars, Leopards, and, historically, potentially Tigers. All black cats are *melanistic*, meaning they have higher amounts of melanin, a pigment in skin, fur, and eyes. The black color in the big cats comes from the MC1R gene, which is not the same gene as the one that makes a house cat black. Melanistic, or black, Jaguars occur about 5% of the time.[15] What people call Florida "Panthers" are actually Pumas, a completely different genus. And they are never black. Fans of the Carolina Panthers football team may be disappointed to learn that the sleek black animal their mascot depicts does not actually exist in the real world.

Tigers are born with stripes that camouflage them in dappled shade (fig. 5.8). They are the only

feline that uses stripes as its primary camouflage pattern. Each Tiger is born with a unique stripe pattern; thus, individuals can be identified throughout their lives. The stripes are on the skin, not only the fur. The standard Tiger design is a burnt-orange coat with bold black stripes and white on the underbelly and parts of the face. At first glance, orange would not seem to be the best coat color for a species that hunts largely in dense green foliage. In fact, no mammal has green fur, but its absence is not a major disadvantage. Like sharks, Tigers often have a high density of rods, retinal cells that allow for better vision in low light conditions. The same is true for many of the Tiger's prey species, including ungulates (hoofed animals). What these species gain in low-light vision, they lose in color perception because they have fewer cone cells. The upshot of this dichromatic vision (seeing in two colors) is that these animals see

FIGURE 5.7 Melanistic Leopard (*Panthera pardus*). Coats of melanistic cats are not uniformly black. In correct lighting their spotted or striped coat pattern is still visible. Myrtle Beach Safari

FIGURE 5.8 Tigers (*Panthera tigris*). The Tiger is the only species of feline that uses stripes as its primary form of camouflage. Each Tiger's stripe pattern is as unique as a human fingerprint. Myrtle Beach Safari

FIGURE 5.9 Dichromatic versus trichromatic vision: how a deer sees a Tiger (*left*); how a human sees a Tiger (*right*). Ondrej Prosicky / Shutterstock

orange and green as the same color (fig. 5.9).[16] Other potential prey animals, including monkeys and birds, view the world in broader color spectra and would therefore more easily spot a Tiger moving in the forest. Although most Tigers come in the standard variety of orange with black stripes, they historically came in many colors, although always with stripes (box 5.4).

In addition to stripes, regardless of fur color all Tigers have white spots on the back of their ears

SORRY, LSU, AUBURN, CLEMSON, ETC.: ALL TIGERS ARE NOT ORANGE

Let's consider some of the historic color variations in Tigers. Why did they exist and what happened to them? And remember, these are colors of Tigers as a single species, *Panthera tigris*, and not different subspecies of Tigers. A standard Bengal Tiger is orange with black stripes. There is the royal white Tiger, a white Tiger with black stripes. White would be good camouflage in snow, and while you probably do not envision snow in India, where the Bengal Tiger originates, remember that India includes the Himalayas, which are certainly snowy. But these animals are also sometimes born in the jungle (natural selection cannot plan ahead!), and it turns out that they likely do okay there too because, as we said above, the grazers they hunt do not see color all that well. In a world of shadows and light, white can hide. Although thought to be extinct in the wild, a form of white Tiger was seen in the Nilgiri Biosphere Reserve in southern India in 2017. This is the first such sighting since 1952. The allele (gene variant) responsible for the royal white coloration has recently been identified by geneticists in China.[17]

STANDARD

ROYAL WHITE

SNOW WHITE

GOLDEN TABBY

Comparison of Tiger (*Panthera tigris*) coat color variation.
Myrtle Beach Safari

(*Continued overleaf*)

BOX 5.4   (Continued from previous page)

A snow-white Bengal Tiger is white with white stripes. You can see these faint stripes, which show up as pale gray as well as slightly darker markings on the face and the end of the tail. Note that these are not albinos, which result from a complete lack of melanin. Albinism is often confused with *leucism*, which is a reduction, but not total absence, of melanin. Snow-white Tigers are not fully white and do not have the pink eyes characteristic of albinos. Like all Tigers, they are born with blue eyes, but unlike many others, they retain the blue color. They were last known to exist in the wild in the 1920s.

A golden tabby Tiger is white with an auburn red saddle about midway up the body and bold red stripes (think of a human with natural red hair). This one has no black stripes, although sometimes there are darker markings at the tip of the tail. These would be well camouflaged in a red clay or amber grass environment. The last known wild tabby was killed in the 1930s in Mysore, India. That is, until a wildlife ranger snapped a photo of a "golden" tiger crossing the road in Kaziranga National Park in Assam, India, in July 2020. Photographer Mayuresh Hendre captured some spectacular images of a golden tabby walking along a rock outcropping in the same park.[18]

Finally, reports of completely black or melanistic Tigers have existed since the early 1700s. There is no photographic evidence, but the number of sightings seems too great to dismiss the claims outright. Heavily striped, marbled, or pseudomelanistic Tigers have rarely been found in modern times, but sightings are increasing. A 2016 census by the Odisha Forest Department revealed at least six "largely-black" wild Tigers living in India's Similipal Reserve. Similarly, two more were recently born at the Nandankanan Zoo.

You may wonder why these colorations are not seen more often. With more than a 99% reduction in the species' population over the past several hundred years, the vast majority of living Tigers have the more common orange and black. Since the alleles responsible for alternative coloration are recessive (orange/black are dominant genes), it is possible for two Tigers carrying those genes to mate and produce a naturally occurring Tiger coloration that has not existed for decades. In fact, the golden tabby Tiger, which was last seen in the wild in the 1930s, reemerged in 1982 when two captive Tigers, both carrying the tabby gene, produced the first golden tabby Tiger cub in nearly fifty years. The same combining of recessive genes from both parents is certainly responsible for the recent appearance of golden tabbies in Assam, pseudomelanistic Tigers in Similipal, and white Tigers in Nilgiri.

The preservation (and breeding) of Tigers that are any color other than the standard orange and black has become a topic of debate. An argument has been made that the emergence of these "other" coat colorations is the product of inbreeding, which exploits recessive genes and results in Tigers with diminished health. The Association of Zoos and Aquariums, the group responsible for bringing the first white Tigers to the United States, has recently prohibited its member zoos from displaying white Tigers. Geneticists note, however, that alternative coat (pelage) coloration is part of the species' natural genetic diversity and is not directly linked to any deleterious health effect.[19] Since genetic diversity is good, a blanket policy of disregarding any unique gene pool that is naturally part of an endangered species seems inadvisable. White coat color, in and of itself, is not associated with health problems; however, if white cats are inbred to produce more white cats, just as with anything else inbred, there may be consequences for cat health.[20]

called *ocelli* (singular: *ocellus*) (fig. 5.10). They are thought to resemble eyes and deter other predators. Not much would attack a big male Tiger, but smaller ones might prove enticing to Lions, Wolves, Leopards, jackals, Cheetahs, and other Tigers. Woodcutters in India working in forests inhabited by Tigers wear masks with white spots on the back of their heads to try to deter Tiger attacks. The white spots may also help mothers and cubs find each other, since they stand out in the forest. Several other cat species have these too.

You have no doubt seen a house cat stalking toys or small animals or insects, followed by a pounce or short sprint to ultimately grab the victim. There are variations on this theme, but all cats take this same general approach to hunting. They are ambush predators, meaning they rely on surprise to catch dinner, whether it is big or small. Their skeletal characteristics are adapted for this behavior. All cats are strong and highly agile, with somewhat lengthened backs and slender, elongated limbs contributing to this strength and agility, although very big cats necessarily have heavier bones. These characteristics endow them all with a good ability to leap and accelerate, at least over a short distance.

Limb length among cats varies depending on their habitat. For example, Clouded Leopards, denizens of dense forests where running is not much use, have relatively short, stocky legs. So do Black-footed Cats (*Felis nigripes*), which were originally forest floor specialists when Africa was mostly forested (6 mya). This probably explains why they are among the worst tree climbers of all cats. Servals (*Leptailurus serval*), which occupy grassy plains, have long legs that serve as stilts, allowing them to see prey over tall grass. They are also exceptionally good at leaping into the air, enabling them to frequently snag birds.[21] Cheetahs also have long legs built for speed, and their claws are only semiretractable, giving them the equivalent

FIGURE 5.10 Ocelli on the ears of a Tiger (*Panthera tigris*). Many animals use eye mimicry to confuse potential predators, including fish, birds, and yes, even cats.
Myrtle Beach Safari

of cleats for improved traction. They also have an exceptionally flexible spine and a long tail with a flattened end, enhancing balance and steering as they make tight, fast turns pursuing prey (fig. 5.11).

Most cats have relatively long tails, providing extra balance during leaps. This is particularly true of some of the cats living in mountainous habitat, such as Pumas and Snow Leopards, where loss of balance will kill you. Pallas's Cats (*Otocolobus manul*), which live precariously in the Himalayas, alternatively have rather short tails. Cats that frequent trees also tend to have long tails. The Margay (*Leopardus wiedii*), or Monkey Cat, is probably the most arboreal—hence the nickname, as it is often found in trees and can catch and eat monkeys (fig. 5.12). In addition to its very long and strong tail, it has highly flexible wrists, allowing it to effectively grasp both branches and prey. Margays can hang upside down by their hind feet while handling prey with their front claws

■ FIGURE 5.11 Cheetah (*Acinonyx jubatus*) on the hunt. Cheetahs rely on speed and maneuverability more than brute strength. Shifting their long, rudder-like tail while running allows them to make quick course corrections to keep up with their quarry.

Jonathan C Photography / Shutterstock

▶ FIGURE 5.12 A wild Margay (*Leopardus wiedii*) in French Guiana, climbing through the forest. The rotating wrists and long tail of the Margay make this cat nearly as agile in the trees as it is on the ground.

Jaagurak/Shutterstock

and can climb headfirst down trees. Jaguarundis (*Herpailurus yagouaroundi*) and Marbled Cats (*Pardofelis marmorata*) both have strikingly long tails, however, and neither is common in arboreal or mountainous habitats. Not every cat needs a long tail. Bobcats (*Lynx rufus*), of course, get their common name from their bobbed tails, along with all three species of lynx.

Leopards are the best tree climbers among the big cats, probably because they are relatively small (fig. 5.13). Leopards will often haul prey up into a tree to keep it from scavengers. A common prey animal in parts of South Africa is the Impala. All Impalas possess a characteristic *M* marking on their

FIGURE 5.13 Leopard (*Panthera pardus*) climbing down a tree in Africa with the assistance of game ranger Bens Marimane (out of frame). The ability of Leopards to climb trees, combined with their increased intelligence, has earned them the nickname "monkeys in cat suits."

hindquarters, which locals describe as standing for "McDonald's" because everybody eats them, humans included (although not *at* McDonald's!). Impalas, like many other large grazers, exhibit synchronized breeding and birthing events. Since there are too few predators to eat all the young, enough survive to maintain the population. Many young Impalas, however, become dinner.

Author Rob witnessed, during the November lambing season, a 12 ft (3.6 m) Rock Python catch and eat a baby Impala, swallowing it whole. The meal created such a huge lump in the python that it immobilized it. A Leopard then caught and killed the overstuffed snake and thought it had two meals in one. When a hyena showed up, the Leopard hauled its Impala-filled python up a tree to eat in peace. After the Leopard draped the dead snake over a branch, the baby Impala fell out of a tear in the snake, and the hyena, still lurking below, grabbed it and took off. That poor little Impala got eaten twice, almost three times, that day! Leopards are not the only big cat climbers, though (see box 5.5).

Cat feet are *digitigrade*, or structured such that they walk on their toes like dogs, rather than

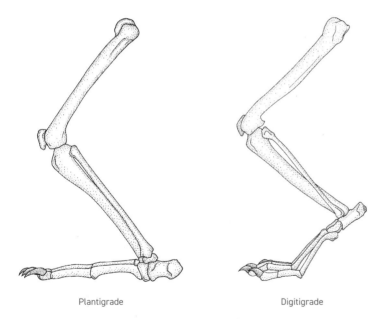

Plantigrade   Digitigrade

◀ FIGURE 5.14 Comparison of plantigrade and digitigrade walking and foot structures.

▼ FIGURE 5.15 Tiger (*Panthera tigris*) claws. Most cat claws are fully retractable, keeping them sharp and allowing them to hold on to prey. Each pivoting claw is fully integrated into the bone structure of the corresponding toe and is important for balance and walking. Declawing a cat can be harmful to the animal, as it requires removing the last phalanx from each toe, like cutting off the last bone from each of your fingers.

FIGURE 5.16 A Jaguar (*Panthera onca*) with a captured Caiman in Brazil's Pantanal. Jaguars are one of the most aquatic cats, allowing them to hunt prey in and around water, including other predators. Gudkov Andrey / Shutterstock

plantigrade, walking on the sole of the foot like you or a bear (fig. 5.14). Apart from Cheetahs, they all have retractable claws (fig. 5.15)—that is, they do not walk on their claws but rather keep them sheathed up in their toes until they need them, so they are always clean and sharp.

Contrary to what you may think, many cats like water habitats. The aptly named Fishing Cat (*Prionailurus viverrinus*) has webbed feet to facilitate swimming and diving after fish. Leopard Cats (*Prionailurus bengalensis*, not Leopards) also have an affinity for water, which might explain why they occupy more islands than any other cat species. Jaguars are excellent swimmers. They sometimes catch aquatic animals like fish or snakes, or even crocodylians (fig. 5.16). Despite their apparent enjoyment of water, Jaguars also live in dry habitats, so they are quite adaptable, like Leopards. Jaguars have been found to eat 85 different animal species, and some evidently enjoy avocados and

BOX 5.5 DON'T CALL THE FIRE DEPARTMENT TO SAVE THESE CATS: THE ISHASHA TREE LIONS

Lions and Tigers are well known for their inability to climb trees. Unlike the nimble Leopard, which can carry twice its body weight into the upper branches of a tree, the bigger cats are built for brute force. Their weight limits their ability to climb and a tree's ability to support them. Of course, biology likes to remind us that the only absolute is that there are no absolutes.

Lions in general are the least dexterous of the big cats. They are built primarily for battle, either with rival predators or their potential prey. This has not prevented a couple of small populations of Lions from adapting to unusual environments. One of these populations lives in Ishasha National Park in western Uganda. This mostly savanna-type environment is unique because of its many fig trees. The large, heavy, horizontal, low-lying branches of these ancient trees can support the immense weight of full-grown Lions. The advantage that resting in such a tree provides to a Lion is an excellent vantage point to spot prey and other predators. The trees also provide ample shade. On hot days in central Africa, a common place to find Ishasha Lions is lounging in the lower branches of a massive fig tree. It is not unusual to find the entire pride, more than a dozen individuals, in a single tree, even cubs.

Male Lion (*Panthera leo*) resting in a fig tree in Ishasha National Park, Uganda. Small populations within a species sometimes develop unique skill sets that help them cope with or exploit their particular environment. This is the case with the tree-climbing Lions of Ishasha.

sometimes grass (although the plant material is thought to help in regurgitating nondigestible material rather than being used for nutrition).

Tigers are one of the most aquatic cat species. They are quite acclimated to the water and are very proficient swimmers, partly thanks to webbing between their toes. They will also lie in the water to keep cool. Wild Tigers have been seen swimming across lakes and down rivers. The Sundarbans of India are a natural mangrove swamp where Tigers are regularly spotted in the water. It is interesting how Tigers seem not to care about temperature. We have seen a captive Tiger on a 14°F (−10°C) day run through the snow and jump into a pond and go for a swim. And we have seen one in a pond with icicles on his chin. Conversely, the heat does not seem to bother them either. They do not sweat but instead pant like a dog to cool down.

As a rule, Lions do not like to swim. In fact, most nonaquatic African animals do not like swimming. Since rivers in Africa are inhabited by large snakes, toothy crocodylians, and Hippos, this avoidance is probably entirely justified. (Hippos? you say. Those mild and clumsy-looking aquatic grazers annually kill more people in Africa than any other large animal: 2,900 on average! And they can kill a crocodile.) However, Lions will also learn to take advantage of a situation. The Lions of the Okavango Delta in Botswana, Africa, a floodplain that is underwater between June and August, are happy to swim. They have learned to take advantage of the water to trap prey. Likewise, Lions are typically not adept climbers, but one population has learned to hang out in trees (box 5.5).

BIOLOGY AND ECOLOGY

Except for females raising cubs, and Lions, which live in groups called *prides*, cats hunt by themselves and lead solitary lives. Bigger cats need to eat a lot (up to 100 lb, or 45.4 kg, per week for a big Tiger), so they need access to sufficient prey. To accomplish this, bigger cats maintain more expansive territories, and males maintain larger territories than females, which provides them access to more females. Bigger territories are necessary when prey is sparse, and if prey is densely distributed, territories will be smaller. Female Lions stay together in a territory, and other feline females often share their territories with their own female offspring, although they do not live and cooperate together the way female Lions do. Males and females use scent to mark their turf, conduct routine patrols, and fight off intruders. As with humans, disputes between big cats tend to arise most often over either property or sex.

Excessive fighting is not good for any cats, though. It wastes energy and can hurt or kill an individual. Like most animals, cats try to communicate before launching into battle. For example, we have talked about roaring in big cats, but there are other methods. Tigers communicate via scent markings like other cats, and they roar, of course, but they also *chuff* at each other to communicate at close range. A chuff means something like "How is everything with you?" or "I am okay." It is a way to reassure each other that there is no threat. Of course, sometimes they "lie" about that. They have been known to chuff, and when their reassured confidant turns his or her back, they pounce. A person can make this sound to captive Tigers to reassure them, too. And the Tigers will do it back. You better hope they are not just kidding!

Male Lions are easily identified by their mane (fig. 5.17). The mane generally starts out matching the rest of the coat but darkens with age and increasing levels of testosterone and thus is a marker of fitness. A strong male in his prime will tend to have a very dark mane. Hence, it serves as an important signal to potential competitors that this guy is not going down easily. That big mane also protects his

▶ FIGURE 5.17 Two Lion (*Panthera leo*) brothers walking across an open area in South Africa's bushveld. Male lions spend more time than most cats engaged in battle. Fighting off other predators or preventing rival males from taking over his pride, a male Lion relies heavily on his mane for battle armor.

Photo taken with the assistance of anti-poaching officer Bruce Missing (out of frame)

▼ FIGURE 5.18 Lion (*Panthera leo*) cubs from three different litters going to greet their father as he returns to the territory. Sharing babysitting duties allows some Lions to hunt while their offspring remain well protected. Eventually everyone gets to share in the kill.

neck and chest in a fight. The opposition might get just a mouthful of fur for his trouble.

Manes matter because males must compete for control of a pride to produce offspring. Generally, a coalition of two or three males, often brothers, will be the leaders of a pride of 10–20 females and cubs (fig. 5.18). The size of the males and their manes makes it more difficult for them to hide and to ambush, chase, and bring down prey, so females do that. The males may then saunter in, taking advantage of their larger size and strength, to deliver the final kill bite, particularly with larger prey. Males protect the territory and members of the pride, though. That territory ranges from 20 to 200 mi^2 (50–500 km^2), often depending on prey density. Other coalitions of males roam around seeking opportunities for a take-over, and the resident males must keep them at bay. If the intruders hear a deep, loud roar and see a coalition of three big, healthy-looking, dark-maned males, they know to just move on. Signals like the manes and the roar reduce the incidence of unnecessary, hopeless fights. A smart coalition will take on only another that looks to be on their level or, better, declining. The best genes win. A coalition generally maintains control of a pride for only two years, on average, very rarely exceeding four years. In the brutal calculus of natural selection, the winners generally kill any cubs (no sense supporting someone else's genes), and this causes the females to go into estrus so the new leaders can father their own brood.

In addition to natural selection sorting out the winning genes, Lions also take advantage of what's known as *kin selection*. Natural selection works on an individual. The organism best suited to survive and reproduce does so, and those good genes get passed along in the population. It has traditionally been thought of as a selfish enterprise. I want *my* genes out there, not yours. So, natural selection does not explain why animals will sometimes help each other. Why waste your energy on somebody else's genes? Kin selection explains this. Consider a pride of Lions. The females in the pride tend to be related to each other because the female offspring tend to stay with the pride when they grow up. So, here you have mothers, but also grandmothers, aunts, sisters, and cousins. On average, they share a percentage of their genes, and it is in their interest to keep those genes in the population. They can do this by raising their own offspring, but if the offspring face better odds in a bigger group, each female can improve the success of her own genes by helping with the extended family. And that's what Lions do. Litters within a pride are often closely synchronized, and cubs born in the same season become a cohort that the whole pride looks after. When some Lions go off on a hunt, some stay back to babysit. Mothers routinely nurse cubs that are not theirs, and if a Lion mother gets killed, her extended family will nurse her cubs and raise them in her absence because those cubs share their genes, too (fig. 5.19). This behavior likely also serves to increase social bonding within the group.

If you are in a group, it is important that everyone bonds with everyone else and remains in contact, and the deafening roar of a Lion allows long-distance communication. Lions also use the same pheromone-laced scent markings as other cats. And they head rub. But be careful if you are friends with a Lion. We know of a Lion caretaker who received a friendly Lion head bump that broke the poor fellow's orbital bone (eye socket)!

If you do not live in a group, how do you find a mate? Solitary cats patrol their territory and home range in search of prey but also keep an eye, or maybe rather a nose, out for potential mates. Associated with a chemical sense similar to a sense of smell, cats exhibit a characteristic grimace called a *flehmen*. They crinkle up their nose and draw back their lips and inhale. You make a similar face when you say "eeewww" in response to something icky. Cats are not grossed out,

FIGURE 5.19 A lioness (*Panthera leo*) with one of her sister's cubs. As with many babysitting duties, the "kids" can get a little annoying at times, but devotion to family means that by protecting your nieces and nephews you are protecting more of your own genes.

however. They are pulling air into the *Jacobson's organ* (also known as the *vomeronasal organ* or *VNO*) (fig. 5.20) at the front of the roof of the mouth. We talked about this in association with snakes, whose flicking tongues similarly draw air into their VNO as a way to "smell" chemicals in the air. Cats seem to use this largely to identify pheromones, which are chemicals released by animals for communication, for example when a female is in estrus and seeking a mate.[22] This is surely not bad news to the apparently grimacing male. Many species possess a VNO, albeit to varying degrees of development (size and functionality). The human VNO is small enough that its existence is still debated among scientists.[23] Whether or not it exists, the VNO is not of great consequence to our sense of smell. Humans, like dogs, rely heavily on *olfactory* smelling—that is, using an *olfactory bulb*. A dog's olfactory bulb (which we will discuss more in the next chapter) is so well developed, the dog can use its sense of smell to track and locate prey. While cats have better olfaction than humans, their sense of smell has evolved to make use of

an enlarged VNO largely for sniffing out rivals and potential mates rather than their next meal. We will discuss the senses they rely on more for hunting in the next section.

In all cats, mating is a rough sport and ovulation is not triggered until copulation occurs (except in Lions, where interaction with pride mates may also stimulate it). Cats will mate repeatedly over the course of a few days. All male cats have penises featuring barbs made of keratin, the same material your fingernails are made of. Ouch! And, in fact, mating probably hurts. The barbs are thought to stimulate ovulation as well as scrape rival semen from the vagina. This latter use is necessary because a female cat in estrus will mate with multiple males, given the opportunity. It is not uncommon for a single litter of cubs or kittens to have more than one father.[24] You have probably heard the process of evolution by natural selection described as "survival of the fittest." "Fittest" in this case means the most successful at producing healthy offspring. It is thus in the male's interest to have only *his* semen in play. And it is

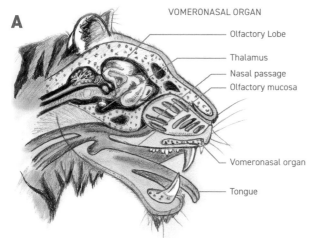

FIGURE 5.20 (A) Vomeronasal organ (VNO) diagram. (B) VNO openings in the mouth of a lioness (photo taken with assistance from wildlife ranger Johnathan Van Zyl, out of frame). The VNO, also known as Jacobson's organ, provides chemoreceptive sensory input and is common in reptiles, amphibians, and many mammals. In cats, the VNO is associated with the detection of pheromones.

in the female's interest to put up with whatever is required to get her eggs fertilized. Mating frequently and with more than one male improves the odds of that happening. Lest you think this is a pretty crazy system, consider that the Human Genome Project has found that early hominids (the family to which humans belong) had similarly spiky penises, a trait that disappeared from our genome about 700,000 years ago.[25] Whew.

In most top predators among the mammals, gestation is short and newborns are small. This allows the female to retain the mobility necessary to hunt even when she is pregnant. Gestation lasts three to three and a half months, and on average two to four cubs are born in some sort of den or shelter. There is not really a breeding season, and if no cubs survive or they die young, the female will immediately go into estrus again. Cubs' eyes are generally closed at birth, opening in a week to 10 days.

Cub coats are woolly and patterned to enhance camouflage. Leopards, Jaguars, Lions, and Pumas are born with spots to help camouflage the

vulnerable cubs. Lion and Puma coats transition to a uniform tawny coloration at about one year old. Cheetah kittens have a ruff of light fur along their backs, over a dark lower half, which makes them resemble Honey Badgers (possibly because not much will mess with a Honey Badger; we will discuss them later). They keep this light fur for about three months, sometimes longer. Cubs develop a set of baby teeth initially, which are replaced with adult teeth by about a year. They first venture from their den at about six weeks.

As is true of most mammals, kitten care is entirely the responsibility of the cat mother, although her territory likely lies within the territory of a male who, in guarding against intruding males, protects the females and cubs. He is not typically a "paws on" helper, although Lion fathers will tolerate some roughhousing. By human standards, female cats are not necessarily that attentive either. If one cub does not have the wherewithal to get to a nipple and eat, mom will just let the cub die. And so, like many predators, cats have a high juvenile mortality rate. Fewer than 10% will survive to reproduce.[26] Natural selection does not pull punches. You must get things right, or get lucky, to survive.

Mothers nurse their cubs for about four months but start feeding them pieces of meat as well at about three months. At first the mother leaves them alone while she hunts, but eventually they come along to learn hunting strategies, which they also learn and practice by playing. Cubs begin learning to hunt and kill at about six months. There is much to learn, so they stay with their mother for over a year, varying by species, with the bigger Lions and Tigers taking longer to mature. Once they mature, the young males disperse, thereby preventing inbreeding, and the females often take over part of their mother's territory or, in the case of Lions, stay with the pride.

All this hunting, mating, parenting, guarding, and defending takes a huge amount of energy, so cats sleep a lot, more than 18–20 hours a day. Most terrestrial predators in general sleep a lot, or at least more than prey do. If you can eat everybody and almost nobody can eat you, you can rest easy. Something like an Impala or rabbit, a hamburger with feet, has no such luxury.

CATS AS PREDATORS

Cats in general are among the most obligate meat eaters of all the Carnivora (meat-eating mammals). They eat only meat, which represents quality food to a cat. Meat is calorie dense, so it provides a lot of energy per mouthful compared to vegetables, and it offers all the essential amino acids an animal needs to build proteins, as well as the necessary vitamins and minerals. But even a mouse is much more of a challenge to catch than a carrot, never mind a Wildebeest. Cats have thus evolved characteristics selected to make them highly efficient predators, regardless of their size (fig. 5.21).

SENSING PREY

First off, like all predators, cats need to find prey. Cats have mostly the same senses as you, but with various strengths adapted to their particular needs.

Cats can use their sense of smell to help them locate potential meals and to "read" communications from other cats left by rubbing or spraying, but as we stated earlier, they do not have the olfactory power of a dog. Instead they rely more on vision, hearing, and touch. Cats are mostly *crepuscular*, meaning they are active at dawn and dusk. Most cats see well, especially in dim light. Many smaller cats have a slit-shaped pupil, rather than a round one like yours (fig. 5.22) or in fact any of the *Panthera*. The advantage to a slit is that it can close more fully. If your eye is designed to see well in low light, at high light levels the problem becomes getting too much light and not being able to see anything, an event you experience temporarily when you exit a darkened movie theater to a sun-drenched parking lot. A sufficiently closing pupil eliminates this problem of excess light.

To see at low light levels, you need a high density of light receptor cells lining the retina. In dim light, colors do not show up anyway, so it is the light- and motion-sensitive rod cells rather than the color-sensitive cone cells that dominate a cat's retina. Cats have the same three types of cone cells you have, sensitive to red, green, and blue, but far fewer of each, so they cannot see all the colors you can. They have six to eight times more rod cells than you, and better peripheral vision, although they are a bit nearsighted, which probably helps them zero in on prey up close. In addition, cats have the same reflecting layer behind the retina that we see in sharks and crocodylians, the tapetum lucidum, which reflects light back onto the retina to enhance vision in low light and results in eyes that seem to glow when light shines on them. All this means that cats are good at spotting

FIGURE 5.21 Black-footed Cat (*Felis nigripes*). This diminutive cat is the smallest and most endangered felid in Africa. What it lacks in size, it makes up for in predatory success. Even at a maximum size of 4.5 lb (2 kg), this feline can make more than a dozen kills per night.

FIGURE 5.22 Feline pupils come in both round and "cat-eyed" elliptical shapes. All big cats have round pupils (*left*). Small cats come in both varieties, depending on the species (*right*).

(*Left*), Myrtle Beach Safari; (*right*), Rebecca L. Bolam / Shutterstock

motion at very low light levels, and motion often means food. The Canada Lynx (*Lynx canadensis*), for some reason, has the reputation of having the best sight of any cat, to the point where some folks think it can see through objects![27] *That* would be *very* cool, but no.

Most cats have large, rounded ears. The *pinna*, or external ear, is cartilaginous and highly mobile. If you watch a cat, you can see it use its pinnae as scanners, like a radar dish, able to focus in separate directions simultaneously, to collect sounds and pinpoint their source. Cats move their pinnae 180° using nearly 30 sets of muscles, compared to your 6. They can hear well over 6.6 octaves of range, down nearly to the low frequencies you can hear, and 1.6 octaves above what you can hear. Those high frequencies are in the range of small rodent sounds, so it is possible that bigger cats are not quite as sensitive to those,[28] but it is apparently difficult to give a Tiger a hearing test, so data are scarce. Like you and other mammals, cats can pinpoint the source of a sound by comparing the timing and intensity of sound waves arriving in one ear versus the other. Caracals (*Caracal caracal*) in particular have huge, mobile ears, complete with impressive tufts of unknown use, similar to those of lynx. Any potential meal will have a hard time being quiet enough to evade detection.

Cats have one other method of detection: touch. If you look closely at the face of a house cat, you will see a bunch of long, prominent whiskers, or *vibrissae*. These are clearly important to small cats in pinpointing when and where to bite small prey at close range. Experiments with blindfolded cats and cats with trimmed vibrissae have demonstrated this.[29] It is harder to see how these would help a Jaguar capture, say, a Peccary (a medium-sized, piglike mammal), but big cats have them, and maybe they use them at close range below their nose where they cannot quite see where they are chomping.

CATCHING AND KILLING PREY

Cats thus have fine-tuned senses that allow them to locate their potential dinner. Now, how do they catch and dispatch it? Cats are all ambush predators, and as such they rely on surprise to catch prey. What features do they have that help them hide during the stalk-and-chase sequence? First, they are all well camouflaged, with coat colors that blend in with the lights and shadows of grasses or scrub. Their patterns help break up their outlines so they disappear into the background. It can be simply maddening to try to locate the napping Jaguar your tour guide has pointed to on the edge of the river as you drift past in a boat in the Brazilian Pantanal. And in that case, you are actively looking!

In addition, cats are patient. All cats rely on getting as close to their prey as possible before trying to leap and grab it. So, they either hide along a trail or near a water hole where they know animals will pass close, or they stalk. Maybe you have seen house cats do this. They crouch down and inch slowly forward, their footsteps silent because of their retracted claws and soft paw pads. When their potential victim looks their way, they freeze. And when they are close enough, they spring into action.

"Springing" is the appropriate description. All cats share a similar skeletal structure that works to make them both quite strong and agile. In general, cat limbs and backbones tend to be longer and more slender than those of other predatory mammals. Most cats can leap far—a large Tiger can cover 30 ft (9 m) in one jump, and other cats jump equally well relative to their size.

Ideally, that first jump lands the predator on its prey, but if not, the cat can accelerate very quickly and run quite fast for at least a short distance. A Cheetah of course is the champion mammalian sprinter, clocked as fast as 64 mph (103 kph), although it can sustain that for only about 1,600 ft (500 m).[30] A big Lioness can very briefly run at nearly 50 mph (80 kph), and Jaguars are almost that fast. Lions do not run down and wear out their prey like, say, a pack of Wolves does, but they work together, coming in from different directions at different times to trap an animal trying to escape. As we discussed previously, most cats have long tails, which provide balance and act as a rudder or counterbalance in running or jumping, so they can aim their landings well. Then it is a matter of knocking the prey off balance and hanging on long enough to get in a strategically placed kill bite.

Cats have two adaptations that help with the final kill, in addition to simply being bigger than their prey and/or hitting hard. One is those retractable claws. Since they are not walking around on those claws all the time, the claws stay long and clean and sharp, and function well as grappling hooks. Cheetahs do not have this advantage, but their exposed claws enhance traction for speed. Cats also have flexible wrists that help them dig those claws in and hang on. And then what? Getting in a prolonged fight with your dinner might get you hurt and will burn a lot of the energy you stand to gain, so cats try to dispatch their prey quickly. They do this with brute strength and sharp, strong teeth.

Comprehensive analysis of felid skull structure and musculature shows a general trend of a bigger skull, a stronger bite, and fewer constraints in prey selection, and the biggest of the felids are, of course, the big cats. Members of *Panthera* have the largest canines of any carnivore except, as we discussed previously, the Clouded Leopard (not in *Panthera*), which has the longest canines relative to its size among the big cats. The teeth of Tigers, Lions, and Jaguars are similarly long (a Tiger's canines extend more than 3 in, or 8 cm!) and conically shaped, with the Leopard's being somewhat shorter and narrower. Tigers have a dental advantage in killing prey. Their canine teeth have a sharp edge on the back side (fig. 5.23). Not only do the teeth provide good puncturing penetration, as they do for other cats, but Tiger teeth also act much like a knife blade in slashing through flesh. A vertical secondary sharp ridge also exists on the inside upper portion of a Tiger's upper canine teeth. It has been suggested that when these teeth penetrate deep into flesh, the small concave area on the inside of the tooth allows blood to flow more freely from the wound, speeding up the kill process. This is not unlike the front-center talons of raptors noted in chapter 4. While Lions may hunt a 1 ton (900 kg) buffalo as a pride, a Tiger can kill something that big all by

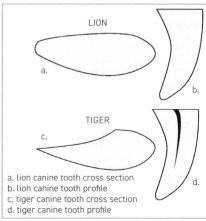

a. lion canine tooth cross section
b. lion canine tooth profile
c. tiger canine tooth cross section
d. tiger canine tooth profile

FIGURE 5.23 Tooth comparison of Lions and Tigers. The canine teeth of Lions are essentially curved, sharpened pegs (a and b). Tiger teeth are similar but have an indented vertical groove and a sharp back edge on the upper canines (c and d, photo). This allows Tiger teeth to both puncture and tear flesh.

itself, thanks largely to superior dentition. As we discussed previously, the Clouded Leopard has these bladelike teeth as well.[31]

The biggest teeth exhibit the strongest bite strength and are hardest to break, although when your prey has big bones, it is still possible to fracture teeth. It is also critical that the teeth be firmly anchored, and the root of the canine teeth in Lions and Tigers extends almost to the eye orbit. Two-thirds of a big cat's canine teeth are firmly rooted inside the skull (look back at fig. 5.5). Likewise, the bite force is highest in the biggest Tigers, decreasing in size from Lions to Jaguars to Leopards and down through the smaller cats. Calculations based on models of skulls and jaw and neck musculature put the bite force of the canine teeth at 1,550 lb (6,895 N) for the Tiger; just slightly less for the Lion mainly because it is smaller; 1,100 lb (4,893 N) for the Jaguar; and 370 lb (1,646 N) for the Leopard.[32] For its size, with huge jaw muscles and a slightly shorter mouth providing more leverage, it is the Jaguar that has the strongest bite.[33] Jaguars do not tend to suffocate their prey but rather bite through the back of the neck or even through the skull, a power that other cats do not quite have. Jaguars pull over larger prey, often breaking the neck of the prey in the fall. Smaller prey are killed with a smack on the head, yet Jaguars like to eat turtles, sometimes rather delicately, without breaking the shell.[34]

Cats thus deploy their speed, size, strength, claws, and teeth to kill prey before it can fight back. Ideally, perhaps, they land on the back of the prey from above and kill it with a sharp bite to the nape, severing the spine. Alternatively, a big cat might grab a bigger animal by the snout, hanging on with its claws, and haul the animal to the ground, either twisting and breaking the neck, or gripping the windpipe and suffocating the unlucky beast (fig. 5.24). A couple of Lionesses might work together to accomplish this. All cats choose easier targets

when they can. They tend to catch animals that are slower because of age or infirmity, serving as agents of natural selection and maintaining the fitness of the prey population. This whole process takes a great deal of energy, and a very good hunter might be successful just 20%–30% of the time. But the prey items are energy rich, often large, and ideally plentiful. And in every case, a cat will grab small items and scavenge when given the opportunity. Nobody turns down an easy meal.

Hunting in a pride also reflects kin selection in action. Why would you want to share your dinner with a bunch of other Lions? Well, if you work with those Lions to catch dinner, your odds of success are better and you can catch bigger things. In fact, the chance a solitary Lion will be successful on any given hunt is approximately 7%–19%. For a pride, that rises to 30%. So, Lions work as a team to either stalk and pounce, or hide by a game trail or water hole and wait to ambush passersby. A team of 300 lb (135 kg) Lions can bring down a Wildebeest, Water Buffalo, zebra, rhino, or even giraffe. They do this by tackling the animal and embedding their rear claws at the shoulder or flank with their forepaws at the neck, chest, or far side. They grab the muzzle and twist the head to drag down the animal. This works best with more than two Lionesses if the prey is big. About the only thing they will not attempt is a mature elephant.

HOW ARE CATS DOING?

Cats make a living by being difficult to see, and they have learned to avoid humans, mostly. So how do we know anything about them? Well, we know about historic population levels based on hunting records, for those species that have been hunted. These records were not precisely kept, but we have some idea of historical populations, and hunting records are still being maintained. In

FIGURE 5.24 Two lionesses (*Panthera leo*) with a captured Cape Buffalo. The Lion is the only species in the Felidae that lives in large groups, has a distinct social hierarchy, and hunts cooperatively. This allows Lions to regularly hunt prey that a single Lion could not handle.

addition, scientists conduct field surveys, in person or via camera traps, to assess populations. Cats are generally solitary, secretive animals, often living in habitats not easily reached or explored, so these are only best estimates. Based on these data, the IUCN establishes the status of species worldwide. The latest reports can be found on the Cat Specialist Group website.[35] Seventeen of the 38 species listed as of 2021 are considered Threatened. Cats require sufficient habitat and prey, but strategies are also needed to deal with conflicts with people and protect cats from poaching and excessive exploitation.

As we have repeatedly stated, because of their position at the top of the ecological trophic pyramid, there are never huge numbers of top predators. In addition, big predators do not reproduce very quickly—that is, they have what scientists call a low reproductive or rebound potential. It takes a while, a year or more, to grow up and learn all they need to know to function as adults. So, not too many offspring are born at once, and cats do not start having offspring until they are a few years old, depending on the species. This means that if habitat destruction or degradation, or some other insult, perturbs a population of predators, it will take a while, perhaps decades, for them to recover their numbers. What is perturbing their populations? Us.

For all the big cats, the biggest problem is habitat loss. Big animals need sufficient space to find enough to eat. Since we began switching from hunting and gathering to more settled cities and farming some 10,000 years ago, humans have taken wild land and turned it into farms and cities. This was okay when there were a few hundred thousand of us, but now that our population is approaching eight billion and growing, there is simply not enough space. We clear land for agriculture and livestock. So sometimes we are adding big grazers to the land, but in no way are we inclined to share them with our neighboring predators! In fact, the quickest way for a big cat or other predator to get itself killed is to target a domestic cow, goat, or sheep.

Aside from lack of space for sufficient food for everybody, however, there is another problem, even when we try to do the right thing. Let's take, for example, national parks, which have been described as "America's best idea." Areas of wild land are set aside for "enjoyment, education, and inspiration" by restricting their use and development. Yellowstone National Park in the United States was the first of these. The United States now has 63 of these protected sites, and other countries have adopted the idea to preserve their own wild lands. All good, right? Well, yes, but as it turns out, not quite good enough for our predators. The problem is that as we have continued to develop land around these parks, the parks have increasingly become, in effect, islands of wild land, isolated and disconnected from other wild lands. This is fine for, say, a salamander or

mouse, which has more than enough room to roam in nearly 2,344 mi^2 (6,070 km^2, or 1.5 million ac) in Everglades National Park. But most parks are not nearly this big, and if you have even a small population of predators like Florida Panthers (a subspecies of Puma) on this large "island," let alone in smaller parks, they are going to run out of room as their protected population grows, and without a connection to habitat elsewhere, there is nowhere else for them.

An average male Florida Panther's territory is 200 mi^2 (518 km^2)! Even if they can still manage to find enough to eat, they will be obliged to start inbreeding, because there are no outside Panthers with access in, and no way for the insiders to get out. Inbreeding tends to depress genetic diversity, usually to the detriment of the health of the animals, as we discussed in our introduction. One solution is to either have such large parks that there is sufficient room, or provide corridors from park to park to facilitate population interactions and therefore genetic exchange. Property owners tend to prefer the second option, but even the most fair-minded conservationist may not want a Panther "greenway" adjacent to his or her kids' swing set. And the Panthers may not want to go down that particular path. Let's look at some problems and solutions specific to some cat species.

Tigers are the largest feline species. They historically ranged from the Middle East to northern Russia to Southeast Asia. It is estimated that as many as 500,000 Tigers existed in Asia before the invention of the firearm. In 1900, there were an estimated 100,000 tigers in the wild. The current population estimate is about 3,000, less than 1% of historic levels, with an IUCN conservation status of Endangered.

Of the big cats, Tigers are most victimized by poaching, largely for the Eastern medicinal trade (discussed in box 5.2), and habitat loss. Tigers now occupy only about 7% of their original range. Tigers are also the big cats arguably most at risk from climate change because a good bit of the habitat they occupy is on small islands along coastal India and Indonesia, land areas likely to slip under the sea as sea levels rise. Tigers are also especially at risk of losing genetic diversity.

Three thousand Tigers is not very many, and they have two additional problems related to genetic diversity. First, loss of habitat has resulted in the remaining Tiger populations being very small and isolated from each other, meaning they are prohibited from interbreeding, which exacerbates loss of genetic diversity. Unfortunately, depending on your definition, Tigers as a species could already be considered *functionally extinct*—that is, they are at a point where the species may not survive without help from humans.

Second, it turns out there is not much genetic diversity there to start with. Tigers have been around for close to two million years, but about 75,000 years ago, a supervolcano eruption in Sumatra (and the subsequent volcanic winter) led to a crash in the Tiger population. Their numbers bottlenecked and the subsequent 75,000 years did not provide sufficient time for them to diversify again. Many zoos and organizations believe that only recognized subspecies are worthy of conservation. Recent genetic research is causing a dispute about how many subspecies actually occur. The chief of genomic diversity at the National Institutes of Health in the United States, Dr. Stephen O'Brien, has led studies of big cat diversity and found that there is more genetic difference between a person from China and a person from Ireland than there is in all Tigers! We do not subdivide humans into subspecies, and assigning Tigers to subspecies and trying to save each subspecies separately does not make much genetic sense (some subspecies have only about a dozen members, not enough to perpetuate the group without inbreeding). Biologically, the animals might

benefit if we mixed them up, thereby enhancing rather than narrowing diversity (see the discussion of Pumas below).

The places where Tigers live also coincide with the areas of highest human population (India and China). Displacing humans to save Tigers has not been tried in earnest, but some efforts are being made. China's "Save China's Tigers" project is attempting to *rewild* captive-bred Tigers for eventual release back into China (fig. 5.25). Because China has a population of more than one billion people and is the single largest consumer of illegal Tiger parts, it is still uncertain whether the country will ever develop a park that could adequately maintain a viable population of wild Tigers and protect them from being poached.[36]

China is the primary consumer of Tiger parts. In October 2018, China legalized the use of both Tiger parts and rhino horns but almost immediately repealed the new law because of international public outcry. The argument is that legalization provides a lawful way for people to acquire Tiger parts, thereby reducing poaching of wild Tigers. The counterargument is, of course, that if trade in Tiger parts is legal, will it create an even larger consumer demand resulting in even more poaching? No one really knows, but Chinese Tiger farms currently have hundreds (if not thousands) of farmed Tigers, euthanized and kept frozen, waiting for legalization.[37]

Tiger conservation reflects the dilemma faced by many species. Captive environments can act as an "ark" for dwindling populations drowning in a sea of humanity. The key is to maintain genetic diversity. Given the low numbers of wild Tigers, odds are that a captive population will be necessary if we are ever going to save or reestablish the wild population. But where are you going to put wild Tigers? It would be nice if all Tigers could be wild, but there is simply not enough suitable habitat because there are so many

FIGURE 5.25 A Tiger (*Panthera tigris*) in South Africa as part of the "Save China's Tigers" rewilding project.

humans. The difficult choice is Tigers in captivity or, eventually, no Tigers at all. Better off extinct than in a zoo? There are those who would argue both sides of this debate.

Lions (fig. 5.26) are the second largest species in *Panthera*. Like the other big cats, they have a life span of about 10 years in the wild, 20 in captivity. Lion and Tiger populations at one time coexisted, with Lions historically distributed all over Africa, through the Middle East, into India, and even in western China. Today their range is limited to a few areas in Africa and in Gir National Park in India. Those last Lions in the Gir Forest are still there only because they were protected as part of a hunting club before the area became a sanctuary. While hunters and conservationists would seem to

have opposing goals, at times they may need each other to save a species.

Lions are considered Vulnerable by the IUCN, with a population of fewer than 20,000, based on a 43% decline in their population between 1993 and 2014. However, this varies widely depending on which population you examine. Lion populations in Botswana, Namibia, South Africa, and Zimbabwe have actually increased by 12%. However, in the rest of Africa, Lion populations show a 60% decline and would be classified as Endangered, and the very few animals remaining in the Gir Forest in India are Critically Endangered. Things may be even worse for populations in African countries in the grip of war such as the Central African Republic and South Sudan. There is no monitoring in these places for safety reasons. A concern for African Lion populations at large is their increasing isolation from other populations. The ability of different populations of Lions to interbreed has bolstered their genetic variability, but this option is being greatly decreased because of habitat loss and urbanization. This could spell genetic trouble even for Lion populations that still appear robust.[38] Once a certain amount of genetic diversity is lost, a population becomes unable to withstand a cataclysmic event like a severe drought or disease outbreak.

Conservation of Lions comes down mainly to good fences and enough management funding to provide for antipoaching teams. The biggest problems for Lions are killing in defense of humans and livestock, prey depletion because of the trade in bushmeat, and habitat loss. There is still trophy hunting, but improvements in its management have moved it toward sustainability. It remains highly controversial, as we saw in 2015 with the outrage over a dentist from the United States killing Cecil, a Lion locally famous in Zimbabwe's Hwange National Park because he was outfitted with a tracking collar for research.[39]

Jaguars (fig. 5.27) are native to North and South America and are mainly tropical. As of 2008, the population was perhaps optimistically estimated to be around 15,000, most concentrated in the Amazon rain forest in Brazil, which makes Jaguars exceedingly hard to find and count. Their current range stretches as far north as the US-Mexico border, making them the largest feline endemic to the United States. Jaguars are the only New World (North and South America) member of *Panthera*. According to the IUCN, the Jaguar's status is Near Threatened. They are the most widespread terrestrial predator besides Wolves—and possibly Red Foxes, as discussed in the next chapter—and

FIGURE 5.26 **A Lioness (***Panthera leo***) with her three-week-old cub. Even a future king of the jungle needs a regular bath.**

are now estimated to have a relatively large world population of around 173,000.

Jaguars once roamed throughout South America and up into the southwestern and central United States but lost habitat and were killed off in the north and east in the 1700s. They are now found mainly in the Amazon basin and prefer wetter areas, so they are most highly concentrated in the Brazilian Pantanal and Bolivian Gran Chaco regions. Settlement and agriculture and Jaguars are mostly incompatible. Jaguar populations were reduced through bounties and fur hunting in the southwestern United States, and the last animals were systematically hunted down by the US Fish and Wildlife Service (USFWS) in the twentieth century, although they reappear sporadically after migrating from Mexico.[40] There have been numerous sightings over the years in Arizona especially, and also in New Mexico and Texas.

Despite this recurring presence of Jaguars in the United States, no recovery plan has been enacted. A proposal was drafted by the USFWS, but the public comment period garnered too much opposition, even from conservationists. Wildlife biologist Alan Rabinowitz ended his comment to the USFWS stating, "The United States does not contain habitat that is critical to the survival of the Jaguar as a species. Designating critical habitat for the Jaguar in the United States is an abuse of the true intent of the Endangered Species Act and a waste of US taxpayer funds." In 2010, the Obama administration agreed to draw up a recovery plan designating critical habitat, but apparently it never happened. Building a physical barrier along the US-Mexico border would certainly impede Jaguar migration, not to mention attenuate the distribution of prey and other species.[41]

The main threat to Jaguars is habitat loss. Brazil, for example, doubled its cattle population between

FIGURE 5.27 A Jaguar (*Panthera onca*) hiding in dense foliage. Jaguars rely on surprise and brute strength to overpower their prey.

FIGURE 5.28 A mother Leopard (*Panthera pardus*) grooming one of her two cubs. Leopards are the most adaptable of the big cats, living in jungles, savannas, forests, and even on the outskirts of some of the world's most populous cities.

1993 and 2013, and an area of Amazon rain forest the size of Italy was clear-cut to accommodate the cattle. This habitat is now unsuitable for Jaguars unless they eat cows, but nobody wants that! However, in 2009, with urging from nongovernmental organizations (NGOs), federal prosecutors began pressuring companies to limit deforestation, and this led to the G4 agreement with the NGO Greenpeace, under which slaughterhouses agreed to buy only from suppliers who had reduced deforestation to zero. It also required ranchers to enroll in a public environmental registry that allowed monitoring. By 2013 60% of suppliers had registered and there was 95% compliance among them, plus 85% compliance from ranchers. This intervention in the supply chain stands as a model for other areas in South America, although current compliance is unknown.[42] Unfortunately, deforestation has again accelerated in the Amazon.

The other approach that shows promise for Jaguars is providing corridors between populations to allow migration and breeding between unrelated groups, to enhance genetic diversity. The NGO Panthera is working on the Jaguar Corridor Initiative throughout Central and South America, which is mapping connecting paths Jaguars could or do use between otherwise isolated patches of known Jaguar habitat. They hope that by protecting and preserving green corridors that Jaguars already use, the small subpopulations will remain connected and able to interbreed, thus maintaining genetic diversity.[43]

Leopards (fig. 5.28) seem to adapt to and survive close to human society much better than most top predators. Leopards range from South Africa to the Middle East and Southeast Asia, and to the Muravyov-Amursky Peninsula in far eastern Russia. The IUCN rates Leopards as Near Threatened, although the population size is somewhat healthier than is the case for most wild cats. If you want to stick around in a world overrun with humans, it helps to live in many different places.

It is still legal to hunt Leopards in some countries, and permits to import Leopard skins are often number one on the USFWS list, although some airlines now refuse to carry such trophies in an effort to avoid controversy that might affect their bottom line.

In Nairobi, Kenya, Nairobi National Park surrounds a portion of the city, and Leopards come and go in the city itself. A population there reportedly lives almost entirely on domestic dogs. A similar thing is happening in India. This ability to adapt to human settlements would seem to help maintain Leopard populations, but it also means they are more likely than other cats to encounter people.

Even though they are the smallest of the *Panthera* species, Leopards kill more people than any other cat. They are very opportunistic predators, feeding on weak and unsuspecting prey, and nearby humans often qualify. Unsuspecting prey also often take the form of livestock, which causes farmers and ranchers to retaliate by shooting, poisoning, and snaring Leopards. Ranchers often hang a cat carcass near their animals as a warning to other cats, an act of desperation that does not work; the Leopards do not care. A more progressive rancher in cooperation with a conservation group might try to translocate problem Leopards to a more natural area. That does not work either, since the animal finds its way back, or you put your problem in someone else's backyard and new animals move in to take its place. Additionally, a translocated Leopard may be displacing an established animal where it is relocated, so somebody still loses. Getting rid of these individual "problem" Leopards might be satisfying for the rancher but does not really help in the long run. What does help?

First, maintaining viable habitat for the cats, with good concentrations of prey. One strategy is to establish watering holes, which attract prey in groups. Like the rest of us, cats will do what is easiest. If wild prey animals concentrate in one

FIGURE 5.29 Livestock-guarding dog in a field with goats (*left*). A mother Cheetah (*Acinonyx jubatus*) (*right*) with three subadult cubs living on the same farm, but without attacking livestock.

(*Left*), Rare Species Fund; (*right*), Cheetah Outreach Trust

place, cats will hunt there and leave livestock and people alone. On the edges of Mumbai in India, for example, people are frequently attacked by Leopards because they regularly cross paths. So how to keep them apart? Well, one main risk arises when people are obliged to hike into the forest at night to relieve themselves because they lack indoor plumbing. The solution is to provide indoor options, which additionally help with hygiene.[44]

Another strategy is to protect livestock with puppies (fig. 5.29). This method is used by a South African group called Cheetah Outreach Trust.[45] How does a puppy take on a big cat? Well, puppies go through a socialization period from 3 to 16 weeks of age, during which they learn what they are and how to behave in society, whether that is a society of other dogs, people, or sheep. If a puppy grows up in a herd of sheep, it will know how to interact with sheep, and here is the key thing when it comes to conservation: it will want to protect its "pack." This has proven to be especially effective in protecting herds in Africa, where large barking dogs are not a natural part of the environment. Not only does this prevent predators from attacking

livestock (meaning the farmer will not dispatch them), but the predators that focus on wild prey are able to stay in their territory. This technique has proven successful in preventing livestock losses from Leopards, Cheetahs, African Wild Dogs, hyenas, and jackals. Farmers are also deploying donkeys, ostriches, and alpacas as guards.

In many places, Pumas are faring quite well. While population estimates are currently unavailable, the IUCN considers the species' conservation status as of Least Concern. In many places, hunting for Pumas is still legal. Of the 15 US states that have a nonthreatened Puma population, only California prohibits hunting them.[46] Like other North American predators, including Wolves and Coyotes, Pumas are enjoying a population rebound after unchecked extermination over the previous two centuries. With their expanding ranges and ever-increasing human expansion into wild territory, it should not come as a surprise that human-predator encounters also seem to be on the rise (fig. 5.30). Occasionally, these encounters go wrong.

Perhaps the most publicized Puma attack in recent history involved the death of 18-year-old

FIGURE 5.30 An adult male Puma (*Puma concolor*). *Puma concolor* means "cat of one color," one of many names this felid is known by. Although large, the Puma is not a member of the *Panthera* group and is technically considered a small cat. Unlike big cats, the Puma purrs and cannot roar. Myrtle Beach Safari

Colorado resident Scott Lancaster, who was killed by a Puma while he was jogging through a forest near his high school (recounted in *The Beast in the Garden* by David Baron). Although Lancaster's death was certainly a tragedy, it nonetheless represents a very rare occurrence. Various estimates of deaths from Puma attacks in the United States and Canada over the past hundred years range from 10 to 27 (or one death every 3.7–10 years).[47] While many people still fear coming face to face with a wild Puma, a recent study shows that you are many hundreds of times more likely to die taking a selfie than when encountering a Puma in the forest.[48] Vanity always seems to get the better of us.

While we humans face very little real threat from Pumas, the opposite unfortunately is not true. Public perception of this predator is becoming more positive, but livestock losses and long-held prejudices are still real problems for the Puma, particularly in South America. And the Florida Panther is still not "out of the woods." During the 1980s, the USFWS estimated that only 20–30 individual Pumas still existed in South Florida.[49] As we discussed regarding Tigers, a small gene pool could spell disaster, and in fact inbreeding and decreased fitness nearly drove the Florida Panther to extinction. Although controversial to some, eight unrelated female Pumas from Texas were introduced into Florida Panther range. The 30-year result was a more than 600% increase in the number of adult and subadult Florida Panthers (120–230 as of 2021). Some people claimed that the Texas Pumas were a separate subspecies, and that the offspring (and eventual full population) are no longer 100% Florida Panther. It was obvious to scientists that the Florida Panther population would not survive without the introduction of unrelated genes. Other species (like Tigers and the Iberian Lynx, *Lynx pardinus*) may also require interbreeding to save the species.

SOME APEX PREDATORS

PUMA
(Puma concolor)

The Puma (fig. 5.31) is also known as Mountain Lion, Cougar, Catamount, and numerous other colloquial names. The Florida Panther is a subspecies of Puma (one of approximately 30 subspecies) that lives in southern Florida. Why does one species have so many names? This is undoubtedly because of its extensive range, including 22 countries throughout North and South America.[50] We will avoid the confusion and call them Puma, which most closely approximates their scientific name (*Puma concolor*). Although not as massive as most of its distant *Panthera* cousins, the Puma is nonetheless one of the top predators in the Western Hemisphere.

Puma concolor means "cat of a single color." This refers to the Puma's nearly uniform tan-gray coat pattern (although like Lions, they are born with spots and eventually lose them). Body size depends heavily on the environment. A Puma living in the steamy jungles of Belize (up to 120 lb, or 54 kg) is much smaller than one living in the Canadian Rockies (up to 220 lb, or 100 kg). It is much easier for a larger body mass to retain heat. Similarly, the list of a Puma's prey species depends on where it lives. As with Leopards, the expansive range of the Puma is a result of its adaptability to different environments. This has contributed to its overall success as a species.

JAGUAR
(Panthera onca)

Jaguars (fig. 5.32) are the biggest cats of the Americas. Although they exist in many of the same countries as Pumas, their range is significantly smaller, particularly because of their propensity

FIGURE 5.31 Hiker coming across a Florida Panther (*Puma concolor*). Encounters with Pumas, like this Florida Panther in Everglades National Park, are becoming increasingly common. Crystal Schnebly

to live only in warmer climates. Jaguars can be found through Central America and even into the southern United States, but their numbers are limited. The largest Jaguar populations exist in tropical and subtropical South America, particularly the Amazon River basin and the Pantanal of Brazil (the world's largest wetland). There may be as many as 173,000 Jaguars in the wild today, although the population is decreasing.[51] Jaguar populations seem to be faring better than those of most other big cats, but their fate will be inextricably linked to conservation of their habitats. As we see the Amazon rain forest being destroyed, it would be unsurprising to see a decline in the Jaguar population. The good news is that we still

have plenty of time to prevent a population crisis for the Jaguar, but there is work to do.

SNOW LEOPARD
(Panthera uncia)

Snow Leopards (fig. 5.33) live at the highest elevation of all the big cats. They occupy a range in central Asia that spans 12 different countries, and they can be found up to 16,400 ft (5,000 m) above sea level.[52] From eastern Afghanistan through the Himalayas, the Tibetan Plateau, and all the way to Mongolia, China, and Russia, Snow Leopards live in mountains and on rocky peaks, typically above the tree line. As you would expect from the name,

FIGURE 5.32 Jaguar (*Panthera onca*) in Brazil's Pantanal.

much of the Snow Leopard's range does indeed experience snow.

Although Snow Leopards have a fairly large range (roughly 780,000 mi^2, 2 million km^2, or the size of Mexico), their population remains relatively small. It is estimated that 4,000 to 7,500 Snow Leopards still exist in the wild. The main theory behind such a wide dispersal is the density of prey. Snow Leopard habitat is typically rocky, cold, and very sparsely vegetated with grasses and small shrubs. This environment can support only limited prey, which results in Snow Leopards needing far larger territories than most cats, sometimes more than 621 mi^2 (1,000 km^2) per cat. The harsh terrain and sparse distribution of Snow Leopards make them incredibly difficult to study, including determining accurate population size.

In 2017 the IUCN downgraded the conservation status of Snow Leopards from Endangered to Vulnerable, although their population is thought to be diminishing. A primary struggle for conservation of the species seems to be a lack

FIGURE 5.33 Snow Leopard (*Panthera uncia*) lounging on a rock ledge.
Eric Kilby from Somerville, MA, USA, CC BY-SA 2.0, via Wikimedia Commons

of data. We simply don't know enough about the species in the wild to draw any firm conclusions. The fate of the Snow Leopard seems almost as elusive as the cat itself.

LEOPARD
(Panthera pardus)

Leopards (fig. 5.34) are the most abundant of all the big cats. They range throughout much of Africa and central Asia. Like many species we've looked at thus far, they have dwindled significantly in numbers over the past few centuries and continue to decline. Although an estimation of 700,000 Leopards in Africa is often quoted, the truth is they have far too large a range for scientists to accurately estimate population size.[53] We simply don't know how many there are.

Leopards are incredibly adaptive, which explains why and how they inhabit such a wide variety of ecosystems. The pressures of human expansion into wild territories are responsible for the decline

FIGURE 5.34 Spotted Leopard (*Panthera pardus*) actively hunting a warthog in South Africa. Author Rob saw this Leopard capture one of the piglets. Instead of immediately dispatching its prey, the Leopard continued to prod the warthog, hoping to lure the mother back with distress calls from her baby, thereby providing a larger meal. This demonstrates an intelligence not often associated with big cats.

in Leopard populations, but some of these crafty felines are adapting to survive near expanding urban development. In large cities (including Mumbai in India and Nairobi in Kenya) Leopards are known to sneak into town at night and feed on stray dogs. Some have become brave enough to casually walk into houses and apartment buildings and kill pet dogs inside people's homes. A quick internet search will show dozens of these predations caught on surveillance cameras. This is a real-world concern for many people. Between January and November 2020, 23 people died from Leopard attacks in the Indian state of Maharashtra.

Author Rob worked with a Uganda Wildlife Authority ranger who recently lost a child to a Leopard attack.[54] The ranger and his family were staying in their lodgings inside Queen Elizabeth National Park, where he was on patrol. A Leopard came through an open window at night, grabbed their sleeping baby, and in an instant escaped back out the window with the dead baby in its jaws. Despite being the smallest of the big cats, the Leopard is responsible for the greatest number of attacks on humans.

The brutal reality of this predator-prey relationship has long put us at odds with species at the top of the food chain (including Leopards). So far, our greatest response has been to exterminate these predators, which we have done largely with ruthless success. Significant efforts are being made to mitigate human-Leopard conflicts and to come up with effective ways for both species to coexist. Some methods (like using dogs or other animals to guard livestock) are producing positive results for both humans and Leopards. These methods show real promise for helping save other predators as well.

6

CANIDS

As with cats, you probably feel that you know a great deal about dogs because many of us live with one, or more. Indeed, our domestic dogs are in same taxonomic group, the family Canidae, as the wild dogs we discuss here. Many canids are quite social animals, and that characteristic has endeared our own Bella and Max (the top names for female and male dogs, respectively, in the United States in 2020) to us. But wild canids are also talented predators, in many cases *top* predators. Probably (hopefully!) your Lucy or Charlie (the second most popular names) does not harbor such intentions much beyond the squirrels in your yard.

Many of the predators featured in this book are members of the order Carnivora. This group emerged about 60 million years ago (mya) into a world already populated by a diversity of other predators, among them carnivorous marsupials like the Pouch Lion (*Thylacoleo*), a large mammal with a strong bite, not closely related to Lions. Around 56 mya, global temperatures rose about 11°F (5°C) in what is known as the Paleocene-Eocene Thermal Maximum (PETM).[1] This change in climate had a drastic effect on ecosystems and biodiversity and offers a cautionary tale about our warming planet. While some species faced insurmountable environmental pressure, others flourished, including early Carnivora. One of the earliest known carnivores was *Dormaalocyon latouri*, a small tree-dwelling mammalian predator that lived during the early Eocene epoch (about

55 mya).[2] This unlikely predator, weighing about 2 lb (907 g), is quite likely an ancestor of modern carnivores.

As the Carnivora group became better established about 55 mya, they began to diverge into two distinct suborders (fig. 6.1), the Feliformia and the Caniformia. The Feliformia (catlike group) eventually diversified into cats, genets, civets, linsangs, mongooses, and hyenas. (Yes, hyenas are more closely related to cats than to dogs. Weird, eh?) The Caniformia (doglike group) diversified to include dogs, bears, raccoons, badgers, weasels, skunks, otters, seals, Walruses, and more.

A variety of canine-like families evolved over time, like the bear-dogs. These were the first known members of the Caniformia to start to reach massive size. The extinct *Pseudocyon* (false dog) was a bear-dog (unrelated to breeds

FIGURE 6.1 **Phylogeny of Carnivora.** The carnivores are broken into two groups, Feliformia (catlike) and Caniformia (doglike). Look closely. There are likely some evolutionary (and genetic) relationships you didn't expect.

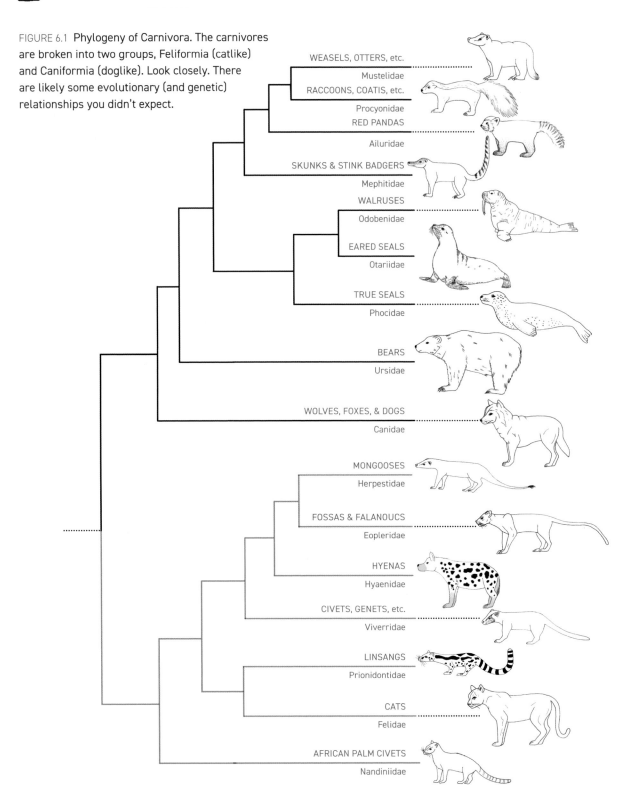

WEASELS, OTTERS, etc.
Mustelidae

RACCOONS, COATIS, etc.
Procyonidae

RED PANDAS
Ailuridae

SKUNKS & STINK BADGERS
Mephitidae

WALRUSES
Odobenidae

EARED SEALS
Otariidae

TRUE SEALS
Phocidae

BEARS
Ursidae

WOLVES, FOXES, & DOGS
Canidae

MONGOOSES
Herpestidae

FOSSAS & FALANOUCS
Eopleridae

HYENAS
Hyaenidae

CIVETS, GENETS, etc.
Viverridae

LINSANGS
Prionidontidae

CATS
Felidae

AFRICAN PALM CIVETS
Nandiniidae

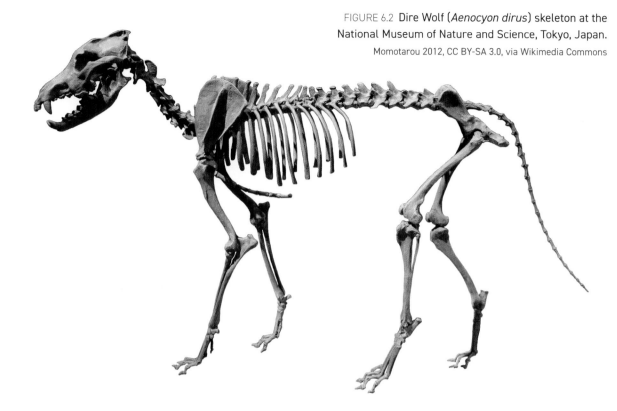

called that today) estimated to have been as large
as 1,704 lb (773 kg). The evolution of modern
canids is typically recognized in three successive
subfamilies: Hesperocyoninae (a group of small,
weasel- and fox-like dogs from 40 to 30 mya),
Borophaginae (more robust "bone-crushing" dogs
from 36 to 2.5 mya), and modern Caninae. Perhaps
one of the most famous extinct modern canids is
the Dire Wolf (*Canis dirus*) (fig. 6.2). While this was
undoubtedly an impressive predator that hunted
large prey, it seems unlikely that it ever reached
the proportions depicted in the television series
Game of Thrones.

Today 37 or so species of wild canids are
distributed among every continent save Antarctica,
and in every variety of habitat, from desert to rain
forest to tundra to cities. Evolutionary geneticists
divide modern canids into four groups, or *clades*:
(1) the Wolf-like canids (Wolves and jackals);

(2) the fox-like canids (including the Red Fox,
Vulpes vulpes; and Raccoon Dog, *Nyctereutes
procyonoides*); (3) the South American canids
(like the Maned Wolf, *Chrysocyon brachyurus*;
and Bush Dog, *Speothos venaticus*); and (4) the
Gray and Island Foxes (*Urocyon cinereoargenteus*
and *littoralis*) of the Americas. Many in these
latter three groups resemble foxes and thus are
commonly called foxes. Raccoon Dogs are the
only true foxes able to climb trees and hibernate.[3]
Wolves (*Canis lupus*) were, for a time, the most
widely distributed terrestrial mammal on Earth.
Now that honor belongs to the Red Fox (fig. 6.3).

The adaptability of canids is in part a result
of their flexible diet. Contrary to the widely
held perception, most of these animals are not
strict carnivores. They assuredly will eat meat
if it is available, but most will also consume just
about anything else edible, from fruits and nuts

FIGURE 6.3 Red Fox (*Vulpes vulpes*) in the forest. With the widespread loss of Wolves to depredation countermeasures, the Red Fox has become the most ubiquitous wild terrestrial mammal species on the planet.

to insects. Canids are not large, so they do not demand as much in the way of resources as larger predators do. In fact, their size generally correlates to prey availability; bigger canids live only where there is larger prey. The Wolf is the largest canid, at up to 175 lb (80 kg), and the smallest is the Fennec Fox (*Vulpes zerda*), weighing less than 2.5 lb (1 kg). Most canids are solitary except in breeding season, but some live in highly structured, cooperative groups, helping each other collect food and raise young.[4]

And, of course, dogs have likely been our own companions for more than 20,000 years.[5] Let us take a look at their wild relatives.

CANIDS AND US

Like many of the other majestically frightening beasts in this book, wild canids figure prominently in the mythology and stories of ancient and modern cultures around the world. Wolves are the most prevalent canid. A she-wolf was said to have suckled and cared for Romulus and Remus in the ancient mythology surrounding the founding of Rome. Wolves are likewise common characters in nearly every native North American culture. The Wolf is associated with courage, strength, loyalty, and hunting prowess. Wolf is a noble creator god for the Shoshone, and for the Anishinabe, a Wolf character is the brother and true best friend of Nanabozho, their culture hero.

In Brazilian folklore, Maned Wolves have various mythical powers. You can supposedly increase your sexual prowess if you remove the right eye of a live Wolf and then release the Wolf (we would rather consume raw oysters). Other myths: A canine tooth tied around a child's neck protects against dental problems. Consuming two small pieces of fresh Wolf heart will keep a snakebite victim alive. And a cough can be cured by drinking hot tea made from dried Wolf feces.[6] There is no scientific evidence that any of these work.

Coyotes (*Canis latrans*) are also common players in the animism of early Native Americans, with attributes varying from tribe to tribe. Sometimes Coyote is a revered culture hero, teaching and helping humans. Sometimes he is a comic trickster, wise without being malevolent, but not exactly helpful either. In fact, the word *coyote* is from the Aztec root *coyotl*, or "trickster."[7] This fellow gets into trouble but is clever enough to get out of it. In modern culture, the Coyote is considered vermin and not exactly wise. You may recall, for example, the creative efforts of Looney Tunes' Wile E. Coyote in trying to catch his nemesis the Road Runner, which he repeatedly failed to do. For the record, Coyotes do eat Road Runners, but Road Runners do not say *beep beep*.

Similarly, foxes are not necessarily out to get you, but they are strictly in it for themselves. You do not want the fox guarding the henhouse, but because he is sly, he very well might convince you to let him do just that. You might want to be "foxy," but you probably do not desire to be considered a vixen, a female fox, but also a sort of femme fatale, quite stunning, but trouble. *Zorro* is not only a swashbuckling vigilante, but also the Spanish word for "fox."

Dingoes (*Canis dingo*) are commonly represented in indigenous Australian mythology in association with The Dreaming, the period when the world was created. In some groups, humans are believed to have a Dingo origin. For example, the Dingo is responsible for giving humans their characteristic shape, in particular their genitals, which are different from those of native marsupial animals in Australia.[8] More menacing, in 1981, Lindy Chamberlain of Australia was accused of murdering her nine-week-old baby while claiming the child Azaria had, in fact, been taken off into the night by a Dingo. She was played by Meryl Streep in the 1988 movie *A Cry in the Dark* (*Evil Angels* in Australia), who made famous the cry "The Dingo's got my baby!" Chamberlain was eventually acquitted. Since then there have been other cases of Dingoes running off with small children.[9]

We searched for examples of jackals in human culture and were startled to find a whole directory of jackals just from Disney. The most famous is Tabaqui the Dish-Licker, the jackal in Rudyard Kipling's *The Jungle Book*, also a popular Disney movie. Tabaqui eked out an existence stealing scraps from the Seeonee Wolf pack and sucking up to the nasty Tiger, Shere Khan. For the record, any canid will scavenge, and any will suck up to the leaders of its own pack. There is no shame in getting along with the bosses.

While Coyotes, foxes, Dingoes, and jackals can be, but are not necessarily, threatening, Wolves have been considered quite scary and are prominent villains in many familiar fairy tales, often symbolic of what happens to misbehaving children. These include *The Three Little Pigs*, *The Boy Who Cried Wolf*, *Little Red Riding Hood*, and *Peter and the Wolf*. But Wolves pursue adults too. The movie *Doctor Zhivago* features a famous and chilling scene where Wolves circle our heroes just beyond the light of their Siberian fire. In the movie *The Grey*, plane crash victims in Alaska are threatened by hungry Wolves. Times are changing, however. Recent children's books include *The Big*

Good Wolf, *Big Bad Wolf Is Good*, and *The Wolf Wilder*. In all of these, the Wolf is a good guy. In the movie *Dances with Wolves*, the Wolf is simply a symbol of wildness.

As is the case with most modern predators, canids have much more to fear from us than we do from them. They are not typically a direct threat to humans since they tend to be fearful of us and avoid contact. Individually, most of them are not nearly big enough to kill an adult human, and a pack is not generally going to bother with something so small and bony—that is, not worth their time nutritionally. As we noted, there have been rare cases of canids killing children, but the operative word is "rare."

There are three primary causes of wolf attacks: rabies, provocation, and defense. Rabies is by far the most common cause. Wolves develop a

BOX 6.1   BEWARE THE FULL MOON

Human lore abounds with fantastic beasts. Monsters lurk in the shadows, plaguing our existence and populating our nightmares. Some of these beasts are dogs. Perhaps the most widely known dog monster legend is that of the werewolf. Shape-shifting wolfman myths appear all over the world, including ancient Greece, China, Iceland, Haiti, and Brazil. The first known printed story of a man transforming into a wolf-beast dates to 2100 BCE Mesopotamia with *The Epic of Gilgamesh*. Ancient Greeks told the story of Zeus turning Lycaon and his sons into Wolves as a punishment. Even in Europe in the Middle Ages, belief in werewolves was held by those in Germany and France. In the 1700s, forces of King Louis XV were deployed to Gévaudan, France, to capture and kill a wolf-creature that claimed nearly 300 victims. There have been more than 150 full-length werewolf movies! Yet there is of course no actual evidence that werewolves exist. Where, then, does

Depiction of a werewolf. Although mythical in origin, werewolf legends are widespread and may have some scientific explanation. Refluo/Shutterstock

particularly nasty "furious" phase of the disease, and there is a documented case of a rabid Wolf biting as many as 30 people. The cause of death, if it occurs, is almost always the disease, not the attack (see box 6.1). A shepherd beating off a Wolf from his sheep may get attacked but is rarely killed. There have been no cases of a predatory wild Wolf attack on a human in North America since the nineteenth century. Bites that have occurred generally came from Wolves, Coyotes, and Dingoes that were habituated to people.[10]

So why have we been out to persecute these animals for millennia? Livestock. They like to eat what we like to eat, so they cost us resources. A pack of Wolves or jackals will take advantage of a handy herd of sheep. In fact, a recent study found that jackals in the Karoo region of South Africa prefer domestic sheep to wild prey. (The small cats

this story originate? There are a few theories. Maybe folks were simply looking to explain serial killers, whose behavior seems beyond anything a normal person would do. Or rabies.[11]

Rabies is a scary disease in and of itself, since it is generally fatal to humans, and it most famously affects dogs (and bats, which have their own public relations issues). A rabid Wolf would likely show little fear of humans. Debilitated by the disease into seeking easy prey, a rabid Wolf would be able to see well on a bright moonlit night and attack humans then. Attacking a human could result in the transmission of rabies from Wolf to person, with the virus incubating for about a month. The victim would appear normal until (yup, you guessed it) around the next full moon. The victim might start behaving oddly, with heightened aggression, foaming at the mouth, and even a fear of water (hydrophobia). In short, it seemed almost as if these humans were becoming the aggressive and foaming dog that bit them, like a werewolf. Another possibility is *hypertrichosis*, or "werewolf syndrome," a rare but widespread disease characterized by abnormal excessive hair growth over parts the body. Its origins are not clear, but it can run in families or show up because of various other disorders.[12] In extreme cases, a person may grow to resemble the descriptions of werewolves.

Latin America is home to another monster dog story: the legend of the Chupacabra. Literally translated, this name means "goat sucker," referring to this monster's propensity for exsanguinating (bleeding dry) livestock. Farmers have reported finding their livestock killed by an unknown creature, while others claim to have seen the monster. Artists' depictions of what witnesses assert they have seen look more like a creature from *The X-Files* than any species known to science. If you are hoping for a supernatural night stalker with glowing red eyes, prepare to be disappointed. Recent photographs have surfaced showing that the elusive Chupacabras may in fact be Coyotes suffering from severe mange. A mangy Coyote is almost unrecognizable, with missing hair and shriveled skin. Its canine teeth are also more prominent, adding to its gruesome visage. Because mange can have severe physical effects, the ability of these Coyotes to appropriately hunt native prey species may be diminished. In a weakened state and fighting for life, these Coyotes would likely resort to killing easier prey, often including livestock. Although it does not make for a scary bedtime story, the Chupacabra appears to be just an unfortunate canid affected by parasites ... but not knowing for sure is enough to keep the legend alive.[13]

in the area, the Caracals, share this preference, so it is not only dogs who are the bad actors.[14]) The colloquialism about the fox guarding the henhouse points to the willingness of foxes to eat chickens and eggs. Humans, of course, also prefer domestic sheep and chickens, so the war began.

WHAT IS A CANID?

Canids, or dogs, belong to the taxonomic suborder Caniformia within the order Carnivora and family Canidae. This suborder also includes bears, seals, sea lions, skunks, raccoons, badgers, otters, weasels, and the Coati and Red Panda.

Like cats, canids are readily able to mix their genes across species, so it can be complicated to precisely define specific canid species as entirely distinct from one another (see box 6.2), but they share some exceptional characteristics. Members of the Carnivora have shearing carnassial teeth just

anterior to their molars, something quite helpful for eating meat. Canids are further characterized by their surprisingly uniform body shape and structure. They are all digitigrade (see fig. 5.14); in other words, they run on their toes. They have four toes on each foot, with naked pads, except species in the desert and Arctic Foxes (*Vulpes lagopus*), which have furry pads to insulate them from the hot or cold ground. The Bush Dogs (*Speothos venaticus*) (fig. 6.4) of South America have webbed toes, suited to their semiaquatic lifestyle.

All canids have relatively long, slender legs, with wrist joints and a locked radius and ulna that allow flexion only forward and backward, not side to side. A dog cannot effectively grasp something in its paws or easily climb a tree because of this. But it is a very efficient design for extended rapid travel over relatively flat ground. This is how all canids hunt, and since natural selection favors efficiency of form, they all share this general shape. We will talk more about this when we get to canid hunting strategies.

 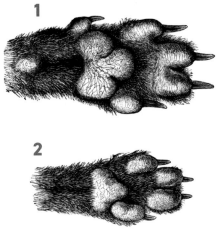

FIGURE 6.4 (*Left*) South American Bush Dog (*Speothos venaticus*) swimming. (*Right*) Depiction of Bush Dog paws. 1. Front right paw. 2. Rear right paw. Notice that the front paw has webbing and is significantly larger than the rear paw, providing greater propulsion. This form of aquatic locomotion is justifiably colloquially known as "doggy-paddling." (*Left*), Steve Wilson, Chester, UK, via Wikimedia Commons

BOX 6.2   HYBRIDIZATION AND DEFINING SPECIES

Taxonomists have a tough job trying to neatly squeeze living organisms into defined units likes species and subspecies. Throughout most of our taxonomic history, we have defined species by their phenotype, or their observable physical traits. This helped scientists decide what is a Red Wolf (*Canis rufus*) and what is a Gray Wolf (a Gray Wolf is a Wolf subspecies). As technology continues to improve, taxonomists have developed new techniques to study living organisms. As our understanding increases, our definition of "what is a species" continues to evolve. A few species (like some Mexican salamanders) look almost identical and are indistinguishable to observers, but molecular analysis shows that they are genetically very different creatures.[15] Conversely, some animals that look very different from each other are closely related.

Perhaps one of the most extreme examples of this is the domestic dog. Although we have different *breeds* of dogs, they are all the same subspecies. In other words, even though they look very different, a Chihuahua is a Greyhound is a Beagle is a Great Dane (box 6.2 fig. 2). While gene sequencing is helping us correctly identify and delineate individual species, it is also providing new taxonomic challenges. Scientists are beginning to recognize that the lack of interbreeding between many related species may be the result of geographic isolation more than genetic isolation; in other words, the populations have been separated long enough to have evolved into and remain as separate species. But what happens when these related species expand or migrate?

Wolf pack (*Canis lupus*) in the woods. Genetic evidence suggests that Wolves inherited the black coat allele through hybridizing with early domesticated dogs. This alternative pelage offered some sort of evolutionary advantage, as nearly half of today's North American Wolves have a black coat. Michael Roeder / Shutterstock

(*Continued overleaf*)

BOX 6.2 (Continued from previous page)

Chihuahua and Great Dane (*Canis lupus familiaris*).
The two dog breeds represent the same subspecies.
While they can genetically breed, their size
difference makes it physically impossible.

   Closely related species sometimes hybridize in the wild. This phenomenon immediately complicates
defining species, because we use this inability or unlikeliness to breed with other species as an identifying
qualification. Very often the hybrid offspring of two parent species is an organism that is less fit for its
environment than either parent. This, along with potential sterility in one sex, reduces the likelihood that
that first-generation hybrid (called F1) will reproduce and pass on its traits. Reduced fitness is not always
the case, though. Sometimes having mixed genes from two species has a neutral effect, rendering the
offspring neither more nor less fit. And sometimes the hybrid develops traits that are superior to those
of its parent species (this is known as *heterosis* or hybrid vigor), endowing it with superior fitness for its
home environment or even allowing it to thrive in areas previously uninhabited by either parent species.
This form of hybridization seems to be a trial-and-error process employed by nature. While those hybrids

that have distinct trait disadvantages do not tend to make it very far in the world, hybrids with increased fitness often thrive and pass on their genes. This is one way a new species may arise, something known in the scientific community as *speciation.*

Speciation through hybridization occurs in both animals and plants, but it is more prominent in the latter. The real challenge we face in putting taxonomic labels on organisms is when *introgressive hybridization* occurs. Introgression takes place when a hybrid organism breeds with a parent species, introducing genes from one species into a different species through the resulting offspring. This *gene flow* leaves the receiving species as a distinct unit, but introgression alters its DNA. We discussed this some with the big cats. Introgression may be one of the most profound methods of evolution, quickly passing adaptive traits from one species to the next. Unlike the evolutionary process of random mutations being selected for or against over eons, introgression can have a rapid effect on a population.

In the canid family, we need look no further than North American Wolves. Thousands of years ago, when humans crossed the Bering Land Bridge (also known as Beringia) from Asia to North America, they likely brought with them domesticated dogs.[16] The process of domesticating Wolves into dogs resulted in some unique traits (including different coat colors) that did not exist among wild Wolves. For example, Wolves were gray but domestic dogs became brown, gray, red, and even black. As humans migrated to America, at least one of their canine companions (and likely more) hybridized with a wild Wolf and introduced the black coat allele (or gene variant). This gene resulted in a few black-coated puppies. Apparently this trait had some advantages because today almost half of North American Wolves have black coats.[17]

There are many examples of wild hybrids. Today what is known as the Eastern Coyote is a mixed genetic stew of Coyote, Wolf, and Dog. As described in box 5.1, big cats have also introgressively hybridized. Even non-African humans are hybrids. You may be aware that *Homo sapiens* hybridized with Neanderthals (*Homo neanderthalensis*) after leaving Africa. Most humans have identifiable genetic markers confirming this, often 2%–4% of their genome (complete set of genetic material). Recent genetic work has shown that there is a specific genetic distance between species that determines whether they can successfully interbreed, and that humans and Neanderthals were more genetically alike than, say, Polar and Brown Bears are.[18] Modern-day Tibetans owe much of their ability to breathe the lower-oxygen air at high altitudes to their ancestors hybridizing with Denisovans (*Denisova hominins*, an extinct species of archaic human). Tibetan Mastiffs recently (within the last few hundred years) inherited a similar adaptation from hybridizing with Tibetan Plateau Wolves.[19] The deeper we dig into genetics, the greater our understanding becomes, but the more blurred the lines between species seem to be. If a plant or animal has signs of introgression, where do we draw the line between "it is a species carrying genes from another species" and "it is a hybrid, not wholly a member of either parent species"? There is no universally accepted answer to this question. Dr. Michael Arnold of the University of Georgia once posited, "What does it matter?"[20] From a wholly scientific perspective, we are inclined to believe it doesn't matter. Species genetics are more fluid than taxonomists would like. Mother Nature will not be confined by our silly little labels.

FIGURE 6.5 Arctic Fox (*Vulpes lagopus*) in snow. With its insulative undercoat making up 70% of the total fur, this species is well equipped for cold climates. Fitawoman / Shutterstock

All wild dogs have tails, often bushy, which aid in balance, communication, and sometimes insulation. An Arctic Fox, for example, has a very bushy tail that it wraps around its face when it curls up to sleep in the snow. The tail of a Bush Dog (fig. 6.4) is rather short, perhaps serving as a rudder for a dog with aquatic predilections. Or maybe it makes less appealing bait for toothy fish or other predators?

All wild dogs are furry, possessing a two-layered coat, with longer protective outer fur, sometimes a bit oily and water resistant, and an insulating undercoat of a density appropriate to their habitat and the season. The Arctic Fox's (fig. 6.5) winter undercoat, for example, makes up fully 70% of its fur in the winter and is so dense and warm that the critical lower temperature for this animal is below −40°F (−40°C). That means the temperature must drop to below −40°F before the brain of the fox notes it is a bit chilly out and begins to crank up the metabolism and fluff the fur. Fat stores also insulate the fox's body. By −95°F (−70°C; approximately the lowest air temperature ever recorded), the fox has raised its basal or resting metabolic rate by only 50% to compensate for its heat loss to the environment. At that point it is maintaining a body temperature over 200°F (110°C) higher than the air temperature![21] This is an amazing feat, particularly for such a small animal. Larger size helps in retaining heat, and if you are bigger, your fur can be longer without you tripping over it, but you also need more to eat and there is not much food to be had in an Arctic winter. Your authors, acclimated to life in sunny South Carolina, USA, start complaining that it is cold before we get to 40°F (4.5°C) above zero! But then we are not especially big, either, or furry.

FIGURE 6.6 (*Top*) Raccoon Dog (*Nyctereutes procyonoides*). (*Bottom*) Bush Dog (*Speothos venaticus*). Both species have small, rounded ears, particularly when compared to the rest of the canids. (*Top*), Ryzhkov Sergey, CC BY-SA 4.0 via Wikimedia Commons; (*bottom*), Ondrej Prosicky / Shutterstock

You can easily recognize the face of a dog. They have the forward-facing eyes typical of a predator, a characteristic long snout, and relatively large, upright, often pointy ears. Exceptions include the Raccoon Dog and Bush Dog (fig. 6.6), which both have small, rounded ears, and of course domestic dogs, whose ear shapes defy classification. Dogs in hot climates, like the Fennec Fox, have extra-large ears to dissipate heat. If it is cold, it is better to have small ears to avoid frostbite, and the Arctic Fox has these.

Regardless of the size and shape, those ears are quite mobile and expressive. In fact, dogs share a common body language they use to communicate with one another. If you are familiar with domestic dogs, you probably understand this language too. When a Wolf jumps forward with front end down, ears up, butt in the air, and tail wagging, she wants to play. Just like your dog. You can even mimic that pose and induce your dog to play. If you are fortunate enough to see a Wolf, jackal, or fox in the wild, however, you should not do that! Those beasts are not going to play with you, and folks will stare. But odds are you will be able to guess that canid's state of mind (see box 6.3).

Finally, all dogs have similar and unusual reproductive cycles. (Domestic dogs are somewhat different from their wild cousins in this regard.) There is only one breeding season annually. The period before estrus (when fertilization can occur) is unusually long, lasting a few weeks. This encourages a close bond between male and female. Each mating session ends with a copulatory tie, a very close bond we will explain below.

All this bonding encourages the male to help with parenting and provide protection and food to both the pups and mother.[22] This is unusual compared to most male mammals.

BOX 6.3   SPEAKING OF DOGS

Of all the top predators, social canines, most notably domestic dogs, are the most familiar to us. Their social behavior offers numerous indicators of their emotional state. This is something you have probably noted. If your dog is wagging her tail, her mouth is open, and she has a silly grin on her face, you can intuit that she is happy. Conversely, if her head is bent, her tail is between her legs, and she is slightly crouched, she probably knows she is in trouble for something, although she may not realize it is bad to eat your favorite pair of shoes. These are two obvious postures that most of us are familiar with, but the variety of body positions and facial gestures social canines use is much more complex.

Tail wagging does not always mean "happy." Sometimes it means discomfort or aggression. Understanding these subtleties is one way that order is maintained within the pack. Ritualized aggression allows dominant members to maintain order without resorting to any real violence. A Wolf lying on the ground, leaking a small amount of urine, is a clear display of submission. An alpha male can often be seen grabbing a subordinate Wolf with his teeth while emitting a threatening growl. If the underling submits, no blood need be shed. Lacking hands, Wolves and wild dogs also use teeth to greet each other, communicate, and play. Some people keep Wolf/dog hybrids as pets. If this has ever crossed your mind, remember that these animals are biologically programmed to be very "mouthy." What they view as a normal form of communication does not always do well with delicate human flesh, especially if there are children in the house.

Body posture and the position of ears, head, and tail can tell us a lot about the current disposition of a canine. The complexity of social canine society and hierarchies is vast, but we have included some of the more recognizable social displays in the figure opposite in case you are ready to give up modern life and join a Wolf pack.

BIOLOGY AND ECOLOGY

As we noted above, wild canids live in nearly every habitat on every continent except Antarctica. The wolf group is found nearly worldwide but is rare in tropical forests. This group includes Wolves (except for the Maned Wolf), Coyotes, jackals, African Wild Dogs (*Lycaon pictus*), and Asiatic Wild Dogs, or Dholes (*Cuon alpinus*). Wolves occupy North America and northern Eurasia. There are African Wolves (*Canis lupaster*) in North and West Africa, and the highly endangered Ethiopian Wolf (*Canis simensis*) is found only in pockets in Ethiopia. Golden and Black-backed Jackals (*Canis aureus* and *mesomelas*) live mostly in Africa. Some studies have now split the Golden Jackal into Eurasian and African Jackal species,[23] with populations of Eurasian Jackals increasing in Asia and in southern and eastern Europe.[24] Coyotes cover North America, with an expanding range like the jackals in Europe, both apparently benefiting from the extermination of Wolves and a tolerance of people. Dholes, or Asiatic Wild Dogs, remain in a few protected places in India. The similar African Wild Dogs live in southern and East Africa. South America has a variety of endemic "foxes" (although they are all more closely related to the wolf group than to the fox group), and the Maned Wolf (fig. 6.7), Bush Dog, and Short-Eared Dog (*Atelocynus microtis*). Red Foxes live just about everywhere except

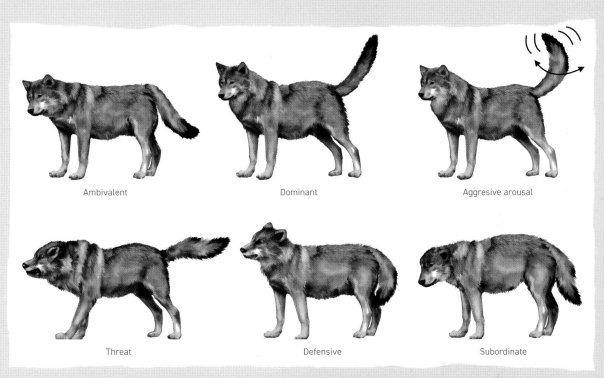

Ambivalent	Dominant	Aggresive arousal
Threat	Defensive	Subordinate

Meanings of Wolf tail positions. Gray Wolves (*Canis lupus*) use extensive visual and behavioral cues to help keep social order among the pack. Knowing one's place within the pack's social hierarchy results in diminished violence.

FIGURE 6.7. Maned Wolf (*Chrysocyon brachyurus*). This species looks somewhat like a Red Fox on stilts. As a member of the isolated South American group of canids, its closest cousin is the unlikely Bush Dog shown in figure 6.6.

mariait/Shutterstock

Africa, including Australia, where they have been introduced and now compete with Dingoes. Dingoes were introduced to Australia, though they have had more time to adapt, so they are considered a *naturalized* species (see box 6.4). Numerous close relatives of Red Foxes are found in Africa, and Gray Foxes (*Urocyon cinereoargenteus*) live in the Americas. The apparently highly adaptable nature of the same basic body design, eating habits, and life history has made these animals successful everywhere they have turned up.

Most people associate Wolves with packs, and this is not wrong, but it might not be what you think, and it is actually atypical of wild canids. Most wild canids are solitary animals except during breeding season and for the time it takes to raise pups to be self-sufficient. Once the pups move on, it is most often every dog for itself again. They roam widely, hunt alone, and pursue prey of a size that a single dog can bring down. So, for example, foxes eat primarily rodents. Coyotes and jackals (fig. 6.8) eat bigger grazers like deer and antelope, but mostly the calves.

Canids will guard the den where pups are born, but other than that they do not defend a territory, although they mark their presence. If a territory exists, it will range in size depending on prey density, so a Red Fox is comfortable using a couple of suburban backyards, while an Arctic Fox can range 1,500 mi^2 (2,500 km^2) looking for enough lemmings to sustain it. Territories may overlap, and individual canids are amenable to that

BOX 6.4   DINGOES: AUSSIES OR NOT?

Dingoes are generally considered to be naturalized in Australia, but they are not marsupials. Most of the other iconic Australian animals—kangaroos, wallabies, Wombats, Koalas, and so forth—have pouches in which they carry their offspring. There were once marsupial lions and wolves too. So why are Dingoes different? It turns out that the Dingo arrived in Australia just 5,000–3,500 years ago and was derived from a protodog population probably originating in Asia before being brought with humans to Oceania (Australia and some of the surrounding islands). The name likely comes from the indigenous Australian word *tingo*, meaning "tamed." They go by other names in Australia: *mirri* and *warrigal* in New South Wales, and *ngupanu*, *papa*, *parrutju*, *tjantu*, *wanaparri*, and *yinura* in Western Australia.

The aboriginals adopted the dogs culturally but did not change them biologically, so although they are considered dogs and not Wolves, they represent a very early branch of "commensal" dogs that have not undergone artificial selection to turn them into breeds. They were often adopted as puppies and taken in, but then they were allowed or encouraged to go back to the wild once they reached sexual maturity.[25] Another interesting example of a protodog is the New Guinea Singing Dog (considered a feral dog by the IUCN), which is present in the wild only on the island of New Guinea, a short canoe ride immediately north of Australia. These "singers" are a sister population to Dingoes but remain completely wild, often living at very high altitudes with very few interactions with humans. They therefore reflect what is most likely ancestral dogness. An interesting video of this ancient and loquacious canine can be found on the internet.[26]

Being dogs, Dingoes adapted quickly to their new land and are now the largest and most widespread predators, *apex* predators, in Australia. A pack of Dingoes can bring down a large kangaroo. They also like sheep. In the 1850s the Dingo Fence—at 3,400 mi (5,500 km) long, longer than the Great Wall of China (but

FIGURE 6.8 Black-backed Jackal (*Canis mesomelas*) in the eastern Kalahari in Botswana. Although they will regularly scavenge and eat insects, it is not uncommon for jackals to prey on young antelopes.

without the tourist draw of the latter)—was erected across Australia to protect the valuable sheep herds of New South Wales from these animals. It also changed the ecology of the herbivores on either side.[27]

Stories about Dingoes historically labeled them as slinking, stinking, and savage. The term *dingo* was a synonym for a corrupt lawyer or money lender.[28] But there is also pressure to conserve them now as iconic Australian fauna, even as Red Foxes, introduced in the 1800s, are offering serious competition.[29] The foxes clearly are not native, but are the Dingoes? Hmm ...

Dingo (*Canis lupus dingo*) on the beach. Dingoes are believed to have originated from domestic dogs that rewilded and became a genetically and geographically isolated population. They still share a close connection and lineage with Wolves and domestic dogs. The three remain different subspecies within the species *Canis lupus*.
Jason Cassidy / Shutterstock

if there is enough food to go around. Regardless of territory size, wild canids generally avoid each other unless they are in the same pack. If you are an alpha male, the alpha female probably lives right next door, but except when it is mating season or there are pups to attend to, you and she may not have much to do with each other. But some dog species tend to be more social than others.

A Wolf pack is generally just a mated alpha pair with the year's pups, plus last year's pups that have not sexually matured yet and are still hanging around. So, a pack may contain from 2 to 12 individuals. This is the case for all Wolf subspecies and the Bush Dogs of South America. Sometimes siblings of one of the alphas will also be present but will not reproduce. The main hunters are the alpha pair, although the others help some and will

carry food back to the pups or stay back to guard them. The alphas are in command in the same way human parents oversee a household. They will stay together until death do them part, and they are the main social unit. The others come and go. And whether they come or go seems to depend on what is available to eat. If there are big, grazing ungulates like Moose to be had, a big pack can be supported and that is how Wolves will organize themselves. Coyotes do likewise. They do not need the pack to hunt, but if the hunting is such that a pack can be both helpful and adequately supported, that is what they do.[30]

The most social dogs are the African (fig. 6.9) and Asiatic Wild Dogs. They are always in a pack, and packs have been sighted containing more than 30 animals, with averages of 8–15 or 20, depending on the subspecies.[31] Like packs of Wolves, African Wild Dog packs consist of an alpha male and female mated pair who do most of the reproducing. If there is enough food to go around, other females may also have pups, but these precarious babies are the first to die if resources get tight. The other adults in the pack are offspring or siblings of the alpha pair. Most often it is the male offspring that stay with their natal pack, serving as dedicated helpers. These African Wild Dogs are in constant communication, helping with the hunt and assisting with minding and feeding the pups. In fact, pups eat first in the African Wild Dog pack, whereas in all other wild canids, pups eat last, and sometimes not at all.

FIGURE 6.9 African Wild Dogs (*Lycaon pictus*) posturing over a kill. African Wild Dogs are one of the most social canids. Atypically for most animals, the young are allowed to eat before the parents. If one kill is not enough for the entire pack, they will hunt again. Depending on the size of prey, African Wild Dogs may make more than five consecutive kills in a night.

Why do these dogs form packs? It has always been assumed that being in a pack leads to better hunting prospects. Yes, you must share your kill, but you are more likely to make a kill as part of a pack, possibly a larger animal than you could handle alone. However, this assumption does not always hold up. Numerous studies of meat supply per capita for Coyotes, for example, in groups versus alone, show that individuals get as much meat or more when alone. This relationship also holds for jackals and Wolves, although for African Wild Dogs, a study that carefully accounted for the energy expenditure associated with pack hunting versus solitary hunting found that Wild Dogs gain more net energy when working together. So even though pack members may get less meat to eat, they expend less energy obtaining it as a member of the pack than they would alone.[32] Are there other reasons to team up besides food? Evidence suggests yes, and in fact defense trumps food as an explanation for the formation of packs. The highly social dog species live where predators are significant threats, especially to their offspring. The threat extends to other predators stealing their catches. Dholes must contend with Tigers and Leopards. African Wild Dogs coexist with Lions, Cheetahs, and hyenas. Bush Dogs live among Jaguars. A pack can better protect its own kills and pups from these formidable predators willing to steal kills and kill pups. In fact, it has been found that African Wild Dog litter size is correlated with pack size, and a pack of about five members is required for the successful raising of any pups at all.[33]

Another puzzle about packs is why a few sexually mature dogs hang around with the alphas if those alphas prevent them from having their own pups. In natural selection, survival of the fittest is the rule, and in evolutionary terms, fitness translates into successful reproduction. He or she who has the most offspring wins—that is, the winner's genes make it to the next generation. Helping somebody else's offspring at the expense of endangering your

own makes no sense evolutionarily. And yet we see helpers in dog packs, Lion prides, bird flocks, and others. Why? Turns out it is again a case of kin selection, as we saw with Lions. If you are an animal that needs to learn and practice to parent successfully, or there is not a territory available this year, or there is not enough food to support another family, your best bet may be to bide your time and help your relatives in the pack.

For dogs that do live in packs, it is vital that all dogs understand the same language, know their role, and coordinate their efforts with minimal conflict. How does that happen? Communication. We already discussed the body language common to Wolves (see box 6.3). Other wild dogs exhibit these same behaviors, but there is much more than body language; dogs communicate via scent and sound as well.

Since we communicate via sound ourselves, we can mostly hear the sounds dogs use to communicate, and in fact those communications are often directed at us by our own canine companions, so let's look at that method first. Canids that are more social use sound much more than the solitary varieties, and it varies from individual to individual, just as there are chatty and quiet people. So, Wolves, jackals, Coyotes, Dholes, African Wild Dogs, and Bush Dogs make all manner of sounds much of the time, while most foxes do not vocalize much at all. All canids growl as a warning or defense, as part of an attack, or simply to assert dominance. Barks are low-pitched, loud, short sounds, usually used to warn of an intruder or identify individuals. No wild dogs are the enthusiastic barkers that most domestic dogs are, though. Foxes emit high-pitched screams in greeting, and Coyotes and jackals yip. Most pups whimper, whine, or squeak when they want or need attention or food, and canids in packs also use these higher-pitched sounds to indicate submission in their interaction with other pack members.

Coyotes and Wolves famously howl. This sound carries over long distances and provides group cohesion when the pack is spread out, identifies individuals and potential mates, and lets neighboring packs know another pack is there. A Wolf howl (fig. 6.10) can be heard 6 mi (10 km) away through forest and 10 mi (16 km) across open tundra.[34] The New Guinea Singing Dog has one of the most unique vocalizations, and it can also carry over great distances and altitudes in its mountainous homeland. These singers are rare canids and the largest land predators on their island. The dogs produce a characteristic harmonic vocalization, described as a Wolf howl with overtones of whale song.

All canids sometimes use their sounds together in combination, but the highly social Dholes, African Wild Dogs, and Bush Dogs, especially, constantly "talk" to each other. Bush Dogs make high-pitched squeaky sounds that carry in the dense forests

FIGURE 6.10 **Howling Gray Wolf** (*Canis lupus*). Wolves make various types of vocalizations, the most well known being the howl. In open areas, a Wolf howl can be heard up to 10 m (16 km) away.

where they live. (A similar original habitat for the modern Basenji dog breed from Africa may explain why it doesn't bark like other dogs.) One study found that Dholes have 11 different sounds made up of low- and high-pitched squeaks, yaps, growls, and barks, including a yap-squeak that seemingly conveys an individual's "name."[35] Another study described the soft, musical contact call, alarm bark, twitter, high-pitched chatter, and sneeze of African Wild Dogs, with the latter expressing the high excitement of the "pep rally" that precedes every hunting expedition.[36] In fact, another study showed that without a quorum of sneezes to confirm enthusiasm from the majority, a hunting party simply will not get going. If a leader sneezes first, this agreement does not take long. But lesser dogs can initiate the sneezing too, and in that case much cajoling is required to get everyone on board.[37]

Canids all have good hearing (we will talk more about that as it relates to hunting), but what they are more famous for is their sense of smell, so it makes sense that this provides another important mechanism for communication. As we will discuss below, pheromones are important in communicating sexual status, but not just that. Dogs have numerous skin glands with secretions that keep the skin supple and protected and provide some cooling, but also deliver scents. People have the same glands. The secretions themselves do not have a scent, but when they interact with body bacteria and pheromones, they provide much information. African Wild Dogs, for example, have a distinct, strong body odor that perhaps lets the group know who is where on a hunt through the underbrush. This type of strong "BO" also occurs in humans, but our relatively weak noses do not allow us to interpret it beyond deciding that a person needs a shower.

A dog can distinguish territory, individuals, gender, reproductive state, and mental state. Dogs really can smell fear. All this information is left behind as they scratch the ground or *mark* via urination and defecation. All that peeing (p-mail?) and sniffing your dog does on a walk tells him who has been by, how they are doing, when that was, and whether there is a threat or the possibility of a sexual liaison. And he leaves his own messages in response for the next passerby. They get the same information from all that butt and crotch sniffing they do when they meet. Wild dogs greet each other this same way. Perhaps it is just as well we do not share this olfactory ability.

In the feline chapter, we detail how cats sense odors largely using their VNO (vomeronasal organ). In contrast, dogs use primarily their olfactory sense. These different types of chemical sensing systems are suited to the way each animal family communicates and functions with others in the family. Evolutionary geneticists have shown that the number of genes used for the vomeronasal system has increased in cats, but decreased in dogs, since they shared a common ancestor over 50 million years ago. The opposite is true for olfactory genes, which have greatly expanded in dogs but significantly reduced in cats.[38] This evolution parallels how they have adapted as predators, since canids rely greatly on olfaction to hunt and felids rely more on sight and movement. Cats also communicate with each other much more through odors and pheromones than dogs, which use these but also rely on behavior and social cues.

Wild canids all follow essentially the same pattern when it comes to mating. They reach sexual maturity at one to two years of age. The single annual mating season lasts about one to three weeks and occurs in the winter so that the pups are born in the spring when there are lots of other newborns around to eat. Males will fight over females, but once a mating pair is established they remain monogamous, mating only with each other until one dies or is driven off. An exception is the Ethiopian Wolf, where alpha females have been seen to mate with males from another pack. This may be a way to avoid inbreeding where there is little opportunity for offspring to disperse to other packs because of scarcity of both habitat and the wolves themselves, as these are the most endangered of the canids. This behavior may also turn out to be more common than we have seen so far.[39]

Monogamy is unusual for a mammal species, let alone a whole group like the canids, and there are several other odd characteristics of dog reproduction. *Proestrus*, or the period before mating, is an unusually long time of bonding for the male and female. It serves to enhance their bond and bring their mating behaviors into sync, probably driven by hormonal signals communicated with pheromones. (Perhaps you are familiar with the phenomenon of the menstrual cycles of human females living together becoming synced. Same idea here.) Once the female reaches estrus, when she is fertile, mating occurs once or twice a day over three to five days. This is what we call "being in heat" in reference to domestic dogs. It is a good thing our male and female like each other by this point because, quite unlike other carnivores or most mammals, most wild dogs undergo a copulatory tie or lock following mating (Bush and Raccoon Dogs do not). The male ejaculates and dismounts from the female, but the bulbous gland at the end of his penis remains engorged and essentially stuck in the contracted vagina, such that the pair must remain attached this way until he deflates, anywhere from 5 to 20 minutes, depending on the species.[40] This likely ensures fertilization during this once-a-year opportunity and prevents fertilization by some other male, so there is no question about paternity, making the male more willing to help with the pups. He knows they are his. Since the couple is reliably monogamous during this period anyway, this seems like overkill to us, not to mention awkward!

Gestation in canids lasts around 50–70 days, and the number of pups varies from 1 to more than 16, depending on the species. Wolves have 1–8 pups, African and Asian Wild Dogs can have as many as 20, and most foxes fall within the range of 3–6. The number of offspring may depend on food availability in any given year. All canid pups are born helpless, with closed eyes and ear canals, and a poor ability to control their body temperature. About all they can do is squirm around to find a nipple and then suckle. The mother must stay with them initially to protect them, feed them, and keep them warm. The father, and other pack members if present, will keep her supplied with food during this time.

The timing of the early life of pups varies according to habitat, food availability, and species, but we will offer what is typical for Gray Wolves as an example of how this would occur for any wild pup. The neonatal period just after birth lasts about two weeks, at which point the pups' eyes and ear canals open and, with improved coordination and strength, they are up and around, although still confined to the den initially. After another week or two, the pups enter a socialization period where they get to know other pack members (if they are part of an extended pack). Pups have sufficient baby teeth at this point to begin to ingest some solid food and start exploring outside the den. The mother will resume hunting now because she is an alpha and good at it. She will leave the pups alone or in the hands (paws?) of another pack member, likely an older sibling of the pups. One study reported that the mother seemed downright gleeful on her first hunt post-pupping, although she returned earlier than the rest of the pack. Most mothers can relate. At 10 weeks the pups have largely moved from suckling to eating solid food regurgitated by any of the pack adults. Wolf pups "ask" for regurgitated food by licking the mouth and muzzle of adults upon their return (fig. 6.11). Think about that when your puppy starts licking your face when you come home. It is not merely a greeting; she is asking you (politely) to vomit.

FIGURE 6.11 Wolf pups greeting an adult member of the pack. The act of a canine licking another's mouth is tantamount to "asking" for food to be regurgitated. Older Wolves may also perform this action as a sign of submissiveness, since it is "puppy" behavior. Domestic dogs may do this to their owners for the same reason, although it may also be a behavioral result of artificial breeding selection for *neoteny* (retention of juvenile features in the adult). Bildagentur Zoonar, GmbH / Shutterstock

Wolf pups play like, well, puppies, learning to socialize and fight and hunt by practicing all these things with their litter- and pack-mates.[41] Note that having a pack, or even just a father to look after you in addition to a mother, is unusual compared to most other mammals except Lions and some primates. Indeed, in most other animals besides birds, babies hatch and are left to their own devices from the moment of birth, or the whole job of raising them falls to the mother. Only bird fathers routinely offer this level of assistance in raising offspring. And, usually, humans.

Wolf pups can join the pack in a hunt at around four months of age and hone their hunting skills this way. They are capable of dispersing from their natal pack at nine months but often hang around for another year or two. Wolf pups are sexually mature at about two years of age. For African Wild Dogs and Dholes (fig. 6.12), males generally stay with their natal pack and females disperse.[42] In the case of the various foxes that do not form packs, pups are on their own after their first summer and everyone disperses, but the females tend to set up small territories near their mother. In this way they can step in and assume the alpha role if something happens to the mother, and they can raise her pups as their own. An Arctic Fox pup, male or female, is chased away by the mother within six months of birth, right as the Arctic winter is descending. That is a tough way to grow up, but pup survival depends largely on the cyclic populations of rodents, simple as that.[43]

FIGURE 6.12 Dhole (*Cuon alpinus*). Unlike with Gray Wolves, male Dholes tend to stay with their natal pack, while females disperse. Vinod V Chandran / Shutterstock

DOGS AS PREDATORS

What do dogs eat? If you live with a dog, the answer to this question will not surprise you: practically anything! The biggest wild canids can be fierce predators, hunting large grazers like deer, antelope, and Moose. Those canids include Wolves, jackals, Coyotes, African Wild Dogs, and Dholes. However, all these will also eat little animals like rabbits, rodents, or birds, or even eggs. The smaller foxes eat mostly rodents, insects, or worms. Bush Dogs prefer rodents. Hoary Foxes (*Lycalopex vetulus*) specialize in termites. Crab-Eating Foxes (*Canis thous*) eat— crabs! They all are known to eat fruits on occasion. All will scavenge anything, including human trash, if the opportunity presents itself. How do they capture this variety of foods?

SENSING

The first task of a predator is finding prey. Canids use smell, hearing, and sight. And a little luck. Remember, they are built to efficiently cover a lot of ground, which they do to find something to eat. As they travel, they sniff, listen, and look. They are good at all these.

The dominant sense for a canid is smell. It is almost impossible for a person to imagine how well a dog can smell. You can smell a teaspoon of sugar in your coffee. Your dog could smell that same teaspoon of sugar in a coffee the size of two Olympic swimming pools! Their sense of smell is 10,000 to 100,000 more acute than ours. How? First, they have up to 300 million olfactory receptors compared to our measly 6 million, and the part of their brain designated for scent analysis is proportionately some 40 times bigger than ours.[44] Then, the architecture of their noses is also better adapted to take in and sort scented molecules (which is apparently most of the molecules out there, if you ask a dog). When we inhale through our noses, that air passes by olfactory tissues for a quick scan as it heads toward the lungs. When a dog inhales, about 12% of the air is directed to a recessed area at the back of the nose consisting of a labyrinth of bony structures called *turbinates*, lined with olfactory sensors that sort and analyze the molecules, sending signals to the brain about the odor (fig. 6.13). On exhalation, this carefully analyzed air swirls out those little slits on the sides of a dog's nose, aiding in pulling in the next sample and allowing the dog to sniff continuously.

Additionally, the nostrils are far enough apart that a dog can distinguish which side a scent is coming from, in the same way that your ears determine the source of a sound. In this way canids can tell the direction prey is traveling since its more recent steps will have slightly stronger odors.[45] Consider that a minute. Plus, they have the VNO, or Jacobson's organ, in the nasal passage, which detects and interprets pheromones, as mentioned earlier in this chapter and in the cat chapter. We should note that these smelling studies originated from domestic dogs, some of which have been bred for their sense of smell, but a variety of breeds have been examined, and while they are not all Bloodhounds of tracking fame, they are all close. And since they are not far removed from their wild ancestors, and wild canids spend a great deal of time sniffing, it seems reasonable to conclude that wild canids share this olfactory prowess.

The dog may thus catch a tasty scent on the breeze or on the ground. She trots along, following the gradient of the scent, which should get stronger the closer she gets to the prey. Maybe she cannot

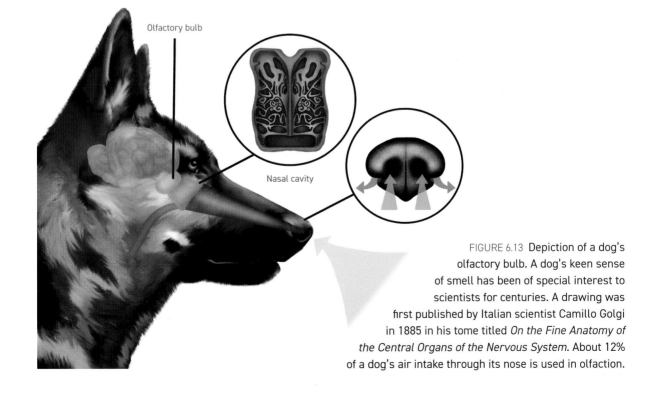

Olfactory bulb

Nasal cavity

FIGURE 6.13 Depiction of a dog's olfactory bulb. A dog's keen sense of smell has been of special interest to scientists for centuries. A drawing was first published by Italian scientist Camillo Golgi in 1885 in his tome titled *On the Fine Anatomy of the Central Organs of the Nervous System*. About 12% of a dog's air intake through its nose is used in olfaction.

BOX 6.5 WE ARE ALL EARS

The Bat-Eared Fox (*Otocyon megalotis*) of South Africa (*left*) has exceptionally large ears, perhaps aiding in finding its preferred food, insects. These little foxes, weighing a mere 9 lb (4 kg), possess the most teeth of any mammal, 46–50, which seems like overkill for a diet of termites,[46] but they are the most ancient of the wild dog lineage, so they probably have not always lived on termites. Perhaps living in the vicinity of larger jackals and Cape Foxes (*Vulpes chama*) has forced them to focus on rather unusual prey for a canid.

Fennec Foxes also have huge ears, but these probably do not appreciably enhance their hearing (*right*). They are the smallest of the canids, 1.5–3.5 lb (0.68–1.6 kg), and they inhabit arid desert habitats where there is not enough water to allow for evaporative cooling like panting. Like jackrabbits and African Elephants, these little foxes dissipate heat over the large surface area of their ears. They can tolerate high levels of urea, a by-product of metabolism, making for concentrated urine so they do not lose much water when they excrete the urea. Fennec Foxes also have fur on the soles of their feet to protect them from hot sand, although they are nocturnal to avoid the hottest part of the day. They spend their days in burrows at the base of sand dunes, well camouflaged and keeping cool.

Bat-Eared Fox (*Otocyon megalotis*). First appearing during the Pleistocene, Bat-Eared Foxes are considered *basal* (primitive) canids and are the only extant species of the *Otocyon* genus.

Yathin S Krishnappa, CC BY-SA 3.0, via Wikimedia Commons

Fennec Fox (*Vulpes zerda*). This diminutive desert-dwelling fox uses its massive ears for both excellent hearing and thermoregulation.

Nicram Sabod / Shutterstock

see the prey through the forest or grass or under the snow or off in the distance, so she listens for it as she trots and sniffs.

You would have a hard time hearing a mouse scurrying around and squeaking at its friends in tunnels beneath your feet. Your dog does not. Dogs have up to 18 muscles to control their ear movement, focusing sound into their ear canals. We have 6 ear muscles and few of us can wiggle our ears at all, let alone aim them. Based on studies of domestic dogs, canines can hear high-frequency sounds well above the upper end of typical human hearing at 20,000 Hz. Those high-pitched sounds are what the mouse produces. We can hear the same lower frequencies that dogs hear, between 3,000 and 12,000 Hz, but dog ears are much more sensitive to those sounds than ours are, meaning they can hear much quieter sounds than we can. What we hear from 20 ft (6 m), they can hear from 80 ft (24 m). Between 12,000 and 20,000 Hz, our ears are about equally sensitive, but dogs are better at discriminating sounds. A dog can hear eight different frequencies between a C and a C-sharp. Our hearing is nowhere near that precise. But we are better at one hearing task: determining where a sound is coming from, because our ears are farther apart relative to head size.[47]

So, the canid hears small prey and can maybe then hunt it. Like other predators, canines have their eyes positioned in the front of their heads, giving them good depth perception and wide, stereoscopic vision. They also have the reflective membrane of sharks, crocodylians, and others, the tapetum lucidum, which helps with night vision. But dogs are not specialized for night vision, as none are strictly nocturnal. They are active at any time of day, though this varies within and between species and individuals, and even with time of year. Dogs have dichromatic vision, meaning they can distinguish, in human terms, blues from longer-wavelength browns, yellows, and greens,

but not reds. This capability is shared by most mammals except primates like us, who have trichromatic vision. Canines also have many more rod cells—receptors that work in dim light but do not discriminate color at all—than cone cells. Thus, like cats, they see motion and see well in dim light, but objects appear a bit blurry, so dogs do not rely heavily on vision (fig. 6.14).

CATCHING AND KILLING

Having sensed prey, how does a canine go about catching it? Well, it depends on what the prey is. A caterpillar or piece of fruit will just sit there for the taking, which explains why wild dogs eat these things. But a caterpillar or piece of fruit does not provide much energy, so many dogs go after bigger items. Since you are reading this book, you probably watch wildlife documentaries and so have seen a fox pursue a mouse. The fox walks along until he suddenly freezes and cocks his head, focusing his ears on those high-pitched sounds to determine the position of the rodent under the snow (dog trainers call this behavior *head-snapping*). He stalks toward the sound, zeroes in above it, then springs straight into the air and lands face first in the snow, possibly getting suspended upside down for a moment. If the fox has judged correctly and has a little luck on his side, his head pops up, he shakes his newly captured prey violently side to side, and bites and swallows it down. Dingoes specialize in rabbits, flushing them out and chasing them down. Hunting rabbits in Australia is helpful since parts of that country are plagued by an overpopulation of introduced rabbits. Short-Eared Dogs (*Atelocynus microtis*) in South America are known to swim after prey. In these scenarios, it is every dog for itself.

Wild canids sometimes pair up with other species. An interesting example of that is Coyotes and badgers. Coyotes have that great dog sense

FIGURE 6.14 A comparison between how humans (A) and dogs (B) see things. The American Kennel Club offers a fascinating webpage demonstrating this phenomenon ("Can Dogs See Color?," https://www.akc.org/expert-advice/lifestyle/see-what-the-world-looks-like-to-a-dog/), and various smartphone apps (like "Dog Vision") show you the world through the eyes of a dog.

of smell, much better than a badger, but a badger can dig. So, the Coyote will locate underground prey and then stand guard as the badger digs it out. When the rabbit or mouse bolts for safety, the Coyote is waiting for it. It shares the catch with its badger partner.[48] Similarly, Wolves and Common Ravens may have a symbiotic relationship. Ravens are a common dinner companion when a Wolf pack makes a large kill, scavenging morsels of meat. The Wolves seem tolerant of the thieving forest denizens, but why? One study suggests that ravens will locate potential prey (like an injured Elk) for Wolves and start vocalizing. By following the raucous caws of the ravens, the Wolves find and kill more prey. The ravens subsequently share in the feast. This strategy seems to be employed more often in winter when food is scarce for everyone.[49]

What about dogs that hunt in packs? Let us look at African Wild Dogs as an example. Their behavior is quite similar to that of Dholes, Wolves, and Bush Dogs, all of whom routinely hunt in packs. Packs range in size from 3 to more than 20 adults, although Wolf packs are smaller. Hunts start with a pep rally of sorts, with intense greetings and excitement in the group. They all head off at a trot of about 6 mph (10 kph). The same lead alpha dog usually heads the pack. Once suitable prey such as a small Impala or Thomson's Gazelle, in the case of African Wild Dogs, is sighted and the pack gets within about 2 mi (3 km) of it, the lead dog takes off, running at about 30 mph (48 kph), with two or more dogs behind at 100 yd (90 m) intervals to head off veering, swerving prey. Unlike cats, which conserve energy by waiting and stalking prey, dogs typically employ a *cursorial* hunting technique. They just keep running and chasing their prey to the point of physical exhaustion (this is similar to the hunting technique of the San Bushmen in Namibia and Botswana).

Dogs, like cats, have claws, but they do not use them for grasping and capturing prey. Rather, their claws are like cleats on an athlete's shoe, used for traction. Because dog claws are nonretractable and are in constant contact with the ground, they are quite dull. It should not surprise you that with a less flexible body and blunted claws, most dogs are unable to climb trees. The one exception to this rule (there's always an exception) is the Gray Fox (fig. 6.15). These nimble little foxes are known to climb up and into the small top branches of juniper

FIGURE 6.15 Gray Fox (*Urocyon cinereoargenteus*) in a tree. The Gray Fox is the only canine species that is adept at climbing and foraging in trees. They can even navigate small outer branches to harvest fruit, which can be seen in this video of a Gray Fox eating juniper berries in the upper branches of a tree: "Grey Fox Climbs for Juniper Berries," https://www.youtube.com/watch?v=INh2JDq7pNA. Danita Delimont / Shutterstock

FIGURE 6.16 Two African Wild Dogs (*Lycaon pictus*) eating a baby Impala. Wild dogs are known for being quite savage in their feeding behavior. Pack individuals often grab various parts of their captured prey, ripping it apart in different directions, often while the prey animal is still alive. Because of their vicious notoriety and their ability to decimate an entire flock of sheep in a night, it is little wonder that they often come into conflict with African farmers.

trees, not to pursue prey but to delicately pick and consume the tiny juniper berries. Gray Fox kits (youngsters) are known to play by chasing each other up and down trees.[50]

Also, unlike cats, dogs do not really deploy a *kill bite* in bringing prey down. The canine teeth of canids are actually less formidable than those of felines. The lead dogs slow down the prey by grabbing and tearing at it, and the rest of the pack catches up and piles on. They basically start eating the unfortunate beast alive, although it quickly dies of shock and blood loss (fig. 6.16). The African Wild Dog packs studied in the Ngorongoro Crater were successful an amazing 85% of the time.[51] In Tanzania, packs were successful with Impala 64% of the time, and 45% of the time overall.[52] Dholes similarly hunt Sambar, a small antelope in India. Wolves in North America pursue larger deer and Elk.

HOW ARE DOGS DOING?

Many, but not all, wild canid species are actually doing okay, if not prospering. One explanation for this positive situation is their small body size and the reduced level of resources (food and space) they require. In addition, as a group, canids are highly adaptable. They all will eat meat, like good carnivores, and there is no shortage of small packages of meat like rodents and rabbits in the world. And insects, other invertebrates, and fruit will all suffice too. In fact, the distributions of Red Foxes, Coyotes, and jackals are actually expanding, although this expansion is partly the result of the misfortune of a major predator and competitor, Wolves, as well as their high tolerance of humans.

Let us look at Wolves first. With the noteworthy exceptions we describe below, most Wolf populations around the world are classified by the IUCN as of Least Concern. In North America, Wolves were listed as endangered when the Endangered Species Act (ESA) was introduced in 1973. The distribution of Wolves has been expanding since then, though, thanks to a concerted effort to reintroduce them in the northwestern United States from populations in Canada. Wolves were initially reintroduced into Yellowstone National Park in the winter of 1994–95. This reintroduction was fraught with controversy, which rages to this day. Livestock ranchers and folks who hunt deer and Elk do not like to compete with Wolves, and they maintain the right to kill them if they seem to be a threat. Currently Wolves are still generally protected by the ESA with exceptions allowing kills if they are a perceived threat to humans or livestock, so usually if a cow gets killed, a Wolf gets killed.[53]

The introduction of these predators to habitats where they had been absent for about 70 years has become an experiment with interesting results. In 2014, an English writer named George Monbiot narrated a short video based on a few studies outlining the effects of the return of Wolves to Yellowstone.[54] The video suggests that Wolves are likely keystone animals in their ecosystems. According to some studies, Yellowstone National Park with Wolves is quite a different place than it is without Wolves. In just a few years after the introduction of Wolves to the park, cottonwood, aspen, and willow trees were found to be growing taller, where they had previously been reduced to chest-high sticks by constant grazing pressure from Elk.[55] Surprisingly, it was not that the Wolves were necessarily reducing the Elk population appreciably; rather, they were forcing the Elk to keep moving. In the presence of Wolves, the Elk lost the luxury of leisurely standing around grazing in one place. Forcing the Elk to experience the stress of potential predation allowed patches of trees to

recover.[56] And this single reintroduction triggered a whole ecosystem cascade of increasing biodiversity. Populations of birds, small mammals, and insects all increased and diversified, taking advantage of the tree cover. Bison numbers increased. And rather than driving the mesopredators like Coyotes and foxes and American Black Bears away, these animals found more prey among the increasing biodiversity and in the form of carcasses left by Wolves. Those carcasses also help scavengers like Bald Eagles and Common Ravens. Beaver numbers have increased because of the increase in trees to browse, and as nature's engineers, they have built pools and raised wetland water tables that support larger and more diverse waterfowl, amphibian, and invertebrate populations. The return of Wolves has maybe even altered the course of the rivers, as regenerating tree roots stabilize the soil and minimize riverbank erosion.[57] In recent years, scientists have reexamined this. They acknowledge that there was a significant recovery for much of the park after Wolves were reintroduced, but they suggest that giving Wolves all the credit is an oversimplification.[58] There are other factors, like the western drought, for example, that likely contributed to the changes in flora composition. But even if Wolves cannot take all the credit, it seems clear that they are an important presence in a balanced ecosystem. The idea that the story is more complex, and there may be some factors that we do not yet fully understand, only demonstrates further the amazing interconnectedness of nature.

Ethiopian Wolves are the most endangered of the wolves that still maintain wild populations, with perhaps only 300–500 individuals left in the mountains of Ethiopia. Ethiopian Wolves exist in small, isolated populations to start with, and loss of habitat isolates them further. There are slightly more Arabian Wolves (*Canis lupus arabs*) left in Saudi Arabia and Oman, but they, too, are critically endangered.

BOX 6.6 THE COYWOLF

You do not need to travel to exotic lands to find wild predators. Sometimes you need look no further than your own backyard. The expansion of the Coyote population in North America, along with Coyotes' ability to adapt to and thrive in a human environment, means that an increasing number of people are likely to encounter these predatory canines. It seems likely that this expansion is largely the result of the reduced eastern Gray Wolf population. But things are not that simple. As mentioned in box 6.2, the Eastern Coyote has hybridized with the Gray Wolf, making what used to be referred to as a *coywolf*.[59]

Genetic research has indicated that populations of "wolves" thought to be distinct are made up of these hybrids. The Red Wolf (*Canis rufus*) is a distinct species that arose because of hybridization between the Gray Wolf and Coyote.[60] In fact, trace Red Wolf genetics have been found in multiple places in the United States, including Texas.[61] Unfortunately, Red Wolves went extinct in the wild in 1980. However, there was a concerted effort to breed them in captivity in the southeastern United States, and eventually several adults were released in North Carolina. This managed population has been doing well, but the recent genetic work identifying them as Coyote-Wolf hybrids has caused some conservationists and wildlife groups to reconsider efforts to preserve their population. Should a population with hybrid ancestry

Coyote (*Canis latrans*) in an urban setting. Coyote territories have been expanding in recent times, largely because of the extirpation of larger predators (Gray Wolves and Pumas) as well as their adaptability to existing in proximity to humans. Matt Knoth / Shutterstock

be conserved? Some say no, but many evolutionary geneticists would argue that this is not a sound approach, and that hybrids are just as important evolutionarily as species that have not hybridized in recent history, if not more so.

When you think about wild canines, your first thought is probably not New York City, but it seems that the Eastern Coyote is making itself at home in the Big Apple. Since the first sighting of a Coyote in NYC in 1995, the population has exploded. Why and how can a medium-large predator live in such an urban jungle? No doubt their stealth has much to do with their ability to remain safe. And their willingness to eat just about anything (including garbage) ensures ample food year-round.[62]

Since the Coyote hybridized with Wolves, it inherited some domestic dog genes along the way (see box 6.8). So, the Eastern Coyote is essentially a coyote/wolf/dog. This might not seem all that important, but just think about the number of Coyotes that might be visiting the backyards of people with dogs. While some people are afraid that Coyotes might eat their faithful pets (which may happen if it is a Chihuahua or Teacup Poodle), many times the Coyotes are curious and seem to enjoy interacting with their canine cousins.

Author Rob had a young male Coyote come to his house every night for a month to play with his white German Shepherd, Hantu (see photo below). The Coyote would sit at the edge of the yard and wait for the dog to be let outside. They would chase each other in and out of the woods for an hour or so, at which point Hantu (14 years old at that point) became tired and went back inside. The Coyote would vanish back into the underbrush and, like clockwork, show up the next evening for another playdate. While it is possible that a Coyote and your playful pooch might breed, the likelihood is fairly low, as Coyotes tend to have a very specific breeding season and your dog does not. A massive invasion of "coydogs" is not something you have to worry about.

Author Rob's German Shepherd playing with a wild Coyote (*Canis latrans*) outside his home in South Carolina. Dogs and Coyotes have much in common and can potentially interbreed. Social interactions between the two species are likely to increase as both humans and Coyotes expand their territories.

Maned Wolves, which look like tall foxes but are neither true foxes nor Wolves and are the largest South American canid, are also not faring well because of habitat loss. They are considered Near Threatened by the IUCN. They are the canid most closely related to the Bush Dogs, which are also Near Threatened and rare, except in Suriname, Guyana, and Peru.

The Asian and African Wild Dogs are faring poorly. These species require space and good-sized prey, and they must compete with the other big predators of Asia and Africa. There are fewer than 2,500 Dholes and fewer than 7,000 African Wild Dogs left in the wild. Both are Endangered. In both cases the problem is habitat loss and, in the case of African Wild Dogs, retaliatory killing, since they will prey on livestock. But African Wild Dogs are doing well enough that finding suitable areas to relocate expanding packs is becoming a challenge.

Coyotes, as we have mentioned, are doing fine and in fact are expanding their populations. They are also apparently hybridizing with Gray Wolves (see boxes 6.2 and 6.6). Likewise, Golden Jackal populations are expanding into Hungary, Serbia, Slovakia, Austria, and Italy.[63] Black-backed and Side-striped Jackal (*Lupulella adjusta*) populations in Africa are of Least Concern according to the IUCN.

Red Foxes benefited from the colonial period of Great Britain. Fox hunting has always been a popular sport in England, and as the English moved to new countries around the world, they brought Red Foxes to hunt. And like any highly adaptable small canine, foxes did well nearly

BOX 6.7 ▸ FOXES AS PETS?

In 1959, a team led by Russian geneticist D. K. Belyaev, working at the Russian Academy of Sciences in Siberia, began trying to selectively breed tame foxes from farmed silver foxes, a color variant of Red Foxes. They initially wanted to breed these foxes to be tame so they could harvest their fur more easily; therefore, foxes that were more tolerant of humans were allowed to breed. Generation after generation of this selection for tameness resulted in several interesting genetic changes in the foxes. Dr. Belyaev was surprised to see new fur patterns appearing in offspring, often manifesting as white patches on the ears, tail, and belly, as well as lightening of some body fur. At around the same generation, the foxes' ears became floppier, their tails curled upward, and they wagged them. They even started to bark and loll their tongues. In short, they began looking and behaving like domestic dogs.

It turns out that selecting for tameness induced a prolonged developmental state called *neoteny*. This essentially means the retention of juvenile-like features, which is accompanied by lower stress hormones and higher relaxing serotonin levels. The genetic changes that accompany this tameness transition are still a focus of study, but the most popular hypothesis is that when humans selected for this behavior, the precursor cells that make up the *neural crest*, an essential developmental region that gives rise to the central nervous system, were affected, leading to cartilage differences (ears and tail) and melanin distribution changes (coat pattern and color), in addition to behavioral changes affecting the brain. Now these animals can be purchased and imported from Russia for several thousand dollars. Be advised, however, that they are not legal in every state, and their urine, which they spray around like cats do, smells pungently of skunk.[64]

FIGURE 6.17 An endangered Channel Island Fox (*Urocyon littoralis*). Although similar in appearance to the Gray Fox (*Urocyon cinereoargenteus*), Channel Island Foxes have been separated from mainland foxes for more than 16,000 years and have been evolving independently.
Shanthanu Bhardwaj, CC BY-SA 2.0, via Wikimedia Commons

and the San Nicolas Island Fox. When an organism moves to an island, it often becomes isolated from its source population and all the genetic diversity that population contained. This *founder effect* means that the island population has only a subset of the genetic variation that the mainland population has. Even if enough other organisms come over to support a population, that population is going to necessarily be rather inbred and, as we already stated, will lack diversity at the outset. This combination can lead to lack of reproductive success over time, and vulnerability to introduced pathogens to which the original population harbors no resistance.

This was in fact the case with the Catalina Island Foxes, whose population was reduced by more than 90% in 1999–2000 by the canine distemper virus.[66] In addition, Golden Eagles have made a comeback in this area since the 1990s, and they are effective predators of the foxes.[67] Captive breeding efforts have helped repopulate the island population, and their numbers are increasing. However, this *population bottleneck* that resulted from canine distemper has caused the Catalina Island Fox population to be susceptible to infection, and it has a high rate of a specific form of cancer in the glands of the ear. Genetic work is ongoing in this population, and the underlying susceptibility to this cancer is being studied.

everywhere they went. Foxes, in fact, have even made it into the pet trade (see box 6.7).[65]

There is a group of foxes living only on the California Channel Islands (fig. 6.17). These populations were separated from the mainland fox population about 16,000 years ago and have been evolving independently on nearly every island since. Individual subspecies exist on most islands, including the Santa Cruz Island Fox, the Catalina Island Fox, the Santa Rosa Island Fox, the San Clemente Island Fox, the San Miguel Island Fox,

SOME APEX PREDATORS

WOLF
(Canis lupus)

Wolves (fig. 6.18) are the largest species of canine and the only one (except Dingoes in their isolated environment) that would be considered an apex predator as an individual. Coyotes, jackals, foxes, and the rest fall in the range of mesopredators,

BOX 6.8 A BIG BAD BEAGLE?

It may not have escaped your notice that in a chapter about dogs we have dedicated very little discussion to actual *dogs*. This is not an oversight. Domestic dogs are included in the section about Wolves because they are in fact the same species! Yes, you read that right. The ominous Wolf of so many fables and your docile lapdog are separate subspecies of the same species: Gray Wolf (*Canis lupus lupus*) and domestic dog (*Canis lupus familiaris*). They are two sides of the same coin (box 6.8 fig.). Although an exact timeline for dog domestication has not been conclusively determined, at least one recent genetic study suggests it started more than 20,000 years ago.[68] It is thought that Wolves began hanging around early humans, scavenging food scraps and feces. (If you have a cat with your dog, this latter food may not be surprising. Yuck! But good for sanitation.) Those Wolves with less fear ate more regularly and produced more offspring with the same disposition. Over the centuries, humans started interacting with and keeping Wolves as "pets," and what a useful relationship it turned out to be. The commensal semidomesticated Wolves offered protection and eliminated our leftovers. Our ancestors started keeping only the puppies they liked (likely eating the rest). This artificial selection skyrocketed the evolutionary process to produce every breed of dog we know today (344 according to the Fédération Cynologique Internationale).[69]

Humans have had an intimate love/hate relationship with Wolves like we have with no other predator (including cats). And yet they evolved to become companions, protect our livestock, and guard us against other predators and even human intruders. In many ways, dogs, and therefore Wolves, are responsible for our success as a species, and we owe them a debt of gratitude. Although visibly different (different phenotypes) from dogs, Wolves are genetically almost identical to them, but there is ample evidence that domestic dogs have a genetic burden their wild ancestors do not share, thanks to us. *Deleterious variation* means that dogs are more likely to carry negative genetic traits than are Wolves or Coyotes.[70] However, because of our love for them and our veterinary attention, dogs' life spans are generally longer than those of their wild counterparts. Although dogs may look different from Wolves, in the end, your pooch is really only a few genetic steps away from being one of the world's top predators. Keep that in mind and show a little respect when you pat your Beagle on the head.

certainly worthy of respect, but not on top unless Wolves are gone. Male Wolves can weigh up to 110 lb (49.9 kg) and, like many species, Wolves tend to be larger in colder climates. Wolves exist throughout much of Asia, the Middle East, Europe, and North America. Once the most widely distributed mammal on the planet, Wolves have experienced significant decreases in their numbers and range, but the Wolf population is considered stable.[71]

AFRICAN WILD DOGS
(Lycaon pictus)

Although not truly a top predator, at least as individuals, African Wild Dogs definitely deserve to be recognized. With a maximum size of 79 lb (46 kg), African Wild Dogs hardly seem a challenge for a Lion or hyena, but hunting in packs, they are a formidable force, a collective apex predator.

Various dog breeds. Dogs (*Canis lupus familiaris*) are actually a subspecies of the Gray Wolf (*Canis lupus*). Although they are genetically similar, the significant visible differences (phenotypes) between various dog breeds and Wolves are the result of hundreds of years of selective breeding. Dora Zett / Shutterstock

Author Rob has spent years in the field following and observing these dogs and the absolute chaos they unleash. He has seen African Wild Dogs chase away individual Lions, battle with hyenas, chase Leopards up trees, and even send herds of elephants scampering into the bush (fig. 6.19). Their almost manic bursts of energy and coordinated attacks are enough to send even the acknowledged top predators packing. African Wild

Dogs have perhaps the highest successful kill rate of any large terrestrial predator (up to 85%) and can annihilate a herd of domestic sheep in a single night (which is why they regularly come into conflict with farmers). Luckily for us, they do not grow to the size of Lions, or we would all be in serious trouble. Despite being smaller than Africa's top predators, the African Wild Dog is one of nature's most finely honed killers.

FIGURE 6.18 Gray Wolf (*Canis lupus*) stalking in snow. Gray Wolves are the largest wild canine and, prior to human efforts to counter their depredation, were the world's most widely distributed species of terrestrial predator.

Holly Kuchera / Shutterstock

FIGURE 6.19 (A) African Wild Dogs chasing a Spotted Hyena away from a kill. (B) An elephant matriarch flaring her ears at an African Wild Dog while her herd moves away from the pack. (C) An African Wild Dog trying to finish off an adult male Impala that it has partially disemboweled.

7

BEARS

You might think of bears as rather cuddly. Teddy bears, hardly a threatening beast, named for Theodore Roosevelt in a nod toward his conservation ethos, remain favorite toys of children everywhere. In fact, except for Polar Bears (*Ursus maritimus*) and sometimes Brown Bears (*Ursus arctos*), most bears do not consider people as possible prey, and they do not even always eat meat. Panda Bears (*Ailuropoda melanoleuca*) consume bamboo almost exclusively, and all bears are omnivores, meaning they include both plants and animals in their diet, just as humans do. And bears are like us in other ways as well. Bear adults are not generally prey. Bears move with a sort of pigeon-toed shuffle on flat feet. They seem easygoing, smart, curious, and dexterous. The cubs are, by any standards, adorable! Humans are drawn to bears. In fact, there is a word for that: an *arctophile* is one who loves bears.

Bears are all easily recognizable as bears. There are only eight species (fig. 7.1), and they share the rounded ears, stocky body, pointed snout, short tail, and thick fur of a typical child's teddy bear. They are in the order Carnivora and are reasonably close relatives of the canids (doglike carnivores) and pinnipeds (seals, sea lions, and Walrus). Fossils from the Miocene epoch, around 20 million years ago, include an animal named *Hemicyon*, or "half dog," which possessed characteristics of both bears and dogs and demonstrates their common ancestry (fig. 6.1). The Sloth Bear (*Melursus ursinus*) was considered the Bear Sloth for a while, and Pandas were thought to be a type of raccoon, but DNA analysis has confirmed that both are in fact bears.[1]

There are bears on all continents except Africa, Antarctica, and Australia. Polar Bears inhabit the Arctic, where they depend on sea ice for hunting grounds. Brown Bears, some of which are known as Grizzly or Kodiak Bears, despite having lost about half their population since the 1800s, are still the most widely distributed bears, roaming the upper reaches of North America, northern Europe, and northern Asia. American Black Bears (*Ursus americanus*) are common through much of North America. Asiatic Black Bears (*Ursus thibetanus*), also called Himalayan or Moon Bears, are found from Japan and southern Asia up into Siberia. Sloth Bears live in and around India. Andean or Spectacled Bears (*Tremarctos ornatus*) are found in the Andes in South America.

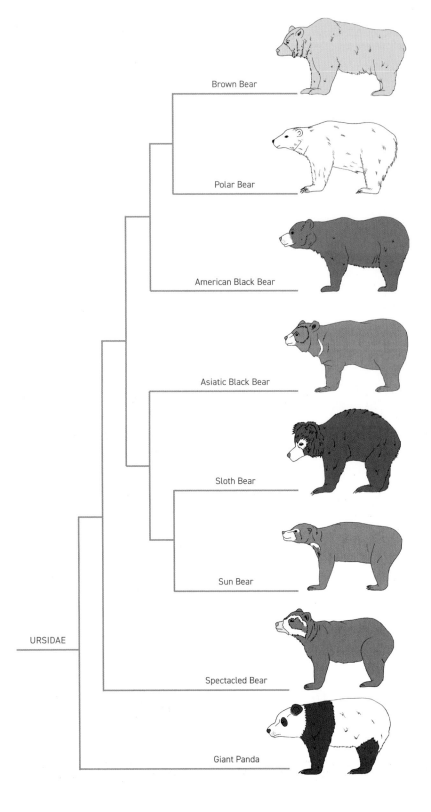

Brown Bear

Polar Bear

American Black Bear

Asiatic Black Bear

Sloth Bear

Sun Bear

Spectacled Bear

Giant Panda

URSIDAE

FIGURE 7.1 **Bear phylogeny.**
Only eight species of bears
currently exist.

Little is known about the Sun Bears (*Helarctos malayanus*) of Malaysia and nearby islands, but they are also called Honey Bears because that is one of their favorite foods. And finally, Giant Pandas are scattered in small forested preserves in the mountains on the eastern rim of the Tibetan Plateau. We will look at each of these, but let us first consider bears in our lives.

BEARS AND US

Humans have been captivated by bears for as long as we have been around. There were Pleistocene bear cults, probably associated with the now extinct Cave Bears (*Ursus spelaeus*). The ability of bears to apparently resurrect, waking from hibernation, endowed them with legendary immortality, and bears show up in numerous early religions as spirits or shamans. Some indigenous peoples of the Americas considered, and still consider, the bear to be a fellow citizen of the forest. The Navajo would not hunt them except in dire circumstances. When bears were hunted, they were treated with respect, and most indigenous tribes feature some type of ceremonial bear dance. Wearing bear claws as ornaments confers power.

Two of the most well-known constellations in the night sky are Ursa Major and Ursa Minor, the Big and Little Bear, also known as the Big and Little Dipper. The Greeks believed Zeus placed them there, or, in another story, Hercules tossed them up by their tails because they were causing trouble on Earth.

February 2 is Groundhog Day in the United States. The groundhog peeks out of his burrow on this day, and legend says that if he sees his shadow, he is scared back underground and we will have six more weeks of winter. If he does not see his shadow, spring is imminent. In Austria, Hungary, and Poland, that legend belongs to

the bear: Bear's Day. Same idea, but we guess that no top-hatted gents haul a bear out of his cave to talk about the weather the way they do with Punxsutawney Phil, the official groundhog prognosticator in Pennsylvania. (Author Sharon is from Pennsylvania, so she was obliged to bring up Punxsutawney Phil at some point.)

A Grizzly Bear is the official animal of California, even though Grizzlies have not lived there in the wild for nearly a century. The Missouri state flag features two Grizzly Bears, representing strength and courage, even though it is not clear whether Grizzlies ever lived in Missouri. More than 50 colleges and universities in the United States have bears as their mascots. Professionally, you can root for the Boston Bruins hockey team, the Chicago Bears football team, the Chicago Cubs baseball team, and the Memphis Grizzlies basketball team.

There are numerous famous bears, including several famous teddy bears. *Winnie the Pooh*, by A. A. Milne, was based on his own son's teddy bear and was named for a popular American Black Bear from Winnipeg, Canada, living at the London Zoo at the time. *Paddington Bear*, by Michael Bond, was about a bear arriving in London from darkest Peru but was based on the tagged child evacuees leaving London by train during World War II. *The Berenstain Bears* is a famous children's book series. Generations of children have found *Goldilocks and the Three Bears* to be "just right." Baloo, the mostly helpful and easygoing bear from Rudyard Kipling's *The Jungle Book*, protected young Mowgli. He was probably a Sloth Bear. *Gentle Ben*, by Walt Morey, describes a friendship between a Brown Bear and a boy, based on real people and bears in Alaska. Then there is Smokey the Bear, who has been helping the US Forest Service convince folks to prevent wildfires since 1944. Bart the Bear was an actual Kodiak Bear (a subspecies of Brown Bear found on Kodiak Island,

Alaska). He was an actor, starring as a cub in the TV show *The Life and Times of Grizzly Adams*. As an adult he costarred in numerous movies with actors including Anthony Hopkins, Robert Redford, John Candy, Dan Aykroyd, Morgan Freeman, and Daryl Hannah, among others, until he died in 2000 at the age of 27. His offspring still appear on-screen, but lately the bears have been computer generated instead of real. These include the ill-mannered Ted, in a movie of the same name, the unnamed bear that nearly eradicated Leonardo DiCaprio's character in *The Revenant*, and the animated Kung Fu Panda. We are all a bunch of arctophiles.

WHAT IS A BEAR?

One of the reasons we humans like bears is that we know a bear when we see one. Bears are mammals, so they are furry and nurse their young. They are in the order Carnivora and are all in one family, Ursidae. All eight species have small eyes, round ears, a pointed snout, a stocky shape, and a short tail. Bears possess large claws and protrusible lips. They sometimes stand upright but mostly walk on all four *plantigrade* feet, meaning on their heels and toes like you. They have five toes on their feet and walk with a pigeon-toed shuffle. In fact, Sun Bears are so pigeon-toed that their front feet often cross. Except for Pandas, however, they are faster than you, running at 25 mph (40 kph) or better, with Grizzlies, a subspecies of Brown Bear, going up to 40 mph (64 kph). They also sometimes sit upright, perhaps leaning against a tree, maybe to engage in a good back scratch.

The flexible, strong forearms of smaller bears enable easy tree climbing. Young American Black, Asiatic Black, and Brown Bears are natural climbers—that is, they do not need to be taught. These species even know to carefully move outward to weaker branches where a bigger predator cannot follow. Heavier adult Black and Brown Bears stop climbing, but Andean Bears routinely climb to find food, escape predators, and even sleep (fig. 7.2). American Black Bears and Asiatic Black Bears have been known to hibernate in trees. Sloth Bears and Sun Bears climb and hang upside down like sloths, using their long claws, but Sloth Bears do not climb to escape, because their primary predator, a Leopard, climbs well too. Polar Bears do not climb trees because, you guessed it, there are no trees in polar regions, and Pandas, ever the oddballs, are poor climbers.

Except for Polar Bears and Pandas, most bears are black, brown, or grizzled, although Spectacled Bears have light-colored circles around their eyes, and Sun Bears and Asiatic Black Bears have a yellow crescent on their chests. And white Black Bears, also called Spirit, Ghost, or Kermode Bears, are prevalent in the coastal rain forests of British Columbia, Canada. Like other white animals, white bears can show up in any population, but whiteness is a rare color morph. It represents only 10%–20% of the population in British Columbia rain forests, where it is relatively common, and it is the result of a single nucleotide replacement in one gene.[2]

Polar Bears are white, of course. No, actually Polar Bears are black. What? Polar Bears have black skin, which helps them absorb heat from the sun, and their insulative undercoat and longer, oily guard hairs are in fact colorless, allowing sunlight and heat right through to the skin and then trapping it there. We made that sound simple, but the physics of it are actually quite, um, "hairy." Suffice it to say that Polar Bears arrived at a suitable compromise between having to absorb heat but also needing camouflage against white ice and snow. They are covered, quite literally head to toe, with the densest fur of the bears. Only the tip of the nose and the footpads lack fur. Fur between the footpads, however, provides both insulation and improved traction. Because the fur is hollow

FIGURE 7.2 **Ecuadorian Spectacled Bear (*Tremarctos ornatus*).** Spectacled Bears are often found foraging in the forest canopy for food as well as making nests and sleeping among the branches. Vladislav T. Jirousek / Shutterstock

and colorless, the bears often take on color from their environment, appearing yellow from accumulated oils and in certain lights, or slightly green if algae are lodged in the fur. As a bonus, those hollow hairs offer some buoyancy when the bears swim, although they are largely useless as insulation when wet.[3]

Another compromise color pattern is that of the Pandas (see box 7.1).

Pandas live in cold, wet environments in the mountains of eastern China and have an undercoat both woolly and oily that resists compaction, keeping them warm and dry as they sit on the damp ground. Sloth Bears look a bit odd, as they lack facial hair but have long hair on their ears

and are otherwise quite shaggy. They have a mane behind their head, a trait they share with Brown Bears and Asiatic Black Bears. Andean or Spectacled Bears have long, thick fur, also somewhat shaggy. American Black Bears have a dense undercoat with long, coarse, thick guard hair, although the density of the coat correlates with their habitat, as some live at warmer latitudes in North America, as far south as Florida and northern Mexico. The more tropical Sun Bears have a short, dense, sleek coat, with cowlicks and whorls around the head. All bears molt annually during spring and summer, replacing all their fur and making them look quite a mess during this period. You have bad hair days. They have bad hair months.

BOX 7.1 IT'S BLACK AND WHITE

Why are Pandas black and white, quite unlike other bears? This question has stymied scientists for some time. There are several suggestions. The black-and-white coloration might be *aposematic*, or a warning of some kind, such as we see in skunks. But Pandas do not excrete anything noxious and will not try to attack or eat you. One study examined this question by comparing coat patterns across carnivores.[4] Could the black-and-white pattern be disruptive to potential predators? Other animals use disruptive coloration to obscure their outlines, making them more difficult to see. Currently Dholes, Leopards, possibly Asiatic Black Bears and Brown Bears, and historically Wolves and Tigers preyed on Pandas, so their color pattern could be a defense. But disruptive coloration is generally arranged to hide features or disrupt outlines. A Panda's black eyes and ears would seem to exaggerate rather than hide those features. Maybe the black around the eyes reduces glare, as in Cheetahs and American football players? But Pandas can be active at night as well as during the day, and there is no correlation between black eye markings and brightness of habitat among carnivores. Do the dark shoulders work as camouflage in shady areas and retain heat? The role of fur in absorbing and retaining heat is more correlated with the thickness of the fur than with its color, but camouflage is a possibility. White is good camouflage against snow, but it is snowy for only about a third of the year in the mountains where Pandas live. In bears as a group, color does not correlate with temperature because bears hibernate in places where it snows a lot, so they are not out and about in it (except Polar Bears). Pandas recognize and remember differences in the facial coloring of other bears, so perhaps it is a form of individual identification. And in carnivores, ears and eyes are often similarly colored.

One study concluded that the coloration is basically a consequence of Pandas' poor diet. They cannot extract sufficient nutrition from bamboo to accumulate enough energy to allow them to hibernate through the snowy season. They are too big to molt quickly to change color, like an Arctic Fox, for example, again because they do not have the energy to do that. Their coat thus perhaps represents a compromise, allowing year-round camouflage in forests and on snow. And they take advantage of the facial coloring to identify each other and communicate. That is the most we can say at this point. Bears have the dichromatic vision and lack of visual acuity common to most mammals other than primates, so we are not seeing them as they see each other. Our conclusion: Who knows?

Giant Panda (*Ailuropoda melanoleuca*) eating bamboo.

BIOLOGY AND ECOLOGY

All bears are solitary animals unless they have collected around a particularly abundant food source, such as Brown Bears gathering along a river when salmon run upstream to spawn, or Polar Bears congregating at a whale carcass. Otherwise the only group you see is a female with cubs, although Sloth Bears, Sun Bears, and Asiatic Black Bears sometimes travel as a family, including the father.

Bears tend to wander, usually in search of food. They are familiar with their ranges, whether that is 1–3 mi^2 (2.6–7.8 km^2) for a Panda, hundreds of square miles for a Brown Bear, or perhaps thousands of square miles for a Polar Bear. They depend on landmarks but also seem to possess an internal compass because they are good at homing, or orienting toward home, in the absence of obvious sensory cues. Polar Bears manage to navigate on ever-shifting, slowly spinning ice. If there is sufficient food, bears are not particularly territorial and will tolerate neighboring bears, although it is a fragile truce.

While bears do not interact much if they can avoid it, they do have a hierarchy, and size is power. Within a species, all members understand where individuals rank in this hierarchy, which helps avoid energetically expensive and dangerous fights. The largest, old males, or females with new cubs, dominate, although new mothers do their best to simply avoid other bears. The roar of a male communicates his size and rank to everybody. Studies have shown that the right roar can immediately double a subordinate's heart rate, sort of like your boss or parent yelling at you. This communication allows fights to be avoided unless a challenge is afoot. Fights to obtain mates or protect cubs can be especially deadly, but they usually end quickly once rank is established. Next in the hierarchy are single subadult males, then other

adult males and females, and then the remaining subadults.

Roaring is a way to communicate a threat. A bear may also emit a low rumbling, or a *woof* or loud jaw-popping sound, both produced by rapidly expelling air. Bears can make a wide variety of sounds but in general do not make a lot of noise. A Polar Bear roar has been described as akin to a bellow from Chewbacca the Wookiee in *Star Wars*.[5] The Sloth Bear makes a rather melodious humming when mating and snores loudly when asleep. The Andean Bear makes an owl-like screech when alarmed. Pandas chirp and yip when greeting. They all grunt, huff, growl, or snort as the situation demands, although it depends on the individual bear. Some bears are simply more talkative than others. And cubs make many of the same sounds as little humans: whimpering, whining, crying, and contented humming.

The facial expressions of bears do not tell you much about what they are thinking. Instead, bears use posturing to communicate, in addition to sounds. Staring is aggressive, while a lowered head is a display of respect with no threat intended. A cranky bear will extend its upper lip or blow bubbles. He will also flatten his ears, although on adult bears this is sometimes hard to distinguish because they have small ears on a big head. In young bears the ears are large relative to the head, so these signals are easier to recognize. Upright ears indicate alertness, and ears are held laterally in play. An open mouth or puckered lips indicate investigation, willingness to play, and sometimes submission, while a gaping or snapping mouth is a threat. Most of this detail comes from a study of American Black Bears, but the same patterns are evident in all bears.[6] A relaxed and happy bear on the one hand, or an angry bear on the other, looks the part. We might point out that this observation is generally true of mammals as a group because we all have similar facial expressions for similar states of mind. Your scared face looks a lot like the scared face of a horse. Your angry snarl looks a lot like the angry snarl of a dog. We primates have more facial muscles so we are more finely tuned facial communicators, but we all follow the same basic patterns.[7]

All bears except the Panda are good swimmers (are you beginning to see why it took a while for scientists to be convinced that Pandas were truly bears?). Pandas will swim if necessary but generally avoid it. Other bears will swim to go after fish or frogs, to cool off, or seemingly just to have fun. Numerous videos online show American Black Bears, often a mother and her cubs, playing in someone's suburban backyard swimming pool.[8] The champion swimmer is the Polar Bear. These bears can swim more than 400 mi (640 km) over 10 days straight in Arctic oceans and so are rightly described as marine mammals.[9] They close their ears and nostrils and can stay underwater for up to two minutes. They swim to get to sea ice (fig. 7.3), where seals, their main prey, abound.

Many people associate bears with hibernation. *True* hibernation is a deep, voluntary state an animal employs to conserve energy during seasonal food shortages and to reduce exposure to winter weather. It involves large decreases in body temperature (e.g., from 99° to 27°F, or 37° to −3° C, in Arctic Ground Squirrels), heart rate, and respiration, and a suite of metabolic adjustments. Hibernation is employed by chipmunks, deer mice, and others, but not bears. Bears enter a similar but less intensive state of energy-saving dormancy called *torpor*. Torpor involves entering a sleep

FIGURE 7.3 Two Polar Bears (*Ursus maritimus*) on an ice floe near Svalbard, Norway. Polar Bears often steal kills from each other, with the larger bear usually victorious. It often requires fewer calories to steal a kill than to make a kill yourself. Chase Dekker / Shutterstock

deeper than normal, but not nearly as deep as true hibernation. Since torpor is more of a sleeplike state, torporous bears can react more quickly to external stimuli or dangers. Torpor, an involuntary state, is generally associated with only a slight decrease in body temperature, and larger drops in respiratory rate and metabolism.[10] Although there are major fundamental differences between torpor and true hibernation, we will use "hibernation" in the more general sense, since "going into torpor" is clumsier than "hibernating" as a catchall term (but we want you to know the difference).

Now that you know the difference, most bear species (including Andean Bears, Sloth Bears, Sun Bears, and Pandas) do not hibernate (including entering torpor) at all. A pregnant Polar Bear, or mother with yearling cubs still dependent on her, will go into an even less intensive dormant state called *denning*, in which she uses her stored energy reserves for as long as three months, but without decreases in heart rate or other functions as occur during torpor. Polar Bears do not den, other than pregnant bears or mothers with young cubs, since winter is when the sea ice is at a maximum and hunting is good. And in some Brown Bear populations, such as Kodiak Bears, some bears neither den nor enter torpor. It is not the cold but rather the lack of food that is the issue. As we discussed in box 7.1, Pandas cannot seem to build up enough fat with their plant-based diet to be able to skip eating for any length of time, so they do not become dormant at all.

American and Asiatic Black Bears and Brown Bears will hibernate, depending on where they live. If resources are highly seasonal and food is mostly unavailable for long periods, it makes sense to conserve energy, and hibernation is an excellent way to do that. Hence, bears that live where there is a long winter all hibernate, except as noted above.

Bears that hibernate do indeed pack on fat in advance. In late summer and fall, hibernating bears enter a period called *hyperphagia*, which means they eat huge quantities of food to build up fat. A Brown Bear consumes 20,000–40,000 calories a day and adds 6–8 in (15–20 cm) of fat. A female may double her weight in anticipation of supporting pregnancy, birth, and then cubs. (Online, check out "Fat Bear Week" from Katmai National Park in Alaska to see some amazing examples![11]) The exception is female Polar Bears, which do not add a great deal of fat, probably because it is impossible. In late summer and fall there is no sea ice to give them access to the calorie-rich seals they prefer to eat.

Brown Bears enter the deepest hibernation of all bears, although because they are so large they cannot slow their metabolism and cool off as much as, say, a bat. It would take more energy than they have available to heat back up from that. But they do slow down all their body processes considerably. The metabolic rate drops by 50%–60% and the breathing rate goes from 10 breaths per minute to only about 1 every 45 seconds. Heart rate and blood pressure correspondingly decline. The body temperature drops from around 100° to about 88°F (38° to 31°C).[12]

American Black Bears that hibernate (fig. 7.4) are considered *semidormant hibernators*, entering a state less extreme than most torpors. Metabolism slows some and heart rate drops from 40 beats per minute to around 8, but body temperature drops only around 11°F (6°C) below normal. (Note that dangerous hypothermia occurs in a human as body temperature drops below just 95°F [35°C], or about 3°F [1.7°C] below normal—a hibernator you are not!) This means that these bears can become fully alert in moments when they wake up, and this rapid reaction lets them take advantage of winter warm spells. You should keep this in mind the next time you consider peeking into what looks like a Black Bear den. The Black Bears living near your authors in the southern United States do not go dormant at all because winter is quite mild, so

FIGURE 7.4 American Black Bear (*Ursus americanus*) eating a salmon near its den entrance. Hibernation requires bears to find a secure location in which to rest, conserve energy, and wait out the lean times.
Glass and Nature / Shutterstock

food never runs out. The one exception is, again, pregnant females. And if there is a particularly nasty cold snap, these Black Bears may nap for a few days at a time. You would probably like to possess that ability during some stretches of cold winter days, too.

Hibernating bears, as we have already established, do not eat, and their body processes necessarily slow to conserve energy. Digestive and kidney function stop almost completely. But their bodies are still metabolizing and therefore should require water and should also be accumulating the toxic by-products of metabolism that are normally excreted. And yet bears neither drink nor excrete for the duration of hibernation. Also, if you essentially slept for a couple of months, your

bones would weaken to the point where you would be unable to stand up when you awoke. Bears lose no bone strength at all. How do they manage these tricks of physiology?

Scientists are not sure. Bears' primary energy source during hibernation is fat, and a by-product of fat metabolism is water, which is apparently enough to meet the body's needs during dormancy. Hibernating bears recycle urea, the main, very toxic waste product of metabolism, back into nitrogen, which is useful for building amino acids to make proteins to maintain muscle and organ tissue.

Calcium is taken from bones for various physiological tasks, as it is in humans, but it, too, is recycled back into bones in a way yours is not when you are not using your bones. For example, astronauts, subject to zero gravity and a consequent lack of skeletal loading, need to make a concerted effort to consume calcium and exercise to avoid loss of bone mass.[13] Once you use calcium for, say, muscle contraction, it does not get reincorporated into your bones, so you must keep consuming calcium to maintain your bones, in addition to making the bones do some work. Studies have shown that the bones of hibernating bears continue to form and resorb calcium just as if they were up and moving around. Bone remodeling slows, but bone mineral density and volume before and after hibernation or in active versus hibernating bears are all the same. Scientists have not quite figured out how this works and would very much like to; it could be a key to solving the problem of osteoporosis and the bone loss that results from inactivity in humans.[14]

Surprisingly, besides these physiological mysteries, bears that hibernate may give birth and lactate during hibernation, allowing a couple of cubs to grow, activities that hardly seem helpful if the goal is to conserve energy!

Reproduction patterns are similar in all bear species. Male and female bears are not easily distinguishable, although males are generally larger. Females produce cubs beginning around four to seven years, with Sun Bears starting a bit younger. Male bears reach sexual maturity around age five, but they cannot successfully compete for mating rights for several years after that. Sun Bears have up to a weeklong estrus (when they are ready to accept a male and mate) at any time of year, like Sloth Bears, except in India and Nepal where Sloth Bears breed from May to July. Brown Bears similarly breed in early summer, with a somewhat wider time frame in Europe than elsewhere. Polar Bear estrus lasts for three weeks, from March to May, about the same as in Pandas. Asiatic Black Bears and Andean Bears have an estrus lasting just a few days at any time through spring and summer,

into October. American Black Bear estrus lasts two to three weeks, from May to July or August. Males will mate with as many females as will accommodate them, and they succeed essentially by fighting off other males and following the females around until they are receptive, then corralling them in place. There is lots of hugging and physical contact, and sometimes quite a racket of roaring, chirping, squealing, and moaning, if one or both partners are talkative to begin with. And there can be some hybridizing (box 7.2).

Once mating and fertilization are accomplished, an embryo speedily grows into a cub in about two months in all bears. Actual gestation varies widely because of variation in implantation of the embryo into the uterine wall, after which it continues to develop. Mating occurs in the summer, but in temperate or Arctic climates it would not make sense for a cub to be born in the fall, right when it is getting cold and food is becoming increasingly scarce. Instead, implantation is delayed so that cubs can be born in the winter months and launched out into the world with their mother in the spring, concurrent with warming temperatures and abundant food. This means, though, that cubs are born during hibernation, for the bears that hibernate. As we said, this is hardly a recipe for energy conservation. How does this work?

First, the female needs to build up sufficient energy reserves prior to hibernating. If she has mated but is in poor condition when she commences hibernation, she will not be able to support cubs, and in that case she typically aborts, and her body absorbs the embryo(s). In fact, this happens with all bears, hibernating or not, except for Sun Bears. The mother's condition when implantation should occur determines whether it will or not.

Whether going into full hibernation or, in the case of expectant mothers, gestating and giving birth to young, bears need a protected spot to do so safely. Bears will den or simply nest on

BOX 7.2 ▶ GROLAR AND PIZZLY BEARS

All bears except Pandas and Andean Bears can interbreed, and many sometimes do. Brown Bears and Polar Bears have been known to interbreed, and because they are closely related, they can produce fertile offspring called *grolars* or *pizzlies*, with the prefix coming from the common name of the male parent. Genetic research has shown that these hybrids have bred back to their parent species.[15] This has resulted in the introduction of Polar Bear genes into the Brown Bear genome through a process called *introgression* (refer back to box 6.2, "Hybridization and Defining Species"). Today, as much as 8.8% of Brown Bear DNA is derived from Polar Bears.[16] Polar Bear DNA, however, does not appear to include any Brown Bear alleles, meaning that while Brown Bears can readily inherit traits from Polar Bears, the opposite does not seem to be true. In recent times, Polar/Brown Bear hybrids have been positively identified in the Canadian Arctic, often bagged by hunters. Typically, Polar Bears and Grizzly Bears share a relatively small amount of overlapping territory. But as the climate warms and Brown Bears can further encroach on Polar Bear habitat, we might expect to see an increased amount of hybridization.

Genomic studies have shown that there is gene flow between other bear species as well; for example, there is a known natural hybrid between Sun and Asiatic Black Bears. Or perhaps the Brown Bear was the intermediary, because it has the widest global distribution and therefore the most opportunity to hybridize with other species. For example, American and Asiatic Black Bear genomes indicate some hybridization, but as far as we know from the fossil record, these two species have never been in direct contact. But both could have hybridized with a Brown Bear, thereby transferring genes from one population to the other. All this swapping may explain why it has been difficult to sort out the evolutionary relationships between bears. There is a complex genetic mosaic of evolutionary history, much like with big cats and dogs. Further genetic research may make those relationships clearer, or even more muddled.[17]

A Pizzly. This is a hybrid between a Polar Bear (*Ursus maritimus*) and a Grizzly Bear (*Ursus arctos*).

the ground. They may excavate a den or take advantage of a natural space under boulders or in trees. American and Asiatic Black Bears often den in trees. A Brown or Polar Bear den is usually on a slope. Generally there is an entrance, tunnel, and chamber, invariably well drained, although sometimes surrounded by water. A Polar Bear den always has a vent as well. Bears will be secretive when building a den in order to hide its location. In fact, Grizzly Bears tend to den during snowstorms so their tracks to the den get covered by snow, which additionally makes an excellent insulator. A Polar Bear den will rarely get below freezing. Some bears build new dens annually and some will reuse a den for a few years. Sometimes a den will be used for decades by different bears.

Bears give birth to two cubs, on average, except Andean (Spectacled) Bears, which almost always have only one. Pandas almost always have two but can really care for only one, so one is usually left to die. Since a Panda doesn't hibernate and lives in a cold mountain habitat, it would be difficult for a mother to hold on to and suckle two cubs for the four to five months it takes for them to become mobile on their own. So, you see how giving birth while hibernating makes sense. The cubs are kept warm and protected with minimal effort on the mother's part while they're all cocooned in a den. Besides Pandas and Andean Bears, other bears often have two or three cubs. Brown Bears have been observed with up to five, and American Black Bears with up to six, although numbers this large are quite rare.

Bear cubs are born small and entirely helpless (fig. 7.5). In fact, bears have the smallest newborns relative to the size of the mother of all placental mammals. They are blind, lack a sense of smell, have only fine hair, and possess virtually no other insulation. About all they can do is find their mother's nipples or complain noisily until they do so. Bear milk is rich in fat and protein, providing three times the energy of cow's milk.

Not surprisingly, Polar Bear milk is the richest. Those little Arctic bears need insulation, and their mother's diet is especially high in fat. Polar Bear milk is also rather fishy, not surprisingly.

Bear cubs can detect temperature changes, which keeps them in contact with their mother so they can eat and grow. Panda cubs are the smallest at birth, at just 3–5 oz (80–140 g), and Polar Bear and Brown Bear cubs are born at 16 oz (450 g). Asiatic Black Bears open their eyes in about a week. Sun Bears take two weeks and Sloth Bears three. The hibernators go four to six weeks and Pandas six to eight weeks. Pandas take 12 weeks to gain the strength and coordination to walk. For Sun Bears, that occurs in just a couple of weeks. The other bears are walking at four to six weeks, except for Polar Bears, which require eight weeks.

A female bear reproduces only once every two or three years, expending a great deal of energy on typically just one or two cubs at a time, so she does her best to ensure they survive. In most cases she keeps them with her for two years, protecting and teaching them. She shows them what to eat, where to find it, and how to catch or harvest it. She sends them scurrying up trees in the face of threats. And she does her best to fight off threats if need be, such as other predatory animals, humans, and most notably, adult male bears. An extremely malnourished mother may kill and eat her cubs, but one assumes that if she is that unhealthy, the cubs are too. If a cub is mortally wounded, however, the mother will likely eat it regardless of her state. The calories invested in a dead cub are precious and cannot be wasted. First-year survival rates are more than 60% for Brown Bears and Polar Bears, and 75% for American Black Bears. Sloth Bear moms have the reputation for being the fiercest defenders of cubs, but also for being the most sensitive and patient teachers. Sloth Bear cubs routinely ride on their mother's back. Most other bears carry cubs in their mouth when needed.

FIGURE 7.5 Brown Bear (*Ursus arctos*) cubs with their mother at a den site. Cubs of hibernating mothers are born and do much of their initial developing inside a secluded den. They must open their eyes and learn to walk before they start to explore the outside world.
Vera Kuttelvaserova Stuchelova / 123RF

BOX 7.3 A BEAR WALKS INTO A BAR ... *REALLY*!

Hiking guides tell hikers to make a lot of noise, try to look large, and back away slowly if they surprise a bear, and to be especially careful of a bear with cubs. With that in mind, consider this dilemma. A friend of author Sharon's was exasperated by Common Raccoons raiding his bird feeders in a housing development on the edge of wetlands in coastal North Carolina. He borrowed a box trap and baited it with peanut butter and birdseed (although note that one of the bear biologists at the South Carolina Department of Natural Resources tells us that Krispy Kreme doughnuts are their go-to bear bait). On two successive nights Ken captured two Raccoons and released them well away from home. On night three, at 3:00 a.m., he woke to hear what sounded like screams coming from outside. Sprinting to the sliding glass doors and flipping on the outside lights, he saw a an American Black Bear cub of about 20 lb (9 kg) stuck in his trap and wailing away! What to do? You have no choice but to release it, clearly, but you can bet its mother is out there looking on and is none too pleased. But nobody wants to listen to a screaming bear cub in the middle of the night. Ken looked carefully for the mother but did not see her, so he ran out and opened the trap, and he and the cub took off in opposite directions. All turned out okay, except maybe for the Raccoons.

Speaking of startling bears, one paper reported that cowboys in Peru claim that if you startle a Spectacled Bear in their mountainous Andean habitat, one defensive strategy is for the bear to take off rolling downhill.[18] We find that a bit hard to believe but wonderful to imagine!

In addition to Ken's tale, author Sharon's second favorite American Black Bear story is from the Pocono Mountains in Pennsylvania. Her parents had a plexiglass bird feeder that fit right into their window frame, such that the birds basically came into the house. Great viewing! Until one day they found themselves viewing a large bear face inside their window, with a big tongue eagerly slurping up birdseed. Might be the only time Josie, their spaniel mix, was rendered speechless.

And finally, as we were writing this chapter in 2019, a story appeared in the local news from Murrells Inlet, South Carolina. A bear walked into a bar. ... Sounds like the start to a joke, but around closing time one night in mid-May, employees gathered in the back of the Inlet Provision Company were surprised when a young male American Black Bear wandered in the front door![19] He poked around for 10 minutes, headed into the VIP room, and then climbed over a wall, finally making his way back out through the same door. No harm, no foul, and no punchline!

After two years, young bears are well equipped for life on their own, although a year-old orphaned bear has a pretty good chance at survival too. Once they leave their mother, siblings often remain together until they reach sexual maturity and seek mates. Growing up as a bear cub can pose some unique challenges, especially in proximity to humans (see box 7.3).

BEARS AS PREDATORS

We're going to look at bears as *eaters* rather than strictly predators, because except for Polar Bears, they are the least obligate of the carnivores and as a rule not solely predators (although they can be). Like us, bears are omnivores. They will eat pretty much whatever food is available (as box

7.3 shows). For example, there was concern that when Spongy Moths moved into Shenandoah National Forest in Virginia, USA, defoliating the trees as they had been doing in New England since their introduction from Europe, few acorns would be produced and American Black Bears would suffer. This had been found to be the case for squirrels, deer, and Wild Boar, which all depend on acorns. Sure enough, from 1985 to 1990, the defoliated oak trees produced almost no acorns. But blueberries, huckleberries, grapes, pokeweed, and spiceweed are also all bear food and did fine, and so did the bears. There was no difference in their reproduction rates or survival through this period.[20] Being omnivorous is a good idea if you have the stomach for it.

Consider Brown Bears. They can be effective predators, like Polar Bears, taking advantage of calorie-rich meat such as carrion or newborn deer or Elk. But in the early 1980s, radio-tagged Grizzlies in Yellowstone National Park were observed excavating something in alpine talus fields, gravelly slopes left behind by retreating glaciers and erosion. It turns out that in late summer, Army Cutthroat Moths metamorphose in large numbers and begin collecting nectar from alpine wildflowers. Being moths, they work at night. During the day they aggregate in large numbers under handy rock formations on talus slopes. Those fearsome bears were gobbling up moths. And those moths are 72% fat and 28% protein. A perfect calorie-rich meal when you are trying to fatten up for hibernation, and one much easier than chasing the small mammals also running around. Big males, females with cubs, subadults, and American Black Bears all take advantage.[21] Sure, meat is the most calorie-dense food, but it's also the most difficult to catch. Grazing on whatever is handy works too.

Bears that hibernate put on the pounds before winter arrives. Some seem to reach their capacity and subsequently slow down their calorie intake. This occasionally leads to some interesting, if not energy-efficient, food choices. Author Rob spent some time in Alaska's Katmai National Park (near the Aleutian Islands) observing Brown Bears. It was toward the end of the salmon spawning season but before cold weather had set in, and most bears seemed to have had their fill. Rob observed several bears lying on their bellies delicately picking individual berries no bigger than a grain of rice from the spongy tundra. He also observed a particularly large male excavating a hole the size of a car into a bluff to catch and eat what the guides suggested might be a wood frog. Walking down to the river, he witnessed very few bears actually fishing. Most were just playing, playfully wrestling, bobbing vertically in the water, or floating down the river with no obvious agenda. It seems that the satiated bears turned their attention from consuming calories to amusing themselves, although this was likely because they were living in an unusually food-rich environment.

Asiatic Black Bears in Japan depend on beechnuts, and if the beechnut crop is down, there are more "nuisance bear" reports, no doubt because the hungry bears are looking for alternatives to the nuts.[22] Andean or Spectacled Bears eat mostly bromeliad (tropical flowering plant) hearts in the winter and ripe fruit in the summer, but also leaves, insects, honey, and rodents and other meat, including deer, cows, and goats, but these bigger animals are likely scavenged, not killed outright.[23] Sloth Bears have especially protrusible lips, long snouts, and long tongues, specialized to slurp up ants and termites, although they also eat fruits, especially figs.[24] Sun Bears (fig. 7.6) also eat a lot of figs and insects.[25] All bears, including Pandas, eat eggs, and all eat honey except Polar Bears, probably only because there's no such thing as Arctic honey. In case you're wondering, nearly all these diets were

FIGURE 7.6 Sun Bear (*Helarctos malayanus*). Sun Bears have long, protruding tongues well suited to excavating honey, pupae, and eggs from active beehives. They must withstand numerous stings to acquire their tasty treats.

Vladimir Wrangel / Shutterstock

determined based on *scat* analysis—in other words, careful investigation of poop. The life of a wildlife biologist is not always as sexy as it might seem.

Of course, Pandas are the weirdos, subsisting almost exclusively on bamboo. But they have evolved to specialize for that, having an extended wrist bone that functions almost like a thumb. This allows them to grip a bamboo stalk while stripping the leaves with their teeth or rolling the leaves into a sort of cigar they can then take big bites of. It also serves as a nice example of natural selection tinkering with a basic design to the benefit of the animal. An engineer might have gone with a true opposable thumb as an even more efficient solution, but only primates have the genetic developmental sequence for that, so evolution works with what it has, adequacy if not perfection.[26]

SENSING

How do bears find food? Like other mammals, they depend on vision, hearing, and smell, although there has not been a great deal of study of the sensory basis of foraging in bears. As you might imagine, they make difficult research subjects.

Let us start with hearing. Bears have a variety of ear types and hearing abilities. Sloth Bears have large, almost floppy, hairy ears, and their hearing, along with that of Pandas, is decent but falls short of that of other bears. Polar Bears and Brown Bears seem to have the most sensitive hearing. Both of these bears hunt prey that live under the substrate, like ground squirrels in burrows, mice under the snow, and seals under the ice, so hearing is an effective way of zeroing in on an unseen target.

Bears, except for Polar Bears, have relatively small eyes. All have round pupils except Pandas, whose pupils are vertical slits like those of a cat, which presumably enhance night vision. They all possess the tapetum lucidum we have seen in nearly all our predators, also for enhancing night vision. Brown Bears and Polar Bears have been shown to have two types of cone cells, at relatively high densities compared to some other mammals. Remember that rod cells are geared to sense dim light and motion and cone cells target bright light and color. This suggests that bears may have dichromatic vision.[27] Recall that primate vision is trichromatic, so this is not as advanced as that, as is characteristic of most mammals. Bears are nearsighted since they feed mostly on things up close, and, because their food is often berries or other fruits where color indicates ripeness, it makes sense that bears should be able to see at least some color.

Polar Bears have the most specialized eyes and the best vision of the bears. Consider that they need to be able to see in the dark Arctic winter, but also in the bright glare of the snow, and underwater. Their eyes are almost as large as a human's, relative to their head size, and have an extra eyelid to filter glare off snow, like sunglasses. They also have a nictitating membrane that protects their eyes when they are open underwater. Yet healthy-looking Polar Bears that are blind have been described, so they must be able to rely on other senses to hunt if need be.

All bears have a pronounced sense of smell. In fact, their olfaction is about seven times more sensitive than that of a Bloodhound. An American Black Bear has 100 times the sensory surface area in its nose as a human. A Grizzly's sense of smell is about 75 times that of a human and it can detect another animal 2 mi (3.2 km) away. The champion smeller is the Polar Bear, though. A Polar Bear can smell the brief breath of a seal poking its nose out of its breathing hole in the ice several miles away. It can detect a seal under 3 ft (1 m) of snow 0.5 mi (0.8 km) distant. There is really no hiding from a bear.

CATCHING AND KILLING

Bears have no great strategy for catching and killing things. Mostly they wander around browsing on whatever they happen to encounter. They are 2.5 to 5 times stronger than a human, so they excel at flipping over rocks, tossing aside dirt, and knocking over trash receptacles to find food. If they discover a cow carcass, good. An anthill, also good. If it is a newborn deer, ripe blueberries, a jar of peanut butter, a hiker's improperly stowed Pop-Tarts—you get the message, all good! A few bears do specialize, though.

Brown and American Black Bears famously gather at streams as salmon head inland to spawn. Fish are slippery and difficult to catch, of course, but in this situation it is a numbers game. The salmon are tightly packed and focused on getting upstream and finding a mate. Brown Bears have long claws and sharp teeth and choose their fishing spots carefully (fig. 7.7). The best method is to simply stand and grab, although the bears sometimes chase the fish. An efficient bear gets a fish in 36% of its tries.[28] Efficiency is higher at night than in daylight, probably because the fish cannot see as well. Or maybe only the most practiced bears fish at night, since a mother with cubs or a young adult avoids the possibility of accidently running into a big male in the dark. It takes practice, but when salmon are plentiful and the bear gets good at catching them, it may eat just the skins, eggs, and maybe the brain—the fattiest parts—and cast aside the meat for some lesser bear or other animal. The bears are really choosy, catching and releasing salmon that have already spawned and so are increasingly devoid of energy-rich tissue. A trained human can tell the spawning status of salmon by looking at them, and apparently bears can do this too.

And bears are important engineers of the terrestrial nutrient cycle.[29] It was found that 15% to almost 19% of the nitrogen in spruce foliage within 1,600 ft (500 m) of a salmon stream in the Kenai Peninsula in Alaska came from salmon, and bears were responsible for distributing about 83% of that just through eating and defecating.[30] The study did not look at the extent to which salmon carcasses themselves got moved ashore after being dispatched and just partially eaten by a bear. Either way, bears are moving nutrients from the ocean to the forest, a rather surprising and likely important example of nutrient cycling.

Since bears are so strong for their size, a single blow of the foreleg to the neck may kill a Moose, Elk, or deer, and then the bear can hoist the carcass in its mouth and carry it off. This strength also allows them to walk off with the kills of others. In Sweden, Brown Bears have been found to be efficient killers of Moose calves up to four months of age.[31] Likewise, Grizzly Bears in Yellowstone National Park in the United States prey on Elk calves. Mostly they search for bedded calves, but sometimes they will ambush or chase a calf, knocking it to the ground and then holding it with those big paws.[32] Sloth Bears have been found to sometimes eat deer, goats, and rodents, but it is not clear whether this is carrion or they catch these animals live.[33]

Of course, Polar Bears are true predators, eating seals almost exclusively if they are available. It has been estimated that up to 44% of annual Ringed Seal pup production in the Canadian Arctic is consumed by Polar Bears.[34] They have also been

FIGURE 7.7 Brown Bear (*Ursus arctos*) hunting salmon. Hunting takes practice. Bear cubs must learn from their mothers as well as through practical experience. Greater skill means more nutrient-rich food.

FIGURE 7.8 Polar Bear (*Ursus maritimus*) eating carrion. Polar Bears can be found scavenging whale carcasses, as well hunting birds, crabs, Reindeer, and more. This is most common when sea ice and seals are scarce.

Danita Delmont / Shutterstock

observed hunting geese, gobbling up gull eggs and nestlings,[35] and eating whale carcasses (fig. 7.8), particularly when sea ice is scarce and seal hunting is impossible, although during the summer, when Polar Bears are obliged to stay on land, they mostly rest and do not eat.[36] Very large bears have also been known to capture the occasional Beluga Whale, Walrus,[37] Narwhal, or Reindeer.[38] As we described above, they can hear, see, and especially smell very well.

Polar Bears have shorter, stouter claws than Brown Bears, but they are longer than those of

American Black Bears and are well designed for gripping both ice and prey (fig. 7.9). The canine teeth are longer (3.15 in, or 8 cm) than in a Brown Bear (2 in, or 5 cm), and the adjacent teeth are reduced, allowing the canines to penetrate more deeply into prey. The prominence of canine teeth means Polar Bears are less well equipped to grind up plant matter than other bears. Polar Bears patrol the ice, swinging their heads side to side, sniffing the air and looking for prey. They stalk seal pups on the ice, staying downwind and hiding behind ridges as they go. Once a Polar Bear gets

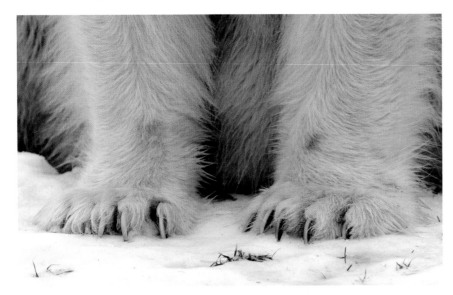

FIGURE 7.9 Polar Bear (*Ursus maritimus*) claws. Polar Bear claws act both as traction on the snow and ice and as a tool to manage prey. They are relatively short for the bear's body size.

Nagel Photography / Shutterstock

close to its prey, it will charge, trying to grab the seal with those big teeth, either from the ice or from the water at its breathing hole. Ringed Seal pups sometimes rest in lairs under the snow in the ice adjacent to the mother's breathing hole, but Polar Bears can sniff them out and will repeatedly pounce on the lair from above, hoping to break through. They are successful in their hunts less than 25% of the time. Perhaps for this reason, Polar Bears routinely steal each other's kills, with smaller, younger bears often falling victim to this thievery. The bears tend to immediately eat the energy-rich blubber layer of a seal. They rarely eat the entire seal, providing the remaining carcass for other animals such as gulls and foxes.[39]

HOW ARE BEARS DOING?

Except for American Black Bears, whose populations are increasing and so are of Least Concern, and Brown Bears, whose populations are globally stable, all bears were listed as Vulnerable by the IUCN as of 2017. Polar Bears, Andean (Spectacled) Bears, and Sloth Bears are particularly

at risk from climate change, while they and other species suffer from habitat loss in competition with human development. Except for Pandas, all Vulnerable bears are decreasing in numbers.[40]

Pandas in the Pleistocene (some 2.5 million years ago) ranged from near Beijing to southern China and into northern Vietnam, Laos, and Thailand but are currently restricted to six isolated populations on the eastern edge of the Tibetan Plateau. Since the 1950s they have been a flagship species for global conservation. They are indeed about as charismatic as megafauna gets. There is a general perception that if an animal is too specialized it cannot easily adapt and is doomed once the population gets small and isolated: an *evolutionary dead end*. Pandas, eating only bamboo and living in scattered, isolated small populations, seem to fit this model. It turns out, however, that the genetic diversity among Pandas is higher than that of other bears. Genetic diversity is one measure of the overall health and stability of a population, so this is rather surprising given the small and specialized niche of Pandas. These bears do not seem to be a genetic dead end at all. Careful management to maintain their genetic diversity, such as the Chinese are doing, should secure these bears a place in the world, provided they also have somewhere suitable left to live.[41]

Bears are still hunted legally in many places, but they are poached even where they are protected. Often this is for sport, but as is true with other powerful predators we have looked at, there is also a thriving trade in bears and bear parts for alleged, and utterly unproven, medicinal uses. This is particularly a problem for Asiatic Black Bears, as well as Brown Bears in southern Russia. Their bones are ground up and included in baby formula so that the child may grow up as strong as a bear. Claws offer good health, good luck, and fertility. Fat is a treatment for rheumatism. Gallbladders are perhaps most in demand. They're about the size of a human thumb, and dried bile has been part of Asian folk medicine for more than 3,000 years, claimed to treat most anything, including asthma, cancer, diabetes, flu, excessive drinking and smoking, parasites, poor appetite, fatigue, fever, and hemorrhoids. Bear bile acid has been found to dissolve gallstones in humans, sparing people surgery, but that is the only proven use for bear bile. Unfortunately, there are over 16,000 "bile bears," mostly Asiatic Black Bears, now kept in bear farms in South Korea, China, and Vietnam. They live miserable lives in small cages, making them easily accessible for gall "milking." This takes some of the pressure off wild populations, but as with big cats, wild parts are more valuable than farmed, so it does not solve the poaching problem.

Since you are interested in predators, you should read the book *Monster of God* by acclaimed science journalist David Quammen.[42] Among other fascinating things, he relays a horrifying tale from Romania about the staged "hunting" of Brown Bears by the dictator Nicolae Ceaușescu. He would be escorted to a high chair in the forest and Brown Bears would be herded toward him by forestry workers, many of whom were simply appalled

by this but did their jobs nonetheless. Ceaușescu would kill more than 20 bears at a time this way, over 400 during his 25-year reign. The Romanian bears are now doing okay and are still legally hunted, for a fee, but compete for space with development and sometimes run into shepherds and their sheep. Still, as one shepherd noted in the book, the bear is a treasure of the forest: "A forest without bears—it's empty." We share that sentiment.

SOME APEX PREDATORS

Among the bears, Polar Bears clearly qualify as apex predators, and we add Brown Bears to the list because although they are omnivores, opportunistically eating berries or moths, they often eat meat. In addition, there is virtually nothing other than another Brown Bear, or a human with a gun, that will take them on.

BROWN BEAR
(Ursus arctos)

Brown Bears (fig. 7.10) are by far the most widely distributed of the bears. Most live in Russia, Alaska in the United States, and Canada, but there are also populations in the northwestern United States outside Alaska, in Europe, and in more southerly parts of Asia, including Iraq and Nepal. Many are considered subspecies, for example Grizzly Bears in the northwestern United States or Kamchatka Brown Bears in Siberia, but we will consider them as a single group.

Brown Bears vary widely in size. An average coastal Alaskan Brown Bear weighs 787 lb (130 kg), while inland in the Yukon in Canada the average is 319 lb (145 kg). The record was a Kodiak Brown Bear over 2,500 lb (1,134 kg). Females are 40%–50% smaller than males. Brown Bears stand

FIGURE 7.10 **Brown Bears** (*Ursus arctos*). Although more omnivorous than other members of Carnivora, Brown Bears are quite capable hunters that can capture prey up to the size of Moose and Reindeer.

3–5 ft (1–1.5 m) at the shoulder and are 5–10 ft (1.5–3 m) nose to tail. Despite this size, they can run up to 35 mph (56 kph) over a short distance. Their color can be nearly black to almost white and they are often silver-tipped or "grizzled," but they are usually brown. Their paws are more than 5 in (13 cm) in diameter, with 4–5 in (10–13 cm) claws on the forepaws. A pronounced hump between the shoulders, often covered with longer hair like a mane, is muscle that contributes to the great strength and strike force of the forepaws.[43]

A single swipe of one of those forepaws can break the neck of prey up to the size of a Moose, but they are also good for snagging salmon out of the water and digging or flipping over rocks to get to insects or roots. They attack prey with their paws rather than their teeth. They have diverse tooth types, like you, with similar wide, flat molars good for grinding plant material like berries and nuts. These bears do not see particularly well, but their hearing is good and they rely mostly on their sense of smell, which is better than a dog's. A Grizzly Bear can detect a human two miles away.[44]

Despite their size and long, thick fur, most Brown Bears hibernate in the winter, although coastal Kodiak Bears, for example, can find food and stay active year-round. In Alaska, Brown Bears will eat to gain up to 3–6 lb (1.4–2.7 kg) of fat per day leading into winter hibernation, but many bears rouse from their torpor to look for food if there is a warm spell.[45] Only gestating females hibernate all winter (and recall that none of this is quite true hibernation—they wake up easily). They breed in the spring or summer and give birth in dens in the winter. There are typically two cubs, although often only one survives. They are tiny at birth, about 6 in (15 cm) long and 16 oz (450 g), but grow quickly. During their first year, Kodiak Bears double in weight every two months. They stay with their mother for about two years, avoiding male bears that might eat them, learning what to eat themselves and how to find or catch it. They reach sexual maturity at four to seven years.[46]

Male Brown Bears require territories of at least 200–500 mi^2 (500–1,300 km^2), but since they do fine in habitat rather inhospitable to established human developments and are widely dispersed, they were listed as of Least Concern by the IUCN as of 2016, with a stable worldwide population of around 110,000 individuals. Brown Bears are legally hunted in Russia, Alaska, Canada, Japan, and parts of eastern and northern Europe. In the continental United States, Grizzly Bears are considered Endangered, but, protected by the Endangered Species Act, their populations are increasing, and efforts are being made to establish *greenways* or safe, undeveloped connections between populations in Yellowstone National Park, Glacier National Park, and the larger bear populations in Canada, for example. Several isolated populations in Europe are considered Critically Endangered. The bears get into trouble, of course, when they encroach on human developments. Loss of sea ice is forcing Polar Bears south, and a warmer climate may be allowing Brown Bears to move farther north.[47] This is causing an increased overlap of territory between the two species.

POLAR BEAR
(Ursus maritimus)

Polar Bears (fig. 7.11) are the undeniable apex predator of the Arctic. They diverged from Brown Bears about 150,000 years ago. They are the generally the largest of the bears, although Alaska Brown Bears can get as big, or bigger. A fully grown male Polar Bear can weigh over 2,000 lb (900 kg), with females being 25%–45% smaller. He can stand over 5 ft (1.5 m) at the shoulder and be over 8 ft (2.4 m) tip to tail. As we discussed, they appear white or yellow, but the fur is translucent, consisting of a water-resistant outer coat and a woolly, insulating undercoat. A big Polar Bear may have 4 in (10 cm) of blubber on his rump and outer back legs. His paws can be up to 1 ft (0.3 m) across, with fur between the pads and partial webbing, functioning as both snowshoes and paddles, with 2–3 in (5–5.5 cm) claws and rough pads for traction on ice. These features allow them to survive the cold Arctic winter (only gestating females hibernate) and even swim for extended periods (up to 10 days, covering up to 400 mi, or 645 km![48]) in the Arctic Ocean. Not for nothing are these animals generally considered marine mammals.

Polar Bears are the most carnivorous bears and will hunt anything that moves, but their preferred prey are energy-rich seals on sea ice. They see well in bright sun, in low light, and even underwater. Polar Bears have the most acute sense of smell of all bears, and they have the longest and sharpest canine teeth. They use stealth to sneak up on seals, staying downwind or in the water to get close. If the seal is under the ice, they use their great weight and strength to break through. Prey are widely distributed, so Polar Bears wander far and wide, in areas ranging from 5,000 mi^2 (13,000 km^2) to over 200,000 mi^2 (500,000 km^2).[49]

Female Polar Bears overwinter in dens when they are pregnant and typically give birth to two cubs. The cubs weigh just a little over 1 lb (0.6 kg) at birth in January, but Polar Bear milk is very rich, 31% fat, and the cubs weigh 22–26.5 lb (10–12 kg) by the time they emerge from the den in March. They stay with their mother for a little over two years.

Polar Bears are currently listed as Vulnerable by the IUCN. Their total population is unknown, but they depend on sea ice for much of their diet, and the extent and season of sea ice decline with every passing year because of climate change. It stands to reason that Polar Bears are suffering from this huge insult to their existence. In addition, as apex predators, they accumulate persistent organic pollutants (we will talk more about this in chapter 8), potentially affecting their health. Finally, developmental pressures in the Arctic including tourism, drilling for fossil fuels and mineral mining, and increasing transportation routes as sea ice declines pose risks to Polar Bears. The international Agreement on the Conservation of Polar Bears was signed by Canada, Denmark, Norway, the Russian Federation, and the United States in 1973, and a Circumpolar Action Plan for Polar Bears was signed by those parties in 2015. This plan is active as we are writing this, encouraging research into the status of the bears and careful monitoring.[50] Reason for hope, but they need their Arctic climate.

FIGURE 7.11 Polar Bear (*Ursus maritimus*) on an ice floe. The largest of the Ursidae, the Polar Bear is also the most aquatic bear, depending largely on sea ice to hunt seals. It has been classified by the US Fish and Wildlife Service as a marine mammal and receives much of the same legal protection as whales and dolphins.

8
MARINE MAMMALS

One of the planet's great delights is seeing a marine mammal in nature. While we are also joyous at seeing a shark, bear, or big cat, we feel a special connection to marine mammals, whales (including dolphins) in particular, that is both visceral and cerebral. They seem to have "knowing eyes," and many say they appear to be looking directly into our souls. As scientists we cannot corroborate that, but as humans we cannot deny it either.

Marine mammals spend most or all of their time in the ocean, and a few also inhabit rivers and estuaries. These include whales, dolphins, and porpoises (order Cetacea); manatees and Dugongs (order Sirenia); and seals, sea lions, Walruses (*Odobenus rosmarus*), Polar Bears, and Sea Otters (order Carnivora) (fig. 8.1). We have already discussed Polar Bears (chapter 6), and Sea Otters are explored later (chapter 9).

Marine mammals are not all predators, but most are. Exceptions are manatees and Dugongs, or sea cows, which are strictly vegetarians, grazing on seagrass and algae in shallow water. When you observe them in aquaria, they are frequently accompanied by floating heads of lettuce. Large herbivores are very common on land but rare in the ocean, a phenomenon explained by the dietary preference of herbivores for large plants and the scarcity of these plants in the ocean. The dominant type of "plant" in the ocean is microscopic algae, or phytoplankton, and most of the herbivores that eat these algae are not much bigger themselves. It is much more common to be a predator if you are big and live in the ocean.

Whales come in *baleen* and *toothed* varieties. Baleen whales, the Mysticeti (*mystax* = mustache), are named for their feeding plates—baleen—which hang in sheets from their jaws where teeth would ordinarily be. Baleen is very untoothlike in that it is made of the structural protein keratin, like your fingernails, but baleen frays along one edge, forming a coarse hair that can filter out plankton and small fish. Baleen whales include the Humpback, Right, Gray, Sei, Blue, Minke, Bowhead, and Fin Whales. They are all large, but some are *huge*. Blue Whales (fig. 8.1), at up to 100 ft (30 m) and 150 tons (13,600 kg), are the largest animal that has ever lived, more massive than the biggest dinosaurs. Remember that in ecosystems a lot of energy is available for growth if you eat low on the food chain.

FIGURE 8.1 Blue Whale (*Balaenoptera musculus*), a cetacean; Manatee (*Trichechus manatus*), a sirenian; and Cape Fur Seal (*Arctocephalus pusillus*), a carnivore.

Baleen whales are plankton eaters (also known as *planktivores*), and since plankton is a mélange of all sorts of creatures both plant and animal (as well as organisms classified as neither), if you consume plankton you are indeed a predator. This makes you reconsider the concept of a predator, no? Planktivores do not populate your nightmares in the same way more toothy beasts do, unless you count Jonah or Pinocchio as ancestors. In just a few instances, scuba divers have inadvertently ended up in a baleen whale's mouth, but none were ever swallowed (at least to our knowledge). In 2020, two kayakers in California unknowingly paddled over a lunge-feeding Humpback Whale. Cell phone video footage shows a school of bait fish exploding out of the water around the unwary paddlers, right before they and their kayak were lifted out of the water inside the whale's mouth. Neither humans nor whale was injured, but one of the kayakers admitted that upon returning to the car, she still had bait fish falling out of her shirt.[1]

The toothed whales (fig. 8.2), the Odontoceti (*dont* = tooth), are also predators. They are perhaps the most widely distributed group of mammals, having adapted to life at most latitudes and in nearly every ocean, sea, bay, and estuary, as well as some Asian and South American rivers. There are over 70 species in 34 different genera. Sperm Whales (*Physeter macrocephalus*), Orcas (*Orcinus orca*), and a couple of beaked whale species are the largest, and dolphins and porpoises are the smallest toothed whales. Belugas (*Delphinapterus leucas*), Narwhals (*Monodon monoceros*), and beaked and pilot whales are all toothed whales.

All pinnipeds ("fin foot" in Latin)—that is, seals, sea lions, and Walruses—are also toothed predators, as suggested by the name "sea lion" or, in Spanish, *lobo marino*, "sea wolf." There are three families of

FIGURE 8.2 **False Killer Whale** (*Pseudorca crassidens*), an odontocete (toothed whale). Despite its common name, this species is the fourth largest of the dolphins. LouieLea/Shutterstock

FIGURE 8.3 Harbor Seal (*Phoca vitulina*) pup and mother. Harbor seals are phocids (true seals) and the most widely distributed species of pinniped.

pinnipeds: the odobenids, otariids, and phocids. The Walrus is the only extant (remaining) member of the Odobenidae. These are massive animals, with adult males weighing in at 1,760–4,400 lb (800–2,000 kg). Males and females both have similar-sized tusks. Walruses are found in Arctic polar regions, with Atlantic and Pacific subspecies.

There are 16 species of Otariidae (*eared seals*), the fur seals and sea lions, distributed through the North Pacific and Southern Hemisphere. Otariids are smaller than a Walrus, with similarly flexible flippers, and are the most land adapted of these groups; they can scale cliff-like, rocky shorelines. They lack tusks but have external ear flaps, which is an easy way to distinguish them from the *true seals*, the Phocidae (fig. 8.3). Phocids have shorter and less flexible flippers and necks, so they are not nearly

as agile on land as the otariids. One phocid, the Southern Elephant Seal (*Mirounga leonina*), is the largest of the pinnipeds, with adult males weighing 4,400–5,950 lb (2,000–2,700 kg). There are 19 species of phocids, and they are found nearly worldwide in oceans, seas, bays, estuaries, and some rivers, but they live only sparsely at tropical latitudes.

MARINE MAMMALS AND US

Among the many amazing insights of the Greek scientist and philosopher Aristotle, who wrote *Historia Animalium* (*The History of Animals*)[2] roughly 2,400 years ago, was the recognition that whales and dolphins are mammals, not fish, because he saw that they nurse their young and breathe air like other mammals. He also figured out that they communicate by sound underwater and described how they sleep. In Greek mythology, dolphins were once humans; ancient Greeks considered them "people of the sea"[3] and accordingly featured them prominently in artwork and pottery. Dolphins today are the national animal of Greece. There is an Amazon legend claiming that Amazon River Dolphins (*Inia geoffrensis*) come to shore at night and morph into handsome men dressed in white, intent on seducing local women.[4]

Early Norse fishermen in Baffin Bay in the Arctic, perhaps with assistance from the native Inuit, discovered Narwhal horns and introduced them to European aristocracy as unicorn horns. Charles the Bold (1433–1477), the last reigning duke of Burgundy, for example, amassed quite a collection, using them in ceremonial displays along with showy items of gold and silver. As unicorn horn was alleged to detect poison, the bold duke always ate meals with a piece of ersatz unicorn horn handy. Since unicorns appear in the Bible, the horns became part of ritzy religious displays throughout Europe right up through the sixteenth century. By then word must have gotten around that these strange tusks were from mere whales, not unicorns. Less magical, perhaps, but no less mysterious, as we are still not entirely sure what Narwhals do with these projections. Only adult males have them, so they are probably involved in sexual selection (i.e., attracting females), but they may have some sensory capabilities and can be used to stun fish.

The Great White Whale, Moby Dick, surely leads the parade of famous cetaceans. The story was apparently based in part on an actual white Sperm Whale, Mocha Dick, who tended to fight back against whaling ships in the Pacific early in the nineteenth century. Sperm Whales in general had a reputation for ramming whaling ships. One was famously responsible for the sinking in 1820 of the *Essex*, a whaling ship out of Nantucket,

Massachusetts, USA. We will consider that ramming behavior a little later.

In modern times, the most well-known dolphin must be Flipper, although if you are not a baby boomer, Flipper might not be so familiar. The movie *Dolphin Tale* recounts the true story of Winter, a Bottlenose Dolphin (*Tursiops truncatus*) that lost her tail after becoming entangled in crabbing gear and was fitted with a prosthetic fluke. For fans of Douglas Adams's *Hitchhiker's Guide to the Galaxy*, dolphins feature prominently as the second most intelligent species on Earth (mice are the first).

Less well known is Carolina Snowball. In our state of South Carolina in the early 1960s, a popular white dolphin was routinely sighted around Beaufort, and she was sought after by the Miami Seaquarium. To protect her, the South Carolina state legislature declared it illegal to harass or capture dolphins in Beaufort County. Not appreciating legal boundaries, in 1962, Carolina Snowball and her gray calf were captured in unprotected waters and carted off to Miami. The mother survived only three years and Sonny Boy, the calf, died in 1973. This story is recounted in *The Prince of Tides*, a novel by famed southern author Pat Conroy, except in his version, Snowball was set free by local kids. As a result of these events, South Carolina became the first state to prohibit the display of dolphins and porpoises, and this prohibition stands.

Similarly, the movie *Free Willy* was based on the story of Keiko the captive Orca, who was eventually released back into the wild in 2002 at 27 years of age and, unlike Willy, died shortly thereafter. Shamu was the first captive Orca at SeaWorld San Diego in the 1960s, launching a series of famous and then infamous captive identically named Orcas, followed by Tilikum, who killed his trainer at SeaWorld Orlando in 2010. For a discussion of some of the controversy surrounding marine mammals in captivity, see box 8.1.

BOX 8.1 MARINE MAMMALS IN CAPTIVITY

Our love of dolphins is universal and owes in part, at least for some, to watching these intelligent creatures performing in shows at aquaria and marine parks, as well as on TV. Likewise, Orcas continue to inspire enthusiasm and drive attendance at SeaWorld and other marine aquaria. Notwithstanding the case of Carolina Snowball, whales and dolphins have been held in captivity for our entertainment with little organized public concern until recent years, when there has been increased scrutiny of the care and well-being of these sentient creatures.

Cetaceans are incredibly intelligent. They are some of the very few species in the world, including elephants, great apes, and humans, that pass the test of self-awareness (although this may be a function of our inability to accurately test other species, and we are sure we all know humans who fail this test).[5] In nature, whales and dolphins are wide roaming and live enriched social lives. For many, the idea of an Orca or dolphin being held in an artificial tank is the moral equivalent of permanently locking your grandmother in the basement.

In the past few years, this sentiment has gained momentum, in part because of the success of the movie *Blackfish*. *Blackfish* is an independent film, acquired by CNN, highlighting the controversy surrounding an Orca named Tilikum, who killed one of the trainers at SeaWorld, Dawn Brancheau. The movie seriously questions the ethics surrounding the keeping of such intelligent beings and describes them as prisoners. The film was so convincing and widespread, it substantially altered public perception of marine mammals in captivity, so much so that in 2017 SeaWorld agreed to a moratorium on breeding any more Orcas at any of its parks.

Blackfish has been criticized by SeaWorld and others for intentional inaccuracies. For example, a "calf" depicted as being taken away from her mother was 12 years old and not a newborn when she was transferred, and the cry of the mother in the movie (actually a male whale at another facility) was not a sound made by any cetacean, but rather a sound effect. The film was more of a political statement than objective investigative journalism.[6] An internet search will reveal the many criticisms of the film.

Although *Blackfish* seems to have misrepresented many relevant facts pertaining to SeaWorld, the authors of this book agree that the underlying issues presented by the film are important and worthy of consideration. For example, is it ethical to keep whales and dolphins in captivity? Given a choice, would they not be happier (as well as better serve their ecological function) swimming freely in their natural environment? How do we objectively measure welfare? Different species differ widely in their responses to a captive environment.[7] Some seem to thrive, but others much less so. Are we merely entertained and amused by their triple flips and playful antics, or is there a deeper value? What about rescued animals or captive-bred animals that cannot be released? A captive dolphin may live 50 years. How can these be cared for without income from human visitors? Many of these questions apply to all captive animals.

Close contact with captive marine mammals over the last half century is one reason the public cares so much about them. If visitors to aquaria never get to see marine mammals up close, will they care enough to support efforts to conserve their populations in the wild? One study shows that visitors to a marine mammal park were more aware of specific environmental concerns than nonvisitors.[8] Are enough students

Orcas (*Orcinus orca*) performing at a marine mammal park. Keeping cetaceans in captivity has become controversial in recent decades. There are philosophical ideologies and scientific evidence to argue both sides of the debate. Unsurprisingly, a majority consensus on the topic is elusive.

going to be inspired to go on to study them? Maybe keeping a few to serve as ambassadors to humans helps save them overall? Do documentaries accomplish the same ends? At least one study suggests they don't.[9]

Alternatively, does keeping marine mammals in captivity lead us to falsely conclude that they are thriving in nature? Does seeing captive marine mammals lead us to change our lifestyles sufficiently to slow the trajectory of their decline?

Not all marine mammal parks offer what many would consider humane treatment of their animals. Some are certainly far better than others, providing a level of care and quality of life that captive marine mammals deserve. A captive environment can possibly be an enriching environment. There are many variables to consider. But after the airing of *Blackfish*, anticaptivity sentiment, even at the best marine mammal parks, seems to have increased. While the debate rages, we continue to ask ourselves whether captivity is an appropriate place to keep marine mammals. Like the dichromatic vision of dolphins, the answer remains gray.

Arguably more friendly to humans than Tilikum was Old Tom (no relation to Two-Toed Tom of chapter 3), an Orca in the 1920s off Australia's southeastern coast that allegedly worked with his pod to assist local whalers by herding baleen whales into Twofold Bay. In return, the Orcas were rewarded with whale tongues, lips, and other scraps. Given that taxonomically, Orcas are big dolphins, it is not surprising that there are also modern examples of Bottlenose and Irrawaddy Dolphins (*Orcaella brevirostris*) assisting fishermen by leading them to fish or herding fish together. Cast-net fishermen in Myanmar attract Irrawaddy Dolphins to their boats. If the dolphins "agree" to engage, they will begin swimming in ever tighter circles around schools of fish, moving them toward the fishermen. When one dolphin gives the sign—a tail fluke pushed in the direction of the boat—it is time to cast the net. Studies show that these fishermen catch more with the assistance of the dolphins than without, and presumably the dolphins benefit from the chaos as the fish try to avoid the net.[10]

WHAT IS A MARINE MAMMAL?

Mammals as a group have fur or hair, suckle their young with milk, and have live births, except for those wacky egg-laying monotreme mammals, the Platypus and echidna. All these features seem inconvenient if one lives underwater, so it is a bit surprising that aquatic mammals even exist. In fact, they did not start out as aquatic at all. Fossil evidence and remnant pelvic and hind limb bones in some modern whales tell us that the ancestors of these animals were terrestrial. Clearly, there were advantages to moving back into the water, and as some did, natural selection worked to make aquatic mammals more hydrodynamic, insulated, and able to breathe and move efficiently in that

environment. Tail flukes and flippers replaced limbs designed for walking, improving aquatic propulsion. This basic bone structure persists, with flippers having shortened "arm" bones and lengthened fingers and toes, compared to yours. Blubber substituted for fur as insulation in cetaceans, and pinnipeds now use both. Young are still born live on land, or ice in the case of pinnipeds. As cetaceans evolved, their nostrils migrated backward on their skulls to the tops of their heads, making breathing easier as they swam, and when they give birth underwater now, the mother quickly assists her new offspring to the surface to breathe until the youngster gets the hang of it. Pinnipeds suckle their youngsters on land, but baby whales and dolphins do just fine underwater.

As mammals, humans share numerous characteristics with marine mammals. How we breathe is not one of them. We have already mentioned that the blowhole of whales, which is the external opening to the respiratory system, is on the top of the head. Here is another: the neutral, or default, position of your nostrils is open. Marine mammals could not suckle underwater if that were the case, and they would drown. The neutral position of the nostrils in pinnipeds and the blowhole of odontocetes is closed, not open (fig. 8.4).

Humans have rather pathetic breath-holding durations, except for trained deep divers and the diving women of Korea and Japan. An elephant seal, however, can dive for two hours, although most marine mammal dives are more in the five-minute range. Natural selection has made several modifications to the physiology and behavior of these animals, allowing them to accommodate to the lack of access to fresh air. There are three facets to this. One is intake. If you can maximize oxygen intake with each breath, you do not need to breathe as often. Second, you can store oxygen internally. Finally, you need to use as little oxygen as possible in satisfying body demands or tolerate

FIGURE 8.4 Common Dolphins (*Delphinus delphis*) exhaling before taking another breath. The blowhole (single nostril) of odontocetes is open only while breathing at the surface. The default position of the blowhole is closed to keep water out. Opening it requires active (rather than passive) behavior.

its absence. Also, there is the problem of pressure underwater on deep dives.

First, consider oxygen intake. One way to maximize oxygen intake is to increase lung capacity. Surprisingly, the lung capacity of marine mammals is within the typical range, relative to body weight, of other mammals, including you. Where marine mammals excel is in air exchange *efficiency*. Consider Bottlenose Dolphins, which have been extensively observed in the wild and studied in captivity, the latter taking advantage of the ability of these animals to be trained by scientists. If you have ever tried to photograph dolphins in the wild, you probably have a lot of photos of rippled water where there used to be a dolphin, because

they breathe very quickly as they swim. In fact, based on a study of seven individuals weighing on average 417 lb (189 kg), dolphins can exhale an extraordinary 29 gal/sec (130 L/sec) and inhale more than 8 gal/sec (30 L/sec) and, remarkably, can exchange as much as 95% of their estimated total lung capacity in a single breath.[11] Humans of average size at rest inhale and exhale a paltry 1.8–2.1 gal/min (7–8 L/min) of air, exchanging only about 10% of the air in the lungs per breath. At any given time, a dolphin has more internal oxygen at its disposal than a human does, even corrected for size.

Does more fresh, oxygenated air in the lungs adequately prepare them to dive? Not necessarily. When you, as a human, are trying to hold your breath, you take as big a breath as possible. Some deep-diving marine mammals *exhale* immediately prior to a dive. How does that make sense, given the need for oxygen during diving? Here is where water pressure enters the picture. Water is, on average, about 800 times denser than air, meaning a volume of water weighs substantially more than the same volume of air. At sea level, marine mammals (and you) are subjected to one atmosphere of pressure, which is the weight of all the air above you pressing down on you. That's about 14.7 lb/in^2 (1 kg/cm^2) and not an issue, right? But for every 33 ft (10 m) you descend, that pressure increases by one more atmosphere. This is the reason your ears might hurt when you dive even just to the bottom of the deep end of a swimming pool. It is that extra pressure acting on the air-filled space inside your eardrums.

Aside from sore ears, as long as you are made mostly of water, like most living things, this pressure is tolerable because one of the many odd characteristics of liquid water is that it is nearly incompressible. Thus, even at depths of 5,000– 6,500 ft (1,500–2,000 m), elephant seals and Sperm Whales, respectively, maintain their shape. But they can't maintain lungs full of easily compressible air at those pressures because their lungs collapse. In fact, you can see the outline of a dolphin's contracted thoracic cavity at about 33–50 ft (10–15 m) down as it dives.[12] A human diver would not recover from such an event, but marine mammals have more flexible rib cages and more elastic lung tissue to accommodate this collapse (fig. 8.5). One advantage of exhaling before diving is that it decreases the buoyancy of the animal, helping it sink, just as when a scuba diver releases air from a buoyancy control vest. And as long as gases under pressure are not diffusing into the muscles or other tissues, as can happen in deep-diving humans using scuba gear, marine mammals do not risk a case of the bends, the painful accumulation and then expansion

of gas in the blood vessels and joints that a human scuba diver experiences if surfacing too quickly.

Next, let us consider oxygen storage. Since the oxygen they need underwater must be stored in their body somewhere other than the lungs, marine mammals have evolved work-arounds. One adaptation to store and transport more oxygen involves *hemoglobin*, the molecule in the red blood cells of vertebrates that binds to oxygen, moving it around the body. This is why nefarious elite bike racers and other athletes transfuse red blood cells into their systems before a big race (a process called *doping*). The added hemoglobin provides them extra oxygen reserves to help fuel their internal power plants.

Marine mammals have naturally higher hemoglobin levels than terrestrial counterparts because the former have both more red blood cells per unit of blood, and more blood, than other mammals, ergo more oxygen storage and transport capacity. For example, human blood volume averages about 7% of body weight (about 1.3 gal, or 5 L, for a typical adult), which is just slightly less than the 7.1% of the short-diving Bottlenose Dolphin[13] (about 3.5 gal, or 13.4 L, for a 417 lb, or 189 kg, specimen), but pales in comparison to the 22%[14] (145 gal, or 550 L, for a 5,500 lb, or 2,500 kg, specimen) of the long-diving elephant seal.

In addition to the blood's oxygen-holding capacity, marine mammals have more *myoglobin*, an oxygen-binding molecule in muscles that is related to hemoglobin. Marine mammals can have as much as 13 times the oxygen in their muscles as in their lungs.[15] Remember that water is nearly incompressible and that blood and body composition is largely water, so blood and muscle are safe places to store oxygen under high pressure. The oxygen is still able to diffuse into muscles and organs that need it to produce energy. Humans also store some oxygen in blood and muscle, but we keep it mostly in our lungs where it

FIGURE 8.5 Sperm Whale (*Physeter macrocephalus*) and free diver. Marine mammals have adaptations for deep diving with quick returns to the surface, something not possible for humans. Shutterstock / Sokolov Alexey

is not as readily accessible to tissues that need it. Marine mammals store most oxygen in blood and muscles, right where it is used.

Our marine mammal thus has maximized oxygen intake and storage. The other side of that equation is to minimize oxygen demand during a dive, a phenomenon that has come to be known as the *dive response*[16] (box 8.2). To a lesser extent, this aspect of the dive response exists in all mammals, including you. Incredible as it may seem, Korean

and Japanese women dive to depths of 80 ft (24 m) and hold their breath for up to two minutes while actively collecting shellfish and seaweed. Even more incredibly, they work for four hours a day with only minimal breaks. In part, this is the result of the dive response, but also lots of practice.

The dive response is a general response to asphyxia, or lack of oxygen, rather than a response to being under water, per se. The hallmark of the dive response is a decrease in heart rate, a

BOX 8.2 ▸ YOU TAKE MY BREATH AWAY

The pioneering work on understanding the dive response was conducted at the Physiological Research Laboratory of the Scripps Institution of Oceanography (author Dan is an alumnus) in the 1930s and 1940s (way before Dan's time) by famed physiologist Dr. Per Scholander. The technology at the time constrained Scholander to study only captive seals. Instrumentation in current use allows mammals to freely swim in the natural environment while data are collected. The data are then transmitted to the scientist via telemetry, or the mammal swims back to the scientist and presents itself so the instrument packet can be removed. Or the animals exhale into inverted cones underwater, and that air can be analyzed.

But back to Scholander's time. To understand the first principles of how seals could dive longer than would be predicted based on what was then known about their anatomy and physiology, Scholander instrumented seals to measure their heart rates and collect exhaled air and strapped them to a tilt board—basically a seesaw. Then, after a control period, he tilted their head into a tank of water, forcibly submerging them while collecting data. This method sounds extreme and would very likely be prohibited by animal use committees at research institutions today, but it did elucidate one of the core physiological adjustments made by diving animals: *bradycardia*, or slowing of the heart (see below).

phenomenon called *bradycardia* (box 8.2). For example, the heart rate of an adult seal can decrease from a normal 55–120 beats per minute (bpm) at the surface to just 4–15 bpm during a dive! A lower heart rate translates into less energy required by the heart and is accompanied by regional *vasoconstriction*, which means that blood vessels supplying tissues that can withstand lack of oxygen are narrowed. Digestive organs, kidneys, and the muscles in extremities get less blood, in favor of the organs most necessary for survival and most sensitive to oxygen deprivation, the brain and heart, which cannot function anaerobically (without oxygen). This also helps conserve heat by keeping blood in the core of the animal rather than near the surface where heat would be lost to the much cooler water. Even with these adaptations, the body temperature of a long-diving animal drops a few degrees. Because the chemical reactions of metabolism (indeed all chemical reactions) occur more slowly at lower temperatures, the temperature drop decreases oxygen and energy demand.

The vast majority of marine mammal dives are of such limited duration that they have enough oxygen to sustain them simply by virtue of the energy-saving mechanisms of the dive response— even up to 18 minutes for a Weddell Seal. And even though blood flow to the brain and heart remains high, both organs can tolerate much lower oxygen levels than is the case for terrestrial animals. For dives of longer duration, even with adaptations to store and deliver oxygen, some mammalian divers need to get additional energy without using oxygen, a phenomenon called *anaerobic metabolism*. Anaerobic metabolism produces energy quickly but inefficiently and leads to the buildup of the mildly toxic by-product *lactic acid*.

You are familiar with anaerobic metabolism if you have ever had sore muscles after intense workouts. Lactic acid lowers blood pH (that is, increases acidity), and a buildup of lactic acid causes soreness. In addition, in terrestrial mammals, including humans, during anaerobic metabolism, carbon dioxide concentrations

A

B

increase in the blood, and this serves as a signal to breathe. Marine mammals can tolerate much higher levels of lactic acid and carbon dioxide than you, in part because vasoconstriction prevents it from circulating everywhere. In addition, the effects of these elevated levels dissipate much more quickly because of marine mammals' efficient overturn of air when they can breathe again, aided by a rapid increase in heart rate, *tachycardia*, upon surfacing, which quickly circulates blood to the lungs to take up oxygen and release carbon dioxide. It would be dangerous if every time a seal went hunting it spent the next day with sore muscles.

What else do marine mammals need to do? How about sleep? Most pinnipeds can and do sleep on land but can also sleep in the water, and of course cetaceans have no choice but to sleep in the water. Here is how cetaceans accomplish this seemingly contradictory feat. If you live underwater but you breathe air, you cannot just doze off. And interestingly, while breathing in humans is automatic, a marine mammal makes a conscious decision to breathe, which it cannot do if it is sound asleep. A solution is to sleep half the brain at a time, a phenomenon called *unihemispheric sleep*, in which one brain hemisphere sleeps and the other ensures that breathing continues. That idea, however, fell out of favor when many cetaceans were observed to routinely just slow down and rest together, without really seeming to sleep. Sperm whales have been observed resting in a pod, suspended vertically in the water in what is referred to as a *tail stand* (fig. 8.6A). In captivity, some Beluga Whales and Amazon River Dolphins rest upside down in the bottom of

FIGURE 8.6 Cetaceans rest in many positions. (A) The *tail standing* behavior of Sperm Whales (*Physeter macrocephalus*); (B) some captive Beluga Whales (*Delphinapterus leucas*) are known to lie upside down on the bottom of their tank.

(A), wildestanimal/Shutterstock; (B), panparinda/Shutterstock

their tanks (fig. 8.6B). That is how a true predator sleeps, without a care in the world! Must have been startling for the keepers the first few times they saw this, as well as concerned visitors.

More recently, though, numerous cetacean species, as well as eared seals (otariids), have in fact been found to sleep unihemispherically. In fact, this is the only way Bottlenose Dolphins sleep. You can observe this in captive Bottlenose Dolphins because each eye stays open alternately. Why would eared seals do this when they could just haul out on land? In fact, they also sleep on land sometimes, but even there, with one eye open. If a fur seal were sound asleep, with both eyes closed, and a Polar Bear pounced, well, you can imagine the outcome. Best to keep an eye on things. In water, eared seals float on their backs, nostrils out of the water, gently circling, literally half asleep while still keeping watch using half their visual field. An Orca or shark might still pursue them, but in the water, an eared seal can possibly outmaneuver those predators. Otariids like sea lions can be immense and are agile even on land, and Walruses are just plain enormous, so they need not worry much about predators on land, but they, too, sleep unihemispherically on land. Likewise, in the water they sleep like most cetaceans, at the surface, moving slowly, with one hemisphere at a time active. Interestingly, the true seals (phocids) have not been found to sleep unihemispherically at all. They float motionless in the water, simply not breathing for long periods, and lift their head to breathe without waking up.

While the physiological adaptations to living in water are similar for pinnipeds and toothed cetaceans, recall that these are two phylogenetically distinct groups. All are intelligent predators in the water, but pinnipeds spend time on land. This leads to differences in ecology and behavior, including hunting behavior, so we will discuss these characteristics of the two groups separately.

PINNIPEDS

Pinnipeds are all streamlined, with flippers for limbs, and blubber and, in most cases, fur for insulation. All visit land or ice to give birth, grow as pups, rest, molt, thermoregulate, and, sometimes, mate. They range in size from the 77 lb (35 kg) Harbor Seal (*Phoca vitulina*) to the 6,000 lb (2,700 kg) Southern Elephant Seal. Both of these species are phocids, or true seals. As we described in our introduction, they are also known as earless seals because they lack external flaps around their ears. This group has a shorter and less flexible neck and flippers and so is less agile on land than the Otariidae, or eared seals (fig. 8.7). The eared seal group includes sea lions, which you might have seen performing in shows. They can be trained to clap and wave and even do handstands using those long, flexible flippers, and with their long, flexible neck and maneuverable whiskers (called vibrissae) they can catch and balance balls on their noses. Eared seals can also walk around reasonably well using their front and back flippers, in comparison to, say, a Harp Seal (*Pagophilus groenlandicus*), which essentially must drag and inchworm itself forward on its belly. The Walrus is in a group by itself, the Odobenidae, and has agile flippers so it can get around like the otariids on land, although it is so gigantic, it is not at all graceful doing so. Walruses are unique among the pinnipeds in having tusks.

BIOLOGY AND ECOLOGY OF PINNIPEDS

All pinnipeds are well designed for efficient life in the water, with flippers and smooth, streamlined bodies. In terrestrial mammals, the vertebrae interlock to provide support to the body against the gravitational pull we all experience, but with water providing support, the interlocking processes in pinnipeds have been reduced as they evolved,

FIGURE 8.7 (A) Baikal Seal (*Pusa sibirica*), a phocid. (B) California Sea Lion (*Zalophus californianus*), an otariid. Phocids are less agile than otariids, both in and out of the water. (A), fibPhoto; (B), Willyam Bradberry / Shutterstock

allowing exceptional spinal flexibility. Their slit-like nostrils close tight underwater, as do their ear canals (though they hear well underwater through another pathway—more about that later). We will discuss their eyes when we consider hunting, since that is the sense they primarily rely on to find prey.

Because they require access to land or ice, all pinnipeds are found mainly in coastal waters rather than the open ocean, although some species, like elephant seals and Northern Fur Seals (*Callorhinus ursinus*), undergo extended foraging migrations into oceanic waters for much of the year. Twelve of the 18 phocid species plus Walruses give birth and rear offspring only on sea ice, an increasingly risky strategy given declines in sea ice associated with anthropogenic (human-caused) climate change. Warming of polar seas is especially a problem for these animals because their superior insulation becomes a dangerous liability as the high latitudes warm: they have more trouble keeping

cool than warm. Not surprisingly, then, not many pinnipeds live where it is warm, like the highly endangered Hawaiian and Mediterranean Monk Seals (*Monachus schauinslandi* and *monachus*). Galapagos Fur Seals (*Arctocephalus galapagoensis*) and Galapagos Sea Lions (*Zalophus wollebaeki*) are equatorial, but the ocean currents around the Galapagos Islands keep the water around 75°F (24°C), cooler than body temperature. Hawaiian Monk Seals can tolerate only limited terrestrial exposure, though; they head for the water when land temperatures reach 93°F (34°C). They, and elephant seals too, are known to dig into the cool sand a bit, where temperatures are a few degrees lower, as well as toss sand onto themselves. As the sand moisture evaporates, their bodies cool slightly.

At sea, where they do most of their foraging, pinnipeds are mostly solitary. Some species stay near familiar, routinely used land or ice masses and haul out frequently, while others stay at sea, coming ashore only for breeding season, to give birth and raise pups, and to molt. Where they reside also depends on seasonal food resources. These behaviors usually co-occur within a population, so, for example, females come onshore at a safe location to give birth and nurse babies around the same time, presumably as a defense against predation. As you will see, males follow.

As with most mammals, breeding season for pinnipeds is short and happens once a year. Some pinnipeds are monogamous, while others are *polygynous*. In the latter, one male guards, fights for, and mates with many females. This is the case with elephant seals, for example, in which the large *beachmaster* (fig. 8.8) rides herd over and mates with a large group of females and fights off rival males. This level of competition with other males explains why male elephant seals have evolved to be so much larger than females. Being large, with lots of stored fat, also allows the males to skip foraging and focus on breeding. There are other mating strategies among the pinnipeds in which females choose males, and there is also *scramble*

competition, in which males and females all mate indiscriminately such that everybody gets to mate, sometimes whether they want to or not. Elephant seals and Gray Seals (*Halichoerus grypus*) are the only phocids that mate on land; however, all otariid species do this.

For true seals, mating occurs immediately following weaning because that is when generally solitary females can be reliably found in a predictable place. Otariids are generally more gregarious and gather on land where males can find females and often fight over them. Gestation lasts about eight months, and given the extreme seasons in the high latitudes where most of these animals live, babies need to be born when the climate is most forgiving, spring into summer. To accomplish this, pinnipeds exhibit *delayed implantation*, which means there is a time lag of 120–160 days between fertilization and the implantation of the embryo in the uterine wall for development. Delayed implantation is fairly common in mammals because offspring need to be born during optimal conditions, regardless of when

FIGURE 8.8 **A dominant male Northern Elephant Seal (*Mirounga angustirostris*) with a much smaller female. Elephant seals are the largest phocids. A dominant male, called the *beachmaster*, ferociously protects a harem of up to 100 females.**

FIGURE 8.9 Hooded Seal (*Cystophora cristata*) and pup. Hooded Seals will nurse pups for only about four days with a super-high-fat milk. During this 96-hour period, the pup will double its weight.
Enrique Aguirre / Shutterstock

mating occurs. Females typically have only one pup at a time.

Pups are born on land or ice, but not in the water. Phocids have very short lactation periods, ranging from 4 to 21 days, and mothers leave to forage only briefly, if at all. This likely occurs because the young are so defenseless against predators that it is a good idea to get them grown and into the water as fast as possible. There are predators there too, of course, but an agile swimmer has a fighting chance. The exceptions to this are the Baikal and Ringed Seals (*Pusa sibirica* and *hispida*), both of which give birth in rocky or icy lairs and lactate for two months. Otariids, which are more agile on land, stay longer, coming and going through four months to a year of lactation.

In general, the role of the mother is to provide the maximum amount of fat-rich, growth-accelerating milk as quickly as possible. Pinniped milk is 50% fat. The milk-fat champion is the female Hooded Seal (*Cystophora cristata*) (fig. 8.9), which provides milk with 61% fat content every 25 minutes for only four days. This fuels a growth of over 15 lb/day (7.1 kg/day), 82% of which is fat! Pups double in weight from 48 lb (22 kg) to 95 lb (43 kg) in four

days. This is an extreme case and is the shortest lactation period for any mammal, but it follows the general pattern. An exception is the Walrus. A Walrus pup suckles on land or in the water for five months but then sticks with its mother for two to three years, during which time the pup may learn foraging techniques from her. Pups reach sexual maturity at four to seven years, depending on the species and environmental conditions.

In cases where the mother leaves to forage during lactation, vocal communication for identifying her own pup when she returns is important (as is sense of smell once close enough). This is important mostly for otariids; vocal communication remains essential for them while the mother teaches her offspring hunting strategies, and they keep talking throughout their lives. One complaint about the California Sea Lions (*Zalophus californianus*) that famously pile up at Pier 39 in San Francisco is that they are irritatingly loud. Phocid mothers, on the other hand, wean offspring so rapidly that they do not need to leave much while still nursing, and then the youngsters are left largely on their own to figure out foraging. An otariid mom almost never knowingly suckles a pup other than her own, but this fostering behavior is fairly common in phocids, particularly among young mothers that have lost their pups.

Males are not directly involved in parenting. In fact, there is some risk to pups onshore from males trampling them to get to females. Males along with females may, however, play a role in defending pups. Galapagos Fur Seal and Galapagos Sea Lion males and females have been seen either mobbing sharks in defense of young or steering the young away from sharks. Hooded and Crabeater Seal (*Lobodon carcinophagus*) males may take up a position near a female and her pup, guarding them until weaning is complete. There is a powerful ulterior motive here, though, because when the mother leaves the pup and heads off, the male goes with her and mates with her.

PINNIPEDS AS PREDATORS

Since pinnipeds hunt underwater, scientists did not know much about their predatory behavior and ecology for a long time. Stomach content and scat (poop) analysis provided some clues about their food sources, but not how, when, or where they were eating. This knowledge deficit changed with the development of time-depth recorders, and then satellite-linked transmitters, allowing scientists to track animal movements for extended periods. We now know that most seals and sea lions dive to depths less than 330 ft (100 m) for only a few minutes. They engage in bouts of diving more than they do single dives, with numerous dives in succession followed by resting periods at the surface. Probably these bouts correspond to when the fishing is good.

Pinnipeds are often crepuscular, foraging morning and evening, but the California Sea Lion, for example, hunts only during the day. This timing may depend simply on prey availability. For example, Antarctic Fur Seals (*Arctocephalus gazella*) eat small, shrimplike krill. Krill migrate

toward the surface only at night when they are less likely to be seen and eaten. Antarctic Fur Seals thus dive at night, too. Even within a species, there are variations. Some Northern Fur Seals follow krill, and some seemingly plow right through the krill layer to deeper depths regardless of the time of day, evidently dining on something other than krill.

Foraging trips vary in both time and space. Galapagos and Steller Sea Lions (*Eumetopias jubatus*) routinely head out from land to hunt and return that same day. Northern and Antarctic Fur Seals may be gone for weeks and travel several hundred miles. With the energy expended by males to fight for mating rights, and by females to raise a baby elephant seal, once those young are weaned, the father and mother head off for foraging trips with nearly continuous diving for 120 and 70 days, respectively. They then come ashore to molt before leaving again to spend an average of 241 and 296 days foraging, respectively.

Pinnipeds eat primarily fish, squid, shrimp, and krill. California Sea Lions cause a bit of havoc by crowding docks and inviting themselves onboard unattended boats (fig. 8.10). Some locals (humans) have started installing life-size plastic Coyotes to keep the pesky pinnipeds away from their yachts (fig. 8.10). California Sea Lions get into more serious trouble for lurking at the base of hydroelectric dams at salmon spawning time, scooping up the salmon waiting to enter the fish ladders that allow them past the dams. Human fishers would also like a shot at catching the salmon or allowing them safe passage so they can make baby salmon, so they are generally not fans of sea lions. In some cases, the salmon themselves are endangered, so here is a complicated conflict between two protected species.

Walruses eat bottom-dwelling invertebrates like crabs and bivalves (clams, scallops). And Leopard Seals (*Hydrurga leptonyx*), probably the most "apex" of the pinniped predators, will eat just about

FIGURE 8.10 California Sea Lions (*Zalophus californianus*) hauled out on a dock (A) and a sailboat (B). This species has made itself at home in some urban ports in California. "Scarecrow" Coyotes (C) are employed by some to keep sea lions from sinking boats and causing property damage.

(C), Arthur Greenberg / Shutterstock

any animals in addition to fish, including other pinnipeds and penguins. How do these predators find and catch their prey?

Sensing

Pinnipeds rely mostly on vision for locating and catching prey, but there's a catch: they need amphibious vision; in other words they must be able to see well both in and out of the water. An eye well adapted to water will tend to be nearsighted in air, and an eye well adapted to air, like yours, will be farsighted and produce blurry images in water. The cornea of your eye, the outermost transparent part, is quite rounded to account for the bending of light rays as they move from air through the optically denser water within the eye. A seal eye has no such issue, as it is mostly looking at light in water already, so almost no bending of light occurs at the cornea underwater. They also have a spherical lens like that of a fish, rather than the disk-shaped lens you have.

But if these adaptations allow them to see well underwater, what happens in air? Their round lens, which is the perfect shape for underwater vision, bends light too much for focused vision in air. It turns out that the slit-shaped pupil of the seal narrows in bright light out of the water, acting as a pinhole camera does, allowing light to enter the eye only though a small hole and thereby maximizing focus at all distances.[17] The large eyes are deep set in protective cushions of blubber, with a tough cover, the *sclera*, for protection. While underwater, a seal's pupil can open much wider than yours to let in as much light as possible in these dimly lit conditions. They also have the tapetum lucidum layer common to sharks, crocodylians, and other mammals that reflects light to the many highly light-sensitive rod cells on the retina. And their eyes, having to transition quickly from darkness below the water to light above, adjust from one to the other much

more quickly than yours do. Hot on the trail of a fish, a seal cannot afford to wait for its eyes to adjust to the dark as it dives. Harbor Seals, at least, can see stars at night, suggesting a possible method of navigating to and from their coastal haul-outs after dark.[18]

As with vision, pinnipeds need amphibious hearing. They must be able to communicate with each other and hear predators both on land and in the water, media with very different sound transmission properties. We will talk more about this with dolphins because their hearing is much better understood. In general, phocids hear better in the water and otariids better in the air, but both can hear in both places. The exception is the deep-diving elephant seal, an otariid, whose ears seem to be much more adapted to water than to air. This is likely because those ears must still work at depths where the pressure is 100 times greater than that at the surface. Even so, the elephant seal must also be able to hear her pup on land. In water, pinnipeds can probably hear predators, either the echolocation clicks of Orcas or the incidental sounds of shark movements, and they may also hear invertebrates and fish communicating with each other or emitting distress signals. They may also use the noise of waves to navigate the coast or find holes in the ice.

Pinnipeds hunt in areas deep enough that light is dim, in water that is sometimes turbid and cloudy, and sometimes they hunt at night. Since they apparently do not echolocate and cannot depend on sight under these conditions, how do they locate their prey? Plump and healthy-looking blind seals have been observed in the wild, so they can apparently find food successfully without relying on vision.

One way to locate prey involves their extensively nerved whiskers, or vibrissae. The innervation of those vibrissae can be up to 10 times that of whiskers in terrestrial mammals, which translates into vastly increased sensitivity. One study using a trained Harbor Seal found that the sensitivity of the whiskers was good enough to feel changes in water velocity of the strength produced by the wake of a swimming fish.[19] Another study showed that a Harbor Seal can discriminate shapes and sizes of items as small as a few centimeters, using only its whiskers.[20] Another showed that a Harbor Seal can navigate a maze using its whiskers,[21] probably a very useful skill when navigating under the ice to find a breathing hole.

Catching and Killing

Pinnipeds are pursuit predators. They are agile and quick in the water, and they chase their prey, unless they need to dig it out of the substrate. Harbor Seals have been observed swimming back and forth around a school of herring, moving to keep the school squarely in their field of vision and singling out individual fish or a small group of fish to attack.[22] Basically, there are four capture strategies: (1) pierce feeding, (2) suction feeding, (3) filter feeding, and (4) grip and tear feeding (fig. 8.11). The architecture of the skull and teeth specializes different species for different strategies, but these intelligent animals are capable of adapting their strategy to the prey at hand and frequently combine strategies. *Pierce feeding* is the most common approach and involves grabbing a prey item small enough to be swallowed whole. Several species of both eared and earless pinnipeds employ a modified version of pierce feeding in which they hold bigger prey with their front flippers, using claws or just the pads, and either chew or shake off smaller pieces.[23] Pierce feeding requires typical carnivore incisors, canine and postcanine teeth (similar to your dog's) that both stab and shear, and a strong bite force. Harbor Seals, Ring Seals, and Spotted Seals (*Phoca largha*), for example, all catch fish via the pierce method.[24]

Suction feeding, a method employed by numerous aquatic animals, also works well. Think about this: if you reach into the water to grab a fish, your hand pushes water ahead of it, and the fish senses the pressure wave, pushing the fish away or enabling it to escape. An open mouth approaching a fish would result in the same escape response. But if you managed to get near a fish and then opened your mouth quickly, water would tend to rush into your mouth, pulling the fish in. This works best if you have a big, gaping mouth or throat, like an angel shark, but it helps pull in the fish even if you have a small gape, so suction feeders take advantage of that. (Note that suction does not work at all in air except with a

FIGURE 8.11 Seal eating a fish. Pinnipeds use the tear feeding method when their prey is too large to swallow whole.

straw. Try to suck something out of the air.) The gape of the mouth and the bite force work against each other, so pierce feeders cannot open their mouths very wide, which is necessary for suction feeding.[25] Among the phocids, Steller Sea Lions combine suction with piercing, while Northern Fur Seals mainly pierce.[26] Bearded Seals (*Erignathus barbatus*) and Walruses are specialized suction feeders, with extra-wide gaping mouths but consequently a weaker bite force. Whichever the process, the vibrissae are directed toward the prey and remain in contact with the prey as it is captured.[27]

What do Walruses do with those tusks? Walruses eat mainly benthic bivalves (two-shelled creatures like clams or scallops), other invertebrates like crabs, and some fish. Seals have been identified in the stomach contents of Walruses and were thought to be fresh kills, but Walruses do not normally eat seals, so the importance of seals to the diet of Walruses is unknown. Female Walruses eat throughout the winter breeding season, but males mostly do not—sex matters more! Whenever males and females are feeding, both come back to land routinely to rest. Among the pinnipeds, Walruses have the most extensive and sensitive vibrissae on their snouts, those famous Walrus mustaches. It is thought they use these to sense prey below the sand as they glide along using their tusks like skis. They then use suction to vacuum the animals from their shells—some serious sucking ability![28]

Filter feeders eat small, shrimplike krill, basically sucking in large mouthfuls of krill-laden water and then closing their mouths and forcing the water out between their lateral high-crowned and intricately cusped postcanine teeth, which are designed like a sieve to contain the krill. Crabeater Seals, the most numerous pinniped, are filter-feeding specialists. Interestingly, though, the other pinniped that filter feeds—the most fearsome predator among the pinnipeds, the Leopard Seal—also employs the final method, the *grip and tear*.

Leopard Seals lurk along the coasts among the pack ice of Antarctica. They have disproportionately large, almost reptilian heads and are the only pinniped known to regularly prey on warm-blooded animals. Not exclusively, though, since as we just pointed out, these animals filter feed on krill, and they also eat fish, like other pinnipeds. But about 35% of their diet is seals, often Crabeater Seals or Antarctic Fur Seals, and 10% is penguins. Easy to see why these seals have a nasty reputation, but a Leopard Seal, like all predators, must eat.

ODONTOCETES

Among cetaceans, the top predators are whales with teeth, or odontocetes. These include the Phocoenidae family, with five or six species of small porpoises, depending on the perilous status of the little Vaquita (*Phocoena sinus*), which may be extinct by the time you read this (we will discuss that later). The Monodontidae family includes the Beluga and Narwhal of the Arctic. The biggest family is the Delphinidae, with 36 or 37 species of dolphins, including pilot whales (fig. 8.12) and Orcas, and a few other smaller whales. The Ziphiidae includes 21 species of deep-diving beaked whales. There is one Sperm Whale, with the Pygmy and Dwarf Sperm Whales (*Kogia breviceps* and *sima*) in a different family. And finally, there are three or four species of river dolphins, depending on whether you count the recently extinct Chinese River Dolphin (*Lipotes vexillifer*). These mammals are found at nearly every latitude and in virtually every ocean, sea, bay, and gulf, plus major river systems of South America and Southeast Asia. Sperm Whales and Killer Whales, or Orcas, are both found nearly worldwide.

FIGURE 8.12 **Pilot whale (*Globicephala*) pod.** Although colloquially called "whales" (mysticetes and odontocetes), pilot whales are part of the family Delphinidae, along with dolphins, Orcas, and others. Andrew Sutton / Shutterstock

BIOLOGY AND ECOLOGY

Odontocetes spend all their time in the water and share a suite of characteristics in addition to their teeth. All have a hydrodynamic fish shape with flippers for front limbs and a tail replacing hind limbs. They have some hair when young but otherwise lose this and depend on a layer of blubber for insulation and streamlining. They all have a blowhole on top of the head, which is essentially a single nostril, but have no sense of smell at all. We have already talked about their various adaptations for diving. Odontocetes communicate with clicks and whistles, and they also use these sounds to echolocate prey.

Bottlenose Dolphins have been found to have unique *signature whistles*, which function like our names do; they use these to identify themselves and each other.

The largest of the odontocetes are the Sperm Whales (fig. 8.13), and they are the most sexually dimorphic in that the male is much bigger than the female, at nearly 60 ft (18 m) for males and less than 39 ft (12 m) for females. The lower jaw is underslung and narrow, and that is where the teeth are. Those teeth are nearly 8 in (20 cm) long and tucked into sockets in the palate. Teeth on the upper jaw never erupt (break out of the tissue that encases them). Sperm Whales eat mostly Giant Squid, captured in what must be epic battles in the

FIGURE 8.13 Sperm Whales (*Physeter macrocephalus*). Sperm Whales are the largest of the Odontocete family, reaching up to 60 ft (18 m) in length. They often show battle scars from their favorite prey, Giant Squid.

Willyam Bradberry / Shutterstock

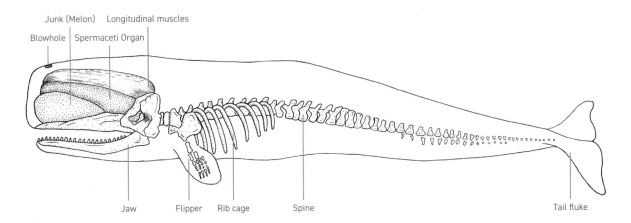

FIGURE 8.14 Diagram of Sperm Whale (*Physeter macrocephalus*) head anatomy. The spermaceti and junk organs were once prized as a source of oil by whalers, which led to a significant decline in the global population of the species.

Redrawn from Spoiled Milk / Shutterstock

deep ocean. Their flanks often feature round scars from the suction cups on squid tentacles. Males have exceptionally large and squared-off heads.

These are the *Moby Dick* whales and, as we noted, had a reputation for ramming whaling ships. Sperm Whales are named for the *spermaceti organ* in their heads (fig. 8.14). Spermaceti oil was much sought after by whalers for use in cosmetics, lubricants, and lamps and apparently resembles sperm, so whalers concluded that it was in fact sperm. (In the head of the whale? Fill in your own joke about the location of the male human's brain.) An adult Sperm Whale may contain up to 500 gal (1,900 L) of the stuff! The demand for spermaceti oil led to a substantial depletion of these whales, so you can maybe sympathize with them for ramming whaling ships.

There are in fact only two documented cases of Sperm Whales ramming ships, and it turns out that from the physics and structural standpoint of both wooden ships and whales, it is possible. The spermaceti organ is used for echolocation, a vital function, so it would not be good for the whale to damage it, but it sits toward the upper back of the giant head, and in front of it is another spermaceti-filled structure called a *junk* (we could not find a reason for that name). That oil was prized too. At any rate, the junk is also involved in echolocation but seemingly also serves as a protective cushion, as its compartments are reinforced by rings of connective tissue. This might allow the whale to ram objects without wrecking either spermaceti organ. Clearly this system did not evolve to ram whaling ships, so what else might it be good for? Well, enhanced echolocation and diving buoyancy, perhaps, and maybe ramming other whales. The whales operate at such depths that we only recently figured out that they eat squid when they are at depth, and only once has one whale been observed ramming another whale. This may be explained by males jousting for the right to mate.[29]

Cetaceans have a range of mating strategies. Some, like Sperm Whales and some beaked whales, are strongly dimorphic, with males bigger than females (fig. 8.15). For these, males may compete to join a group of females, defend a harem for a period, and subsequently mate with multiple females. At the other end of the spectrum, most dolphin males are only slightly bigger than females (like humans), and there is some competition for females but a single male does not dominate a group. In the *promiscuous* mating system of Bottlenose Dolphins, both sexes may mate with multiple partners, though in many populations, two or three males will form an alliance and may accompany and/or isolate a reproductive female for a period of hours or days. It is not clear whether the displays of the males trigger estrus or vice versa, but the female will go through this routine multiple times, mating with multiple males and presumably thus ensuring fertilization. In Orcas and pilot whales, there are long-term matriarchal pods made up of multiple generations. Genetic study of Orcas has shown, however, that there is not a great deal of inbreeding. Instead breeding occurs between different pods, though often within a larger population of frequently interacting pods. It is described as somewhat like human society where marriages might occur within one town or two adjacent towns, but not within one family.[30]

Muscular contraction makes an odontocete penis erect, and it can be maneuvered at will. There are thus some interesting examples of misplaced sexual activity in Bottlenose Dolphins. Males have attempted copulation with the underside of boats, inserted a penis into a pipe, and explored undoubtedly startled human swimmers with the organ. This leads to speculation that the penis is indeed a sensing organ as well as a sexual one.[31] Alternatively, studies of animal behavior in general show that males, eager to get their sperm out there and thus their genes into the population as

FIGURE 8.15 Sperm Whales (*Physeter macrocephalus*) mating in the Indian Ocean. Sperm Whales demonstrate strong sexual dimorphism, with males growing significantly larger than females. Martin Prochazkacz / Shutterstock

stipulated by Darwinian fitness, are somewhat predisposed to make mistakes. Or maybe it just feels good?

Gestation ranges from 10 to 12 months for most odontocetes, though pilot whales, False Killer Whales (*Pseudorca crassidens*), and Sperm Whales take longer at 14–16 months, and Orca gestation is up to 17 months. The life span of whales generally correlates with size, such that small porpoises and river dolphins may live 25 years, Bottlenose Dolphins may live 50 years, and Orcas and Sperm Whales longer still. Given these long life spans, particularly for larger animals, and given that calves are born large and well developed, whales do not typically give birth annually, although smaller species such as porpoises will do so if there is adequate food to support it.

Odontocete calves are born tail first and are well developed because they must be able to swim immediately. The mother may be tended by other females, and she or these "aunts" push the newborn to the surface for those initial breaths. We have seen newborn Bottlenose Dolphins on surveys in our coastal waters in South Carolina. These babies are adorably wrinkled and have furled fins from being folded up in utero, and not surprisingly, this has also been observed in newborn Sperm Whale and Orca calves. The mother provides milk with a fat content of 14%–53%, depending on the species, to facilitate the quick development of the insulative blubber layer and rapid growth. Lactation again roughly correlates with body size, with smaller porpoises providing milk for eight months, Bottlenose

SAY "CHEESE": USING FIN IDENTIFICATION IN DOLPHIN RESEARCH

How do scientists figure out migratory movements, family relationships, and who is in which pod, for animals that surface only to breathe? Most of the time you cannot see them! It turns out these animals can often be identified by their dorsal fins (dolphins and some whales), tail flukes (Humpback Whales), color patterns and markings (Orcas), and other distinctive features (Right Whales). We will consider Bottlenose Dolphins as an example, because these are studied at Coastal Carolina University (CCU), where your authors work.

Since 1997, our colleague Dr. Rob Young and his students have been conducting photo surveys of Bottlenose Dolphins inshore along the estuaries and northern coast of South Carolina, USA. To do this, they venture out on relatively calm days (it is difficult to see dolphin fins in waves) in the study areas and take photos of dolphin fins. Many fins have distinct notches and markings. Sometimes one has a bite out of it. They are sufficiently variable that individuals can be identified (see figure below).

Examples of dorsal fins. Eve (A), seen here with her calf Dodger, was the first confirmed female in the Coastal Carolina University catalog. A resident of the North Inlet salt marsh system for over 20 years (and counting), Eve typically has a calf every two years, which is faster than Marge (B), who follows a more typical three-year cycle. Knobby (C) is a large male who is often seen with females during mating periods in May and June—who knows how many calves have been his over the years? Dorsal fins, which have no bones and are composed of fibrous tissue, can sustain substantial damage (D), though many remain undamaged and are not easily identified in photos.

Rob Young, Elizabeth Moses; all photos taken under NOAA/NMFS Permit 976–1582

(*Continued overleaf*)

BOX 8.3 (Continued)

CCU has a library of fin photos, so if you were to get a good photo of a dolphin fin around here, you could probably find the animal in the library and learn a bit about him or her. (But it is illegal to follow and harass dolphins, even unintentionally, for photos without a federal permit!) To make it easy to refer to these dolphins and keep track of who is who, the researchers name them, sometimes basing names on their fin characteristics, and sometimes just making them up.

With over 20 years of sightings recorded, we know a lot about the populations and movements of these animals around our coast and elsewhere through collaborations with other researchers. Photo ID can be used to identify seasonal movements, population boundaries, and social groupings, and in combination with DNA analysis from skin biopsies, it can help identify populations of interacting and interbreeding dolphins. In fact, bottlenose dolphins along the US East Coast can be classified into over a dozen different coastal migratory, coastal resident, and estuarine stocks (population-based management units). These dolphin stocks, ranging from hundreds to thousands of dolphins each, have distinct impacts on their ecosystems and are vulnerable to similar threats to their shared habitats, allowing us to manage activities that may impact their own health and the health of the coastal ecosystem.

Dolphins nursing for around a year and a half, and Sperm Whales and pilot whales nursing for more than two years. Orcas may nurse longer than that, perhaps allowing time for learning within the advanced social structure of the Orca pod.

The mother-calf bond is strong. Bottlenose Dolphin mothers have been observed attending to dead calves, helping them to the surface over prolonged periods. In July 2018, an Orca from the long-studied J Pod of Southern Resident Killer Whales off the northwest coast of the United States and southern Canada pushed her dead calf along for 17 days and 1,000 miles before finally letting it go. The baby had lived for only half an hour.[32]

ODONTOCETES AS PREDATORS

Sensing

How do toothed whales find prey? We noted earlier that they have no sense of smell. They also lack the vibrissae of pinnipeds. Their eyes are relatively small compared to those of pinnipeds, but they are structured similarly. Odontocetes have the same high density of photoreceptors, mostly low-light-sensitive rod cells, along the retina. They also have a well-developed tapetum lucidum layer to reflect dim light back at the retina. They can see well in and out of water (fig. 8.16), accomplishing this with the same nearly spherical lens as in the pinniped eye. They can also adjust quickly to dramatically changing light levels from above the water to deep below the surface by tightly constricting their pupils in bright light, as pinnipeds do. These animals can see their prey, but studies have demonstrated that their main way of "seeing" is actually "hearing."

The precise echolocation with which a dolphin "sees" its environment is so different from anything humans are capable of that it is challenging for us to imagine (fig. 8.17). You can tell when something producing a sound is coming toward you by that sound getting louder. You can also tell which direction it is coming from. If you do not recognize the sound, though, you cannot identify the source. An odontocete can. It can "see"

FIGURE 8.16. Orca (*Orcinus orca*) hunting a seal on an ice floe. Orcas locate potential prey on top of icebergs by thrusting their head above the water in a behavior known as spy hopping. Odontocetes have good visual acuity both above and below the water. vladsilver/Shutterstock

with sound. How does this work? First, they all make sounds themselves. Odontocetes vocalize to communicate and identify each other, as we discussed, but they also use sound to hunt. Each species has a spectrum of click pulses at particular frequencies. The higher the frequency (and they can produce and hear sounds much higher than humans can hear), the smaller the object that can be detected and the more detailed the "picture" will be. Thus, inshore Harbor Porpoises (*Phocoena phocoena*) and river dolphins use the highest frequencies to detect small fish prey in their cluttered habitat. These animals do not produce other sounds, as far as scientists know. Nearshore and offshore, their habitat is less busy and the animals living there tend to stay in larger groups.

Thus, they need to both communicate and detect relatively large, sometimes distant objects, which requires lower-frequency sounds.[33]

All odontocetes have tiny ear canals that lead to a tiny pinhole on the side of their heads, but it seems they do not really use that pathway to hear. Sound in water is more efficiently transmitted through the vibration of bones and specialized fats than through their ear canals. There is also a pocket of oil tucked inside the lower jawbone. The whale produces clicks by forcing air over *phonic lips* in its nasal passages. These sound waves travel through the *melon*, or forehead area, which concentrates the sounds into beams and amplifies them. This is the reason toothed whales all have such pronounced foreheads, with the Sperm Whale

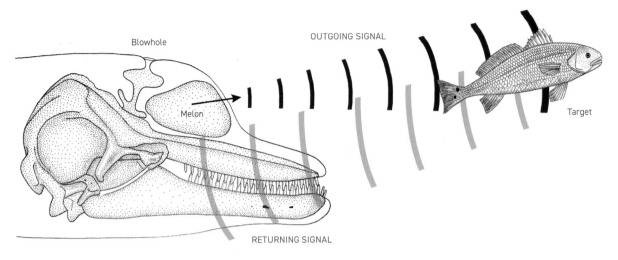

FIGURE 8.17 Anatomy of echolocation in toothed whales. Echolocation allows odontocetes to "see" with sound, a concept foreign to human perception. The melon amplifies sound to target prey. Redrawn from Wikimedia Commons Dolphin Anatomy figure, CC BY-SA 3.0

forehead and its combined spermaceti and junk being an extreme example. In other odontocetes, such as the Bottlenose Dolphin and Beluga Whale, their bulbous forehead is the similarly oily *melon*. If you could feel it, you would find the melon somewhat squishy, since it is basically a ball of oily fat. Beams of sound leave the whale, travel speedily through the water, hit objects like fish or rocks or other whales, and bounce off back to the original whale. The sound waves enter the lower jawbone, travel quickly back through that oil to the inner ear, and voilà, the clever whale brain translates the sound to a picture almost immediately! Fast enough that our cruising whale does not slam into a head of coral in the dark and can grab a moving fish.

Catching and Killing

Odontocetes feed on individual fish or squid, except for Orcas and False Killer Whales, which eat other mammals, including other cetaceans. Toothed whales are *homodonts*, meaning all their teeth are the same conical shape and are not specialized for various tasks like yours or those of pinnipeds.

Deep-diving beaked whales possess only two teeth, which are used mostly for gouging other males during battles for mating rights, so beaked whales are suction feeders, probably eating mainly soft-bodied squid. Sperm Whales also dive deep and eat squid. Some dolphins have more than 40 teeth per row. These teeth help grab, hang on to, and manipulate often slippery prey. Smaller teeth are specialized for smaller prey, and larger teeth are necessary for larger prey, so it is not surprising that Orcas have the biggest teeth among dolphins.

Toothed whales forage year-round, usually in groups. These can be family groups called *pods*, smaller groups of bachelor males, or huge schools, depending on the species or the specific population. They forage where prey is abundant, which sometimes requires migrations. Migrations may depend on older, experienced members of the group leading the way, as we see in similarly long-lived elephants, for example, which share a similar matriarchal social structure. Beluga Whales and Narwhals travel in groups around the Arctic, following seasonal movements of prey and sea ice. Up to 30 subgroups of 8–10 Dusky

FIGURE 8.18 Common Dolphins (*Delphinus delphis*) corral fish into a bait ball. This cooperative hunting technique allows some dolphins to hunt from a condensed group of fish while other dolphins keep the fish corralled. The dolphins take turns corralling and feeding, resulting in better success for all dolphins involved. wildestanimal/Shutterstock

Dolphins (*Lagenorhynchus obscurus*) forage a few kilometers apart off Argentina but evidently maintain contact, because when one of them finds a big school of Southern Anchovies, they converge and cooperate. Some dolphins force the school upward and herd it into a *bait ball* (fig. 8.18), while others plow through the dense concentration of fish and chow down, swapping places so that all members participate in the buffet. If the dolphins can successfully force a few schools of anchovies together, they can feed for hours. Common Dolphins (*Delphinus delphis*) in the Gulf of California may be joined in this effort by California Sea Lions and diving birds such as Brown Pelicans. This type of cooperative feeding has been observed

in Atlantic Spotted Dolphins (*Stenella frontalis*), Fraser's Dolphins (*Lagenodelphis hosei*), and Pacific White-sided Dolphins (*Lagenorhynchus obliquidens*), among others.

Long-snouted Spinner Dolphins (*Stenella longirostris*) off Hawai'i feed at night on the animals in the deep scattering layer (DSL), so named because this large aggregation of marine life shows up on depth finders. This nearly worldwide assemblage of fish and invertebrates takes part in the largest migration in the animal kingdom, from the darkness at 755–2,300 ft (230–700 m) deep up to the surface each night to feed under the relative protection of the night. But with echolocation, there is no need to have light to "see" them, and dolphins take advantage of their superior prey-finding abilities. The diving patterns of both Dusky Dolphins and Common Dolphins also correlate with the movements of the DSL.

Dolphins and porpoises living in coastal and river habitats tend to be smaller and slower moving, speed not being as critical to prey capture, although the well-studied Bottlenose Dolphins are an exception as they inhabit both coastal and open waters and are big *and* fast. In these turbid waters, echolocation can be critical, but some dolphins instead focus mainly on noisy fish and listen for them, avoiding giving themselves away with their own clicks. Likewise, they deploy specific strategies

BOX 8.4 BREAKFAST ON THE BEACH

Toothed whales as a group are intelligent animals, so you would expect them to have some tricks for catching prey. Below we will mention several hunting strategies deployed by Orcas, for example. But we have a neat example of dolphin intelligence right in our home state of South Carolina. A few inshore populations here are known to *strand feed*.

In a few salt marshes in South Carolina and Georgia, certain populations of Bottlenose Dolphins routinely wait until two to three hours on either side of low tide and, in a coordinated group, line up and rush the muddy creek bank. Their bow wave pushes fish onto the bank, and the dolphins, always lying on their right sides and revealing to observers how large they are, slide up after the fish and catch them as they frantically flop on the bank to get back to the water (see photo below). With the dolphins right next to each other, if one

Bottlenose Dolphins strand feeding in a South Carolina estuary. Dolphins take advantage of confined shallow water by using their bow wave as they swim to wash fish up onto shore. They then slide onto the "pluff mud," always on their right side, to capture and consume the stranded fish. Rob Young/Adam Fox; photo taken under NOAA NMFS Permit No. 976–1816

in specific populations. In the seagrass beds of Florida, dolphins use *fish whacking*, where they smack fish with their tail flukes, stunning them into submission. In *kerplunking* they flap their tails to produce a wall of bubbles or stirred-up mud that disorients fish. In the Bahama Banks, dolphins have been seen digging into the sandy substrate after buried fish.[34] In our backyard in South Carolina, coastal Bottlenose Dolphins engage in a unique feeding strategy called *strand feeding* (see box 8.4).

All odontocetes are active predators, but not for nothing is the biggest of them named the Killer Whale. Orcas are clearly the *most* apex, if you will permit us to use that superlative, of all marine

misses a fish squirming in the mud, odds are the unlucky fish gets picked off by the next one. The dolphins then wriggle themselves off the bank and back into deeper water to regroup for the next round.

In an interesting example of why apex predators matter, it turns out that wading birds, in particular Great Egrets (see photo below), are attracted to these events. They assemble on the bank above the maneuvering dolphins and gobble up the stranded fish that the dolphins do not reach. The birds that are attracted to strand-feeding events obtain more food than other wading birds in the marsh; they forage more efficiently. Remember that energy efficiency matters greatly in natural selection, and based on the energy requirements of these egrets, it seems they get all they need by taking advantage of the behavior of the dolphins.[35] A win-win proposition, except for the fish.

Great Egret fishing alongside a strand-feeding dolphin. Shorebirds often follow strand-feeding dolphins and pick up fish they miss. This has proven to be a successful foraging strategy for these clever birds.

Rob Young/Adam Fox; photo taken under NOAA NMFS Permit No. 976–1816

predators. They eat sea turtles, penguins, fish including sharks and rays, and marine mammals including pinnipeds, otters, and other whales. Their prey species include the largest marine predators, such as Sperm Whales, Blue Whales, and White Sharks. They deploy various techniques for these. Finding fish involves basic echolocation, herding, and grabbing, much as with other odontocetes. They also have some tricks. For example, if you flip a shark or ray upside down, it relaxes and stops moving, a phenomenon called *tonic immobility*. Clever Orcas off New Zealand use this response to their advantage. Swimming upside down, they approach a stingray. Avoiding the venomous barb at the base of the tail, they grab the stingray and flip themselves over, thereby spinning the stingray upside down and rendering it immobile, after which the Orca eats it![36] Turnabout is fair play, though, and sometimes an Orca is found dead, suspected of being stabbed by a stingray barb.[37] The predator-prey evolutionary arms race continues.

Off the North Pacific coast of North America, Orcas exist in distinct populations with different cultures of prey preference and strategies.

FIGURE 8.19 An Orca (*Orcinus orca*) hunting sea lion pups in the shallow waters off Patagonia, Argentina. Sea lions are less agile on land than they are in the water. Rushing to capture them where ocean meets land is an effective method of attack. Foto 4440 / Shutterstock

Resident pods focus on aggregations of salmon that migrate toward their spawning rivers in the summer. *Transient* pods move from place to place here year-round, specializing on marine mammal prey, in particular Harbor Seals. Transient pods make less noise than their salmon-hunting cousins, instead deploying a quiet ambush approach to catching seals or other, larger prey that passes by. This same pattern of resident Orcas eating salmon and transients eating mammals is also found in Prince William Sound in Alaska.[38]

At high latitudes, Orcas have been seen working in a coordinated fashion to create waves to wash several species off ice floes, including Crabeater, Leopard, and Weddell Seals (*Leptonychotes weddellii*), plus Adelie Penguins.[39] They also work together to harass and eventually kill other types of whales, usually juveniles,[40] such as Gray Whale (*Eschrichtius robustus*) or Humpback Whale (*Megaptera novaeangliae*) calves. You have perhaps seen videos of Orcas riding waves ashore in Argentina, nearly beaching themselves to snatch young sea lions and elephant seals venturing into the water from the rookery onto the beach (fig. 8.19). Pinniped horror films must feature such clips. Only certain whales in a group participate in this surfing saga, and then they share what they catch with the rest of the pod. Juveniles watch and learn the hunting strategies of their pod—in effect, their culture.[41]

HOW ARE MARINE MAMMALS DOING?

Marine mammals are charismatic predators that do not pose much risk to humans. They seem uninterested in killing and eating us. They do, however, compete with us, as is the case with California Sea Lions eating salmon along the northwest coast of the United States, plucking

them up as they gather to make their way through fish ladders around hydroelectric dams, or even jumping into boats to help themselves to the catch. South American Sea Lions (*Otaria flavescens*), Antarctic Fur Seals, and Orcas steal Patagonian Toothfish from commercial fishers' longlines near South Georgia.[42] Sometimes the "helpful" river dolphins in the Amazon that we mentioned earlier eat more than their share and hopelessly tangle fishing gear in the process, and Bottlenose Dolphins do the same thing in several artisanal fisheries in the Mediterranean Sea.[43] Marine mammals also compete with us indirectly. For example, cetaceans account for 36% of all fish predation off the northeastern US coast, and 20% of all primary production in the Gulf of Maine goes to supporting the cetacean food chain. These mammals may not always be eating fish we would like to eat too, but they are gobbling up a lot of the food for those fish.

We do not, mostly, hunt these animals ourselves, for food or trophy, although we certainly did do that for a while, quite successfully. The Marine Mammal Protection Act in the United States has been around since 1972, and the United States works collaboratively with other countries to share this successful model of marine mammal protection.[44] If a population or species is threatened with extinction, it is theoretically protected under the auspices of the Endangered Species Act and/or CITES. In 1986 the International Whaling Commission banned nearly all whaling. In 1994 the Southern Ocean Whale Sanctuary was established. So why are we still worrying about marine mammal conservation, since these national laws and international agreements must protect them?

Well, many are indeed thriving, especially comparatively speaking. Prior to the end of commercial whaling in the United States, the great whale populations were heavily depleted. The whales targeted by whalers were the large

FIGURE 8.21 (*Left*) A dead Minke Whale (*Balaenoptera acutorostrata*) entangled in a net washed ashore in Maine, USA. (*Right*) A seal entangled in a net and debris still managing to survive (for the time being).
(*Left*), DejaVuDesigns; (*right*), Ian Dyball / Shutterstock

us) if these harmful chemicals were not still in our environment, but while we have banned manufacture of some of these products, they linger a long time (and are thus also known as *persistent organic pollutants*, or *POPs*), a problem for decades, even centuries, to come.

Trash or marine debris, mostly in the form of plastics, is now ubiquitous in the oceans. Plastic fishing nets and other gear, long lost to its owners, continues to entangle and kill marine mammals and other animals (fig. 8.21). Bits of plastic are increasingly consumed and accumulate up the food web. In spring of 2019, a young Sperm Whale washed up in Sicily and a Cuvier's Beaked Whale (*Ziphius cavirostris*) washed up in the Philippines. The guts of both were stuffed with plastic. The beaked whale was found to have died of dehydration and starvation; it contained 88 lb (40 kg) of plastic bags![54]

Noise pollution is yet another issue of concern, one that has reached the public consciousness as we have become aware of the dangers of loud noises associated with seismic mapping of areas potentially suitable as sources of offshore oil and natural gas. As we have discussed, sound travels fast in the ocean, and with increasing boat traffic, sonar, seismic surveys of the seabed, and industrial noise, it is increasingly dangerously noisy in our oceans these days. If you depend on hearing to find your food and to communicate with your fellow whales and dolphins, this is potentially a problem. Controversy exists about the dangers of noise to marine mammals because effects can be subtle and not all populations appear to be affected, but there is evidence of acoustically induced strandings of beaked whales (necropsies revealed damage to their acoustic organs). Decreased foraging efficiency, reproduction, and group cohesion have been observed in some species. Beluga Whales call more often and at different frequencies, or avoid areas altogether, when boats are around. Orcas, pilot whales, and Bottlenose Dolphins change their sounds in the presence of sonar. Fin Whales (*Balaenoptera physalus*) and Sperm Whales are known to fall silent for extended periods around seismic surveys. Noise is thought to contribute to declines or lack of recovery in some populations of Orcas and Gray Whales. Elephant seals respond to noise by diving more quickly and surfacing more slowly, an evasive behavior.[55]

Marine mammals, like all the rest of us on Earth, are impacted by humans changing the environment. Marine mammals from temperate zones might move poleward to cool off and then wind up competing with the mammals already there, and most marine mammals already live at higher latitudes. The decline in annual sea ice area is taking a toll on pinnipeds, in particular, which depend on this as a nursery ground, mostly safe from predators.[56] These changes also affect the habits of prey. For example, krill populations in the Antarctic appear to be declining, possibly related to the decline in sea ice as well as overfishing. The importance of krill in the Antarctic food web cannot be overstated. Everything from fish to penguins to whales relies on it.[57] Marine mammals depend on us to prevent these dramatic changes from happening or to reduce the trajectory of the more dangerous impacts. Maybe they are not so intelligent after all.

SOME APEX PREDATORS

ORCA
(Orcinus orca)

Orcas (fig. 8.22), also called "wolves of the sea" as we stated earlier, are the undisputed apex predators of the oceans. No predators eat them, and they can, and sometimes do, eat just about

FIGURE 8.22 Orcas (*Orcinus orca*), a.k.a. Killer Whales. These odontocetes are surely apex predators, as they are known to prey on Great White Sharks. Tory Kallman / Shutterstock

any other marine mammal (at least the younger ones, in the case of other whales) as well as fish, cephalopods (octopus and squid), and birds. Orcas have a worldwide distribution, although mostly in cooler waters, have a striking black-and-white coloration, and are the largest of the dolphin family. Males are slightly bigger than females, getting up to 32 ft (9.6 m) and 8–9 tons (7,000–8,000 kg), with a 6 ft (1.8 m) dorsal fin. Orcas normally swim around 6–8 mph (10–13 kph) but can reach 28 mph (45 kph). Dives typically last 1–4 min but have been known to go as long as 21, and they can dive as deep as 3,289 ft (1,000 m). Orcas have a life history like that of humans. Studies of populations along the coast of British Columbia, Canada, indicate that females reach sexual maturity around 12–16 years and produce about five calves over the course of the next 25 years. They then seem to go into something like human menopause, no longer reproducing but remaining active in support of their pod and living up to 80–90 years.

Orcas generally prefer coastal areas, and populations are organized into a society of matrilineal pods. These populations may be resident or transient, meaning they either stay within the same general area or travel widely. Resident and transient pods mostly do not interact; resident pods focus on eating fish, while transients eat whatever they find, often other marine mammals. They communicate with clicks associated with echolocation, whistles, and pulsed calls, with "dialects" specific to each pod.

Because of their wide distribution, Orcas are difficult to study, but while individual populations such as those off Washington State, USA, appear to be declining, in general these whales seem to be stable. As we discussed, they are at high risk from human-produced chemical contamination. They may also be at risk as human fisheries and climate change disrupt the availability of fish, with impacts up the food chain.

BOTTLENOSE DOLPHIN
(*Tursiops truncatus*)

Like Orcas, the other dolphins probably all qualify as apex predators in their subdivisions of marine or riverine worlds, so we have chosen the one we know the most about to represent the marine group: Bottlenose Dolphins (fig. 8.23). There are distinct coastal and offshore populations of these highly intelligent and social animals, weighing 300–1,400 lb (140–635 kg) and measuring 6–13 ft (1.8–4 m). Offshore populations are generally larger than coastal populations. They often travel in groups, or pods, communicating via touch, whistles, and the clicks that they also use to echolocate. They are known to have signature whistles, the equivalent of a human name.

Bottlenose Dolphins eat fish and squid, and inshore populations also eat crabs and shrimp. They use various methods to catch fish. We discussed strand feeding, fish whacking, and kerplunking (see box 8.4). In open water, they herd schools of fish into bait balls, driving the fish ever closer together and toward the surface, and then take turns charging through, chomping as they go. As we saw with the egrets taking advantage of the strand-feeding dolphins, seabirds take advantage of these bait balls, as do other predatory fish. Bottlenose Dolphins (and probably other odontocetes) can use sound to stun fish.

Bottlenose Dolphins are at risk from the environmental contaminants we have already mentioned, but also from oil spills, such as the 2010 Deepwater Horizon spill in the Gulf of Mexico. There have also been die-offs of Bottlenose Dolphins linked to biotoxins from harmful algae blooms along the Atlantic and Gulf of Mexico coasts around Florida and up the East Coast of the United States. One of the main threats to them is entanglement or capture in commercial fishing gear, and Bottlenose Dolphins are hunted in some

FIGURE 8.23 Bottlenose Dolphins (*Tursiops truncatus*) playing in the wake of a boat. Tory Kallman / Shutterstock

places in the world, most famously Japan, for food and for protection of other fisheries. As of 2018, however, worldwide they were of Least Concern according to the IUCN.

AMAZON RIVER DOLPHIN
(Inia geoffrensis)

Amazon River Dolphins or Pink Dolphins (fig. 8.24) are apex predators of the South American rivers they inhabit, and you may not have even been aware there was such a thing as a river dolphin, so we are including them here. These *botos*, as they are called in Spanish, are the largest of the river dolphins, found in the Orinoco and Amazon River basins, and the Araguaia River in Brazil. Males are larger than females and can be 5.6–10 ft (1.7–3 m)

and 350 lb (160 kg). Their dorsal fin is long and low, and they have a highly rounded head and a long, highly sensitive and slender beak, equipped with bristles. Echolocation and touch enable them to find food in the murky river water, and although their eyes are small, they also see well. Their neck vertebrae are unfused, giving them a great deal of flexibility to look around, and their fins can circle independently, allowing them to swim every which way, including backward and upside down. They are usually solitary except for mother and calf pairs, or small pods of three or four individuals, and are generally quieter than other dolphins.

Amazon River Dolphins are the only odontocetes to have teeth akin to your molars, allowing them to chew their food. And they will eat just about any animal: over 40 species of fish, shrimp, crabs,

FIGURE 8.24 Amazon River Dolphin (*Inia geoffrensis*). These are the largest of the riverine dolphins and are sometimes called Pink Dolphins for obvious reasons. COULANGES/Shutterstock

and turtles. In fact, to some extent, Amazon River Dolphins are apex predators of the rain forests too! During the rainy season when the rivers flood, the dolphins are adept at traveling well into the flooded forests and submerged grasslands in search of food.

The IUCN lists Amazon River Dolphins as Endangered, and their population is decreasing. Considering that the similar Chinese River Dolphin recently went extinct, this cautionary listing is not surprising. These animals have a relatively small distribution to start with, and the rain forest on which they depend for clean rivers is one of the most imperiled habitats on Earth. Their rivers are increasingly surrounded by development, energy production activities, and agriculture, all of which diminish water quality. Dams isolate populations and impede water flow. While we saw that the dolphins sometimes assist fishermen, they are also prone to entanglement in gear and may be killed if they are viewed as competition.

SPERM WHALE
(*Physeter macrocephalus*)

Sperm Whales (fig. 8.25) are distributed throughout oceans worldwide. They are both the largest of the toothed mammals on Earth and apex predators of the deep sea. We already discussed their large heads and the source of their name. Males are much larger than females, by about a third, getting up to 60 ft (18 m) and 38–55 tons (35,000–50,000 kg). Males do not reach maximum size until they are about 50 years old, and the whales can live to 70. They can swim 22–28 mph (35–45 kph). They stay in stable social units of 20 or so individuals, with females more likely to stay with the same group than males, who come and go and are more solitary. Female groups generally stay in temperate or tropical waters, while males may venture to higher latitudes.

Sperm Whales make a variety of sounds including groans, whistles, chirps, pings, squeaks, yelps, and wheezes, some of which can be heard many miles away. Their echolocation clicks are the loudest animal-generated sounds in the world, and it is possible they use sound to stun prey like dolphins do. They use this echolocation to hunt primarily Giant Squid, which are found in waters at least 3,300 ft (1,000 m) deep, perhaps consuming 3% of their weight in squid per day. They are known to dive as deep as 1.2 mi (2 km) for nearly 90 minutes, but 20–75 minutes is more common. These whales often carry round scars from the suckers of Giant Squid. These squid can get to 43 ft

FIGURE 8.25 Sperm Whale (*Physeter macrocephalus*) breaching. Sperm Whales are the largest toothed mammal in the world. They were renowned for their aggressive behavior toward sailing vessels from the 1600s to 1700s. This is likely because of the prevalence of whaling ships that hunted these whales for their spermaceti organ. Their proclivity to fight back was the basis for the popular novel *Moby Dick*. wildestanimal/Shutterstock

(13 m) and weigh nearly a ton (1,000 kg). Imagine that battle! These whales will also herd fish into bait balls like dolphins do.

Sperm Whale populations are stable as far as we know, but the IUCN lists these animals as Vulnerable. They are vulnerable to entanglement in fishing gear, collisions with boats, bioaccumulation of human-produced chemicals, consumption of plastic, possibly the increasing sound pollution in the oceans, and decreasing prey availability because of overfishing and shifts related to climate change.

LEOPARD SEAL
(*Hydrurga leptonyx*)

Leopard Seals (fig. 8.26), or Sea Leopards, are apex predators around Antarctica and associated southern islands, adults especially frequenting the outer edge of the pack ice. Only an Orca might eat them. These phocids (true seals) are the largest seal and are very toothy. Females are generally a bit larger than males, reaching up to 11.4 ft (3.5 m) and 1,100 lb (500 kg). They have disproportionately large, almost reptilian heads with long canine teeth, and long front flippers. They eat krill, other seals, penguins, fish, cephalopods, crustaceans, and, in one account, a platypus.[58]

Leopard Seals are solitary animals, rarely hauling out of the water, and gathering only to breed. Out of the water they will growl at whomever comes too close, and their turned-up mouths give them a sort of menacing "smile." They are known to make noise underwater during breeding season, but scientists do not know much about what is going on. It is risky to study these beasts; one scientist is known to have been dragged and drowned, although a National Geographic photographer, Paul Nicklen, also had a female Leopard Seal repeatedly bring him penguins in various states of disrepair, apparently mistaking him for a rather clueless fellow seal in need of assistance.[59] Leopard Seals also earn their nasty reputation from their penchant for playing with their food. They will torment penguins, chasing them around ice floes and preventing them from getting to shore in a game of cat and mouse, often not even bothering to eat them at the end of it all.

As far as is known, Leopard Seal populations are stable. They have never been exploited. They likely risk carrying high levels of human-produced chemicals because of bioaccumulation, but this should be less of a problem in their remote home—although remember that all the oceans are connected. The prospect of krill farming may pose a risk since that can make up 45% of their diet.

FIGURE 8.26 Leopard Seal (*Hydrurga leptonyx*). Leopard Seals are the second largest true seal, and arguably the *most* apex of the pinnipeds. Their diet ranges from tiny krill to penguins and other seals. Only Orcas are known to regularly prey on them. Marcos Amend / Shutterstock

9

SOME OTHER TOP PREDATORS

We have discussed the major groups of top predators (except humans, next up), but there are a few that fall outside those groups, and we would feel remiss to ignore them. They include a couple of representative bony fish, several mustelids (weasel family), and three peculiar mammals that also fall outside the major mammal groups we have discussed: Fossas, Tasmanian Devils, and Spotted Hyenas. Certainly, there are additional top predators worthy of inclusion, but space limits are prohibitive.

BONY FISH

Author Dan is a shark biologist, so sharks got full (and first!) billing in this book, but we are really giving short shrift to bony fish, probably unfairly so. With about 33,000 species, this is the most diverse group of vertebrates, and a few are top predators in their ecosystems. But we could not include them all. So, to mention at least some big, toothy fish that are not sharks, we will introduce you to a marine and a freshwater species as representatives of this big group of predators.

First, how do these bony fish relate to sharks? Sharks and their kin are cartilaginous fish, Chondrichthyes (*chondr* = cartilage; *ichthyes* = fish). The other group of jawed fish, representing 97% of all fish living now, are bony fish, Osteichthyes (*osteo* = bone). Within that group are the *ray-finned fish* or Actinopterygii, and the *lobe-*

finned fish or Sarcopterygii. This latter group now includes only the coelacanths and lungfish, but it used to be much more diverse, and it is ancestral members of this group, with their fleshy fins, that took the first steps onto land. In fact, we are more closely related to lobe-finned fish than lobe-finned fish are to sharks!

The Actinopterygii includes two groups: primitive ray-finned fish, Chondrostei (yes, cartilaginous bony fish, 30 species that include sturgeon and paddlefish), and advanced ray-finned fish, Neopterygii. Ninety-five percent of that latter group are *teleosts*. We will talk about only two of these 33,000 species. Short shrift, indeed.

How are these different from sharks? Instead of cartilaginous skeletal support, teleosts have ossified (rather heavily calcified) bones, although they are not as robust as yours. Why? Because bony fish have water to support them, so they do not need their bones to hold them off the ground against the pull of gravity. The bones are highly calcified but still relatively flexible except for the vertebrae, which support the muscles that propel the fish side to side as it moves forward, or hold it stiff as the *homocercal*, or symmetrical tail (different from a shark's heterocercal tail), whips back and forth. (Our two predators each use one of these approaches to swimming, so we will discuss this further.)

Most have *cycloid* and *ctenoid* scales, both of which are a very thin outer bone layer made of concentric ridges representing growth increments, although they do not often work for accurately determining the age of a fish. The scales overlap like roof shingles over the fish's body. Cycloid scales are smooth edged and ctenoid scales have a fringed edge. These provide protection against parasites, predators, abrasion, and so forth while still allowing flexibility. The skulls of these fish are often compressed laterally—that is, their faces are rather tall and thin, although not always (the Muskellunge we discuss below has a rather flat head). The gills on either side just behind the head are covered by a bony *operculum*, which protects them and provides the capacity to pump water over them if the fish is still. Recall that the external openings of shark gills consist of from five to seven gill slits. That is one easy way to distinguish a bony fish from a shark.

What makes these fish good predators? We will dive into specifics below, but here are some general fish features. The front of their jaws extends outward (make a fish face and note how you position your mouth) while the rear drops, providing good suction as they seize prey (or graze, in some cases, but we are talking predators here). Both upper and lower jaws are toothed. We have heard of fish (especially sharks) being described as "swimming noses," and we talked about the impressive chemoreception of sharks, but the olfactory lobes also make up most of the brain of a teleost, so olfactory acuity is similar in both groups. In addition to being able to smell prey, fish can see well, particularly if they are denizens of generally clear water. As we explained regarding hearing in marine mammals, sound travels well in water, and fish can listen for prey. And fish can feel water pressure via pressure-sensitive cells on their bodies. You can usually see these concentrated in the *lateral line* that extends down the body on either flank, and if you have ever tried to catch a fish in a dip net, you know how well it works. The fish always knows the net is coming and seems to predict where it is going! Our two top predators do not have to worry too much about nets (except, for one, commercial fishing gear), or many predators. They are too big and fast.

ATLANTIC BLUEFIN TUNA
(*Thunnus thynnus*)

Atlantic Bluefin Tuna (fig. 9.1) can grow to more than 10 ft (3 m) long and nearly 2,000 lb (900 kg). They are found throughout the entire North Atlantic Ocean and Mediterranean Sea, from the equator to north of Norway and from the Black Sea to the Gulf of Mexico. This is the widest geographic distribution of any fish (other than a few sharks), and this is the only large pelagic (open-ocean) fish that lives permanently in temperate and subtropical waters. Atlantic Bluefins can reside in temperatures from nearly freezing, 37.4°F (3°C), to 86°F (30°C), maintaining a relatively stable body temperature, an unusual feat for fish, a cold-blooded, or *ectothermic*, group.

We saw the same phenomenon in a small group of sharks that includes the Shortfin Mako, White Shark, and thresher sharks, which have heat exchangers that retain some of the body heat produced during swimming and other activities. And it is easier to stay warm in water if you are big. A bluefin's massiveness makes keeping its body core warmer than its environment much

FIGURE 9.1 Southern Bluefin Tuna (*Thunnus maccoyii*), a close relative of the Atlantic, or Northern Bluefin (*Thunnus thynnus*). Al McGlashan / Shutterstock

easier.[1] Bluefins thus break the rules by which the overwhelming majority of their relatives abide, exhibiting a form of *endothermy*. Endothermy equates with, and perhaps is a prerequisite for, *high performance*, a term that seems to apply to most of the predators in this book. Refer to chapter 2 for a review of the basic principles of endothermy in fish.

Atlantic Bluefin Tuna in the eastern Atlantic reach sexual maturity at about age 4 and can live to at least 35 if we let them. They spawn once annually and in warmer water, in May and June in the Mediterranean Sea and Gulf of Mexico. As is common with marine bony fish (but not sharks!), females release huge numbers of tiny (0.04 in, 1 mm) translucent eggs, correlated to the size and age of the female. A 5-year-old female might release 5 million eggs. A 15- to 20-year-old might release 45 million! These tiny eggs hatch a mere two days following fertilization, after which the defenseless 0.12–0.16 in (3–4 mm) long larval fish are on their own.[2] You see the need for the big numbers, since obviously there is no parental care.

Adult Atlantic Bluefin Tuna feed on schooling fish, including herring, sardines, anchovies, mackerel, sand lance, and Bluefish, as well as squid. Many of these prey items eat at low to intermediate levels on the food pyramid, so you might challenge designating this species as an apex predator and selecting it as one of only two fish for this chapter. We have done so because adult Atlantic Bluefin Tuna have few natural predators (just larger billfish and sharks, as well as toothed whales). Additionally, compared to the terrestrial environment, the marine environment has fewer full-time, true apex predators. For example, in some ecosystems Tiger Sharks eat lower in the food chain than other sharks because the former eat herbivores, thus challenging our perception of what an apex predator is. Would you take us seriously if we omitted Tiger Sharks from this book for that reason? Finally, Atlantic Bluefin Tuna are such magnificent beasts and consummate predators that learning about them is a magical experience. Author Dan was fortunate enough to swim in an offshore pen with schooling Atlantic Bluefin Tuna weighing 400–1,000 lb (180–450 kg) in the Mediterranean Sea near Malta. Ignoring the warning to feed the fish only by dropping herring into the water, he held one out. He vaguely recalls a monster tuna nearly instantaneously snatching it from his hand, and the resulting *whoosh* as the leviathan's transit sent him into a full somersault.

The prey of Atlantic Bluefins are typically fast-moving, schooling animals. How does the tuna catch them? First, note the *countershading* we saw in many sharks, where the fish is light underneath and dark on top, hiding in plain sight against either the lighted sky or the dark depths.

In addition, tuna and their kin, the billfish and Swordfish (*Xiphias gladius*), are the fastest fish in the ocean. If you were to design a fish to maximize hydrodynamic efficiency based on physics, a tuna is what you would devise. Their length-to-width ratio and shape are physically optimal for moving through water. To smooth that shape, tuna fins tuck neatly into grooves along their sides and back, a feat that their shark counterparts, such as the Shortfin Mako, are incapable of. A row of *finlets* runs from the dorsal fin (top) and anal fin (underside) to the base of the tail (fig. 9.2). These act like spoilers on cars, minimizing turbulence from the friction of the water against the fish that would slow the fish down. And they swim with maximum efficiency. That big body contains strong muscles attached to tendons that run through the narrow base of the tail, the *caudal peduncle*. Those muscles are kept warm by trapping instead of losing heat. The *lunate* tail (shaped like a crescent moon) is tall and stiff and is whipped side to side by these tendons while the body is held relatively straight. The more a fish tail looks like this, the faster a swimmer it is. These fish can travel at up to 40 mph (60 kph)! Only an

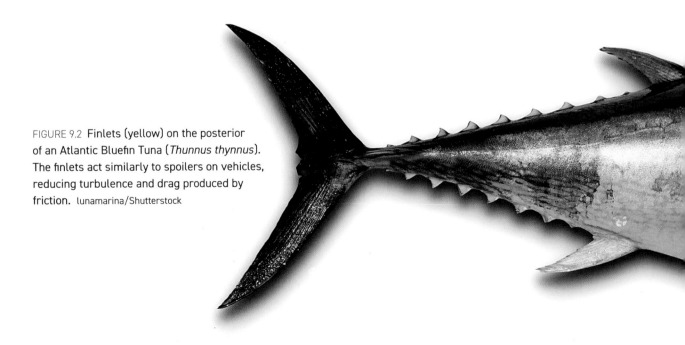

FIGURE 9.2 Finlets (yellow) on the posterior of an Atlantic Bluefin Tuna (*Thunnus thynnus*). The finlets act similarly to spoilers on vehicles, reducing turbulence and drag produced by friction. lunamarina/Shutterstock

Orca and a few big sharks can catch a bluefin at top speed. And alas, so can we.

Bluefin tuna are caught via longlines, traps, and purse seines. Purse-seine nets are deployed in the open ocean over deep water. Those used for tuna are typically 3,300–6,500 ft (1,000–2,000 m) long, about 100–2,100 ft (300–650 m) in diameter, and as much as 650 ft (200 m) in depth. The seine has a float line and a lead line and is positioned to encircle a school of fish, with one or two speedboats pulling it from the larger harvest vessel. Once the float line circle is closed, the opening at the bottom of the net is closed, or pursed, by cinching the lead line and thus preventing fish from escaping, and the catch is hauled aboard the harvest vessel and placed in the ship's hold. Drift netting has been banned, although large-scale illegal operations persist.

Unfortunately for bluefins, they are quite tasty and very much in demand, particularly for high-quality sushi in Japan. In 2019, at the first fish auction of the Toyosu Fish Market in Japan, the so-called "King of Tuna," owner of the Sushi Zanmai restaurant chain, paid more than US$3 million for a 612 lb (278 kg) Pacific Bluefin Tuna (*Thunnus orientalis*). This works out to about $4,900 per pound.[3] This price was largely a marketing ploy, but a quality bluefin tuna commands a high price.

Despite the establishment of the International Commission for the Conservation of Atlantic Tunas in 1967, fishery quotas have been in place only since 1982, and in 2007 a recovery plan was adopted. Atlantic Bluefin Tuna are rated Endangered by the IUCN. Those exorbitant market prices, however, are too much to resist, and thus bluefins are still fished illegally. Suggestions to ban longlining to allow safe spawning in the Gulf of Mexico were derailed by the Deepwater Horizon oil spill in 2010, which happened right in the middle of spawning season there.[4] And so the fish caught are ever smaller and harder to find. The frustrating thing with these marine fish is that with the huge number of eggs they produce, provided they have food and space, their population could make a comeback if we would just let it.

MUSKELLUNGE
(Esox masquinongy)

Muskellunge are a member of the pike family of large, freshwater fish, top predators in the rivers and lakes in the Great Lakes region of North America. Muskies, as they are sometimes called, can live 30 years and can exceed 60 in (1.5 m) and 60 lb (23 kg). Our colleague Dr. Derek Crane, a Muskie expert, tells us that any "records" of Muskies much bigger than that are fish tales! Their native range runs from western Quebec to northwestern Ontario and south to Tennessee and western North Carolina in North America; however, they have been introduced in other temperate waters. Like other top predators, they occur at low densities, for example one adult for every 2–10 ac (0.8–4 ha) is common. These fish are dark colored and can be barred or spotted. These patterns offer good camouflage in the weedy or wooded waters they prefer, although they can also be found in clear water. In one study in Wisconsin, Muskies were found to eat 31 different species of fish, including suckers and Yellow Perch, as well as other predators such as Largemouth Bass, Walleye, and Northern Pike, ranging in size from 6% to 47% of the total length of the Muskie.[5] That's a big meal! How do they catch a dinner like that?

As you can see in figure 9.3, a Muskellunge is quite different in both color and design from a tuna. You immediately deduce that this fish does not chase down its prey like a tuna does but rather is an ambush predator. The first clue is the dark coloring; notice there is no countershading in this species. Not necessary, since nothing but the bottom is under this fish, which looks up

FIGURE 9.3 Muskellunge (*Esox masquinongy*). The Muskie is one of the largest predatory fish in the northern half of North America, often residing in relatively small landlocked freshwater lakes. Engbretson, Eric / US Fish and Wildlife Service

while lurking among vegetation, mud, or rocks. Also notice the streamlined shape of the fish, but with the dorsal and anal fin set back and a rather wide tail with a muscular base, or peduncle. This is an arrangement built for strength, not endurance. The Muskellunge accelerates quickly as it lunges up after prey that it does not chase (and is physiologically incapable of chasing) for long, although it can swim at speeds approaching 30 mph (48 kph) over a short distance.[6] You see a tail like this on salmon, and as you are probably aware, those fish swim from the ocean upriver to spawn, against the current and up waterfalls! There is serious power in a tail like that. Finally, of course, look in that large mouth and you will discover lots of teeth! Once the Muskie has a grip, the unlucky fish in its jaws is there to stay.

Muskellunge spawn when water temperatures are in the 50s F (10s C). A 40 lb (18 kg) female fish might release 200,000 eggs—a lot, but not compared to the millions of a tuna. This phenomenon is typical of freshwater versus marine fish. Freshwater fish release far fewer eggs, but those eggs are bigger. Bigger is better because you are bite sized for fewer creatures with every incremental increase in size. And freshwater fish can sacrifice numbers for size because there are more places to hide in a freshwater habitat than in the open ocean.

Muskies are classified as Least Concern by the IUCN. They were once commercially fished in western Lake Erie but are rare there now, and habitat loss and degradation have led to population declines, particularly in the southern portions of their range. Careful management and stocking maintain populations in some places where they were once native, but other native populations are stable even though they are a hugely popular game fish. Muskies are big and strong and they fight hard. So why are they prospering in areas where they are prized catches? Management and stocking help, but importantly, a firm catch-and-release ethos has been embraced by Muskie fishers since the 1990s. More than 95% of recreationally caught Muskies are now enthusiastically returned to the water by their fishers to allow them to reproduce and grow for the next battle. As a result of capture-release practices, the average size of fish caught and the number longer than 50 in (127 cm) have increased.[7] This is a success story for which conscientious fishing enthusiasts deserve real credit.

MUSTELIDS

The Mustelidae (*mustela* = weasel) is the largest and most diverse family in the order Carnivora (see fig. 6.1), with 59 different species. This family includes weasels, stoats, otters, badgers, ferrets, minks, martens, and Wolverines. All are predators but most are small, about the size of pet ferrets. Least Weasels are the smallest of the Carnivora at about 1 oz (25 g) and 4–10 in (11–26 cm) long (fig. 9.4), and the Carnivora range up to the nearly thousand times bigger Sea Otters, at 20–100 lb (9–46 kg) and 3–5 ft (1–1.5 m) long. Regardless of size, mustelids have long, slender bodies; short legs; short, round ears; and highly odiferous anal glands. They are found on every continent except Australia and Antarctica.

FIGURE 9.4 The Least Weasel (*Mustela nivalis*), the smallest of the Mustelidae. This species hunts primarily mice. With its diminutive size, rounded ears, and small black eyes, it looks much like a mouse itself. Wim Hoek / Shutterstock

Mustelids and procyonids (raccoons are the most familiar of those) diverged from the common ancestor they shared some 29 million years ago (mya). Today the largest mustelid is the Wolverine, discussed below, but as recently as 22.7–18.5 mya, a giant mustelid, *Megalictis ferox*, prowled the central plains of North America. This animal was about the size of a modern Jaguar, 100–150 lb (45–54 kg), but with an even wider head, supporting larger jaw muscles and bone-crushing teeth. Based on the jaw and tooth structure, it is thought to have been a *durophage*, like a hyena, able to quickly consume an entire organism—bones, hooves, teeth, and all.[8] Those saber-tooths we mentioned in the cat chapter were prowling around then too. Not a good time to be prey!

Mammals generally have anal glands, small paired sacs tucked inside the anus that supply a chemical cocktail excreted with feces. As with other mammals, for mustelids, these deposits communicate information on sex, age, reproductive status, and individual and group identity. The compounds are sulfurous and quite smelly in mustelids.[9]

You may not notice these deposits from your dog, for example, unless, like author Sharon, you have a dog who gets so stressed by the attentions of the vet that she releases her glands at every visit. You indeed notice that! This excretion is why your dog sniffs at routine spots on a walk. He is reading the messages left behind. Mammals also have ventral glands in various locations on the body, such that objects can be marked just by rubbing against them. Mustelids are no exception. All this marking communication within and between territories along regular routes of travel is like Facebook for mammals, mustelids included.

All mustelids are carnivores and therefore predators, but only the biggest could be considered top predators: Giant River Otters, Sea Otters, Honey Badgers, and Wolverines.

GIANT RIVER OTTER
(*Pteronura brasiliensis*)

Giant River Otters (fig. 9.5) occur in slow-moving rivers in densely forested areas throughout central South America, mainly in the Brazilian Amazon. They are up to 6 ft (2 m) long but are slender, weighing up to 70 lb (32 kg), making them the second largest of the otters. They have distinct long necks, and flattened tails that enhance swimming speed. Giant River Otters are social animals, living in family groups of 3–20 individuals, generally consisting of a breeding pair of animals and several generations of their offspring.

In addition to communicating by strong scent like other mustelids, Giant River Otters are the noisiest of the group, communicating with almost constant loud chatter, including underwater, and frequent physical contact. They mark their territory along a river, called a *campsite*. Here each group member spends considerable time trampling and tearing out vegetation, scent marking, and rubbing mud in its fur. Little is understood about the messages they are sending, but explorers and scientists note that it is immediately obvious when one approaches such an area because it stinks.[10]

These otters are active in the daylight and sleep in holes along the riverbank at night, also using these as nests for pups. The dominant male and female in the group breed, and they will typically have two to five pups. Unlike with other otters, the father sticks around, and he and the rest of the family group play with, protect, and provide food for the new family members. In the land of Jaguars, caimans, anacondas, and even lone adult male Giant River Otters, such protection is critical. An adult Giant River Otter is safe, but pups are snackable.

In a zoo setting, a mother otter was observed to take pups at about one month of age out into their pool where she would initially swim with a pup in her mouth (fig. 9.6) and then release it, swimming

▲ FIGURE 9.5 A Giant River Otter (*Pteronura brasiliensis*) with a captured fish in the Amazon. Giant River Otters are very social and quite noisy, with nearly two dozen distinct types of vocalizations. A *raft* (group) of these otters can be heard for quite some distance in the Amazon jungle. jo Crebbin / Shutterstock

▼ FIGURE 9.6 A Giant River Otter (*Pteronura brasiliensis*) in Brazil's Pantanal swimming with a young pup in her mouth. Mother otters swim with their pups, often releasing them into the water and then pushing them toward the surface, encouraging the pups to learn to swim. Uwe Bergwitz / Shutterstock

along with it against her chest and pushing it up for air, apparently to introduce it to swimming. After a month of this practice, pups were confidently swimming on their own.[11] Pups begin traveling and hunting with the group as early as 3–4 months and are weaned by 5–6 months. In about 10 months the pups are grown but stay with the family group until age two.

Giant River Otters eat fish, about 10% of their body weight per day. They take advantage of large, 4–16 in (10–40 cm), slow fish species that hunker down in shallow water during the day. These include *trahiras*, or wolf fish, and similarly slow-moving catfish and cichlids. They are also known to sometimes, but rarely, eat small mammals, amphibians, birds, snakes, caimans, and turtles. While they forage as a group, they do not appear to hunt cooperatively. Their large eyes and their own overpowering scent suggest that they probably rely on sight when catching fish, although they probably also use their facial vibrissae when the water is murky.[12] Long, sharp claws, and sharp teeth at the end of the long, flexible neck help them grab and hold on to slippery fish. Otters catch fish by repeated short (10–20 sec) dives. For smaller fish they tread water with their webbed rear feet (fig. 9.7) and feed the fish headfirst into their mouths with their forepaws. When otters catch a big fish, they haul it out onto the bank or a tree branch to eat.[13]

Historically Giant River Otters were hunted for fur, but since 1973 they have been protected by CITES. Despite this, as of 2014, the IUCN rated Giant River Otters as Endangered and decreasing as a result of habitat loss, deterioration of river water quality, canine diseases from domestic animals, and some level of persecution by fishers because of perceived competition with fisheries.[14]

SEA OTTER
(Enhydra lutris)

Sea Otters (fig. 9.8) are the biggest of the mustelids, and only an Orca or large shark will eat an adult, so they are top predators. They are also a keystone

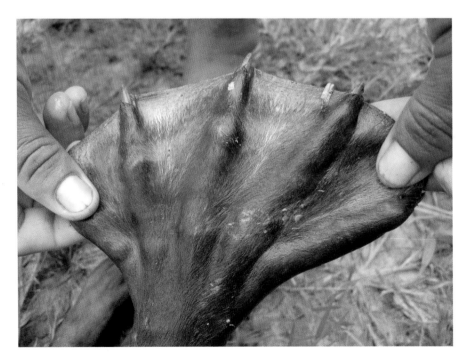

FIGURE 9.7 Webbed paw of Giant River Otter (*Pteronura brasiliensis*). These act like flippers, giving otters substantial thrust and maneuverability through the water.

guentermanaus/Shutterstock

FIGURE 9.10 Mother and baby Sea Otter (*Enhydra lutris*). Unlike Giant River Otters, Sea Otters do not spend much time on land. Since so much of their lives is spent in the water, Sea Otters find interesting ways to do even the most routine activities, like resting or sleeping. This mother is acting as a floating mattress for her offspring. worldswildlifewonders/Shutterstock

Because they eat so much, Sea Otters have been found to have an outsized effect on their habitat: they are *keystone species*. We talked about Wolves serving this role and maintaining terrestrial forest habitats. Sea Otters likewise help maintain kelp forests. Ecologists learned this when Sea Otter populations were depleted by hunting in California and Alaska. The coastal kelp forests also nearly disappeared. It turns out that without Sea Otters to keep the urchin population under control by eating them, the urchins become numerous enough to gobble up the kelp from the base, faster than it can grow. Kelp famously grows 1.5 ft (50 cm) a day to reach 60 ft (18 m) when conditions are right,[19] and all that productivity, along with hiding space, provides habitat for a huge diversity of species including the invertebrates already mentioned, plus numerous bony fish. Seals, sea lions, and even Gray Whales use the kelp forest for a bit of protection and as a place to find food. Without Sea Otters, the whole ecosystem becomes dysfunctional, although wave action, nutrient availability, and climate have also been found to play key roles in the eastern North Pacific.[20]

From the mid-eighteenth century until the early twentieth century, Sea Otter populations were devastated by hunting for their pelts. An estimated worldwide population of 150,000–300,000 was reduced to a paltry 1,000–2,000 individuals. The population off California decreased from 16,000–20,000 individuals to a mere 50. With protection from the Marine Mammal Protection Act and repatriations from robust remnant populations in central Alaska and the Aleutian Islands, California's population is now estimated at about 3,000, and more northern populations have largely recovered to historic numbers, although Sea Otters are still listed as Endangered by the IUCN. The population bottleneck has reduced the genetic diversity of Sea Otters, potentially putting them at risk for issues related to inbreeding.[21] They are also at risk from oil spills, and diseases resulting from toxins and pathogens in runoff from nearby land.[22]

HONEY BADGER
(Mellivora capensis)

If you spend time on social media, the Honey Badger brings one thing to mind: "honey badger don't care." This was a popular meme from the video "The Crazy Nastyass Honey Badger," the most-viewed YouTube video of 2011 (with nearly 100 million views at the time of this writing). Honey Badgers have a reputation for fearlessness (fig. 9.11). In 2014, there was an episode of the Public Broadcasting System

FIGURE 9.11 Honey Badger (*Mellivora capensis*) chasing a Black-backed Jackal. Honey Badgers have a reputation of being aggressive and fearless, which seems to be the case here. Dirk.D.Theron/Shutterstock

Nature series titled "Honey Badgers: Masters of Mayhem."[23] Are these depictions of this animal justified? What is a Honey Badger, really?

Honey Badgers are relatively large mustelids, weighing 13–30 lb (6–14 kg). They are 24–30 in (60–77 cm) long, and slender, with short legs like other mustelids. Their shoulder height is about 10–12 in (25–30 cm). Males are larger than females. They range through sub-Saharan Africa, extending through Arabia, Iran, and western Asia to the Indian Peninsula. They are the only species in the genus *Mellivora*, so they are not sister species to other badgers. Honey Badgers are solitary carnivores, other than mothers with typically a single cub. Like other mustelids, they communicate via scent marking, but unlike their relatives, Honey Badgers have large, reversible anal glands and they will spray a foul-smelling warning scent when threatened, as a skunk does. Their loose-fitting skin gives them flexibility and helps them resist serious damage from bites.[24]

Reproduction seems to happen any time of year. Scent marking helps Honey Badgers find one another to breed, despite their normally solitary existence. Kits are born nearly hairless but get their characteristic black-and-white coat in three to five weeks. For nursing, the mother positions the kit on her belly with its tail near her head and wraps her forearms around it. By 2–3 months, the kit is foraging with its mother and is weaned within this period, but it stays with its mother for 12–16 months. Kits depend on her for food for a year before they are proficient enough at digging, climbing, and hunting to reliably get their own. This long period is comparable to that of other carnivores that hunt, probably because practice is required to become a successful hunter.[25]

Honey Badgers probably rely on their sense of smell to find food. Practice hunting may be especially important for Honey Badgers, given that they will eat a wide variety of prey, some at rather high risk. They consume small rodents, often digging them up after first blocking potential escape holes from the burrows. They similarly dig up spiny lizards. They climb after nests of juvenile raptors and other birds. The dig up or chase down both venomous and nonvenomous snakes, scorpions, insects, and, not surprisingly, honey and bee larvae. Honey Badgers also scavenge when there is opportunity.[26]

Contributing to their reputation for ferocity is that Honey Badgers do not seem to avoid the territories of apex predators in Serengeti National Park: Lions, Leopards, and Spotted Hyenas. Honey Badgers are sparsely distributed, with male territories ranging up to 209 mi^2 (541 km^2) and females 49 mi^2 (126 km^2), but they are found in the same places as these apex predators. While Honey Badgers are sometimes killed by these predators, their aggressive threat display apparently works much of the time.[27] Author Rob witnessed this bravado firsthand while tracking a pride of Lions in the South African bushveld. The Lions came across a Honey Badger in the tall grass. Six two-year-old Lions circled the Honey Badger and started batting their paws at it. The Honey Badger let out a guttural growl and bounced back and forth between the Lions, biting at their feet with no sign of backing down. Bewildered by the aggressive behavior of such a small animal, the young Lions gave up and wandered off to catch up with the rest of the pride. The Honey Badger's success seemed to be a result more of bluster, and the inexperience of the Lions, than of physical dominance.

Honey Badgers even show aggression toward nonpredators far larger. Figure 9.12 shows a Honey Badger in a battle with a large male Oryx (a large antelope) at a watering hole in Namibia. Why exactly a Honey Badger would attack an Oryx is unknown, but when the Honey Badger attacked the Oryx's face, the Oryx flung the Honey Badger through the air like a rag doll. Honey Badgers are

FIGURE 9.12 Honey Badger (*Mellivora capensis*) being tossed by an adult Oryx in Namibia. It seems that these overzealous mustelids sometimes don't know when to quit. Dirk.D.Theron/Shutterstock

FIGURE 9.13 (*Left*) Honey Badger (*Mellivora capensis*). (*Right*) Cheetah (*Acinonyx jubatus*) cub. Because most animals will leave an angry Honey Badger alone, it is believed that the markings and fur of young Cheetah cubs resemble those of a Honey Badger as a form of protection from other predators. (left), Maggy Meyer; right, Lauren Pretorius / Shutterstock

so, shall we say, "respected" in the animal kingdom that Cheetah cubs have long white fur on their backs, making them look something like Honey Badgers (fig. 9.13). This may serve to dissuade potential predators.

Honey Badgers are resistant to some snake venoms. They are known to have survived bites from Puff Adders, Black Mambas, and Cape Cobras. One study looked into the genetics of Honey Badgers compared to venom-susceptible mustelids, as well as domestic and wild pigs, hedgehogs, and mongooses, all of which are also resistant to venom, to find the source of this resistance. A similar genetic sequence was found that accounted for the resistance in the pigs, hedgehogs, and Honey Badgers. This is an example of convergent evolution, where a similar trait arises separately in distantly related organisms in response to a similar selective pressure, in this case snakebites. You can see the advantage of this trait: these animals have access to prey that is simply too risky for anyone else to go after.[28]

Honey Badgers are of Least Concern according to the IUCN Red List but are declining, and since they are solitary and widely dispersed across their habitat, their populations are difficult to assess. Historically, the main conservation concern for Honey Badgers has been their ongoing conflict with beekeepers. Somewhere between 100 and 200 Honey Badgers are killed annually by beekeepers in South Africa, for example. They are also subject to hunting for parts for traditional medicine and for use as protective charms. In 2002, the Wildlife and Environment Society of South Africa launched an effort in support of "badger-friendly" bee products, and sales of these are rising, so currently, in South Africa at least, Honey Badgers are considered a conservation success story.[29] It is not clear what is happening elsewhere.

So, given their penchant for secrecy, fearlessness in the face of apex predators and venomous snakes, an ability to throw their stink at you, and no problems eating just about whatever there is to be had, maybe, in fact, Honey Badgers don't need to care. We hope so.

WOLVERINE
(Gulo gulo)

Wolverines (fig. 9.14) are the largest terrestrial mammal among the mustelids, although they are generally just 24–40 lb (11–18 kg) and around 3 ft (1 m) long, plus a 7–10 in (18–25 cm) bushy tail. They have short legs like the rest of the group, and relatively large, flat feet, so they look somewhat like small bears. In fact, they are sometimes called Skunk Bears. Since you have read about "eau de mustelid" several times now, we bet you know why that is: these, too, can raise a stink. They have a reputation for strength and ferocity and can carry away Moose and Caribou carcasses and rip apart steel traps.[30] This has earned them the monikers Devil Bear and Wood Devil, and their scientific name, Gulo gulo, means "glutton."[31] Wolverines inhabit the northern latitudes of North America, Europe, and Asia, on tundra and in boreal forests where it is snowy and cold. Those big feet make good snowshoes and they have heavy fur.

Wolverines are omnivores, as you would expect in the harsh climate they call home, since it is best to eat whatever is available under those conditions. Their claws are semiretractable and help them dig and climb, and their robust teeth and jaw muscles allow them to tear into frozen flesh and bone. In spring and summer, Wolverines are active predators day and night, eating mainly smaller mammals like rodents, Roe Deer, birds, and hares as well as berries, nuts, and invertebrates. They have also been observed in Alaska chasing down Caribou and are able to kill them after exhausting them with a long pursuit.[32] Much like their cousins the Honey Badgers, Wolverines appear to take

FIGURE 9.14 Wolverine (*Gulo gulo*). Wolverines are the largest of the terrestrial Mustelidae. Like Honey Badgers, Wolverines are known for their ferocity and are typically left alone, even by larger predators. Michal Ninger / Shutterstock

advantage of proximity to apex predators. A study in Norway showed that the restoration of Wolves to boreal habitat there led to an increase in Wolverine populations as well, because the Wolves provided them carcasses to scavenge over the winter. Moose was shown to be the Wolverines' main winter meal in this case, but they were never observed attempting to kill a Moose, scavenging instead. Wolverines seem to avoid direct contact with Wolves, though, sticking to higher terrain and avoiding Wolf trails. They also take advantage of leftovers from kills by lynx and Brown Bears.[33] This steep habitat also allows Wolverines to take advantage of avalanche kills of large mammals.[34] Wolverines are known to cache food year-round among boulders, in snow, or in bogs in their rugged terrain, hiding it from other predators for future use, short and long term.[35]

Wolverines are solitary and widely distributed, with perhaps just five in 400 mi^2 (1,000 km^2).[36] In Scandinavia, for example, one male may maintain a territory over 400 mi^2 (1,000 km^2), averaging 258.3 mi^2 (669 km^2) and including multiple female territories about two-thirds smaller within this, averaging 65.6 mi^2 (170 km^2).[37] Both males and females patrol their territories, marking the boundaries and behaving aggressively if they encounter another Wolverine (fig. 9.15). These inhospitable and expansive territories make Wolverines difficult to study, but we know some things based on tracking animals with radio collars.

Female Wolverines reach sexual maturity at 15 months, but they typically do not successfully raise kits before age three. Beyond that age they den annually but are successful at raising kits on average only half the time and are not likely

FIGURE 9.15 Wolverine (*Gulo gulo*) baring its teeth. Wolverines are known to be both territorial and aggressive and have purportedly even fought with Wolves over a dead carcass. DenisaPro/Shutterstock

to reproduce, or even survive, much past age eight. They typically have two or three kits, raised exclusively by the mother. This is a relatively low reproductive rate for a carnivore, likely reflecting the high energetic cost of reproduction in general, coupled with the unreliable food availability in the cold climate of Wolverine habitats and the high energetic demand of living there.[38] Kits are weaned in 9–10 weeks and reach adult size at seven months but stay with their mother until sexual maturity.[39]

Wolverine kits are born in dens that are tucked between boulders or downed trees or tunneled under snow drifts, either way insulated by at least 3 ft (1 m) of snow for most of the winter. Kits are born in the late fall or early winter, and the female may move them a few times over the course of the

winter. The family abandons the last den with the spring thaw. Wolves and Golden Eagles are known to kill Wolverine kits, but the location of dens in steep, difficult terrain with deep snow helps protect them early on, and the family moves to even higher terrain in the summer, partly to avoid predation but also because there is more small prey—rodents and birds—at higher altitudes.[40] Kits stay with their mother for two years, and then usually the female kits find territory nearby, while males disperse. Genetic studies have shown that males can disperse as far as 300 mi (500 km) in Scandinavia; however, one male traveled over 800 mi (1,300 km) from Montana to North Dakota in the United States.[41]

Historically, Wolverines were hunted for their fur and were nearly extirpated in Scandinavia until they were protected in the 1970s. The population in Sweden is rebounding into its historical range but has lost genetic diversity compared to the North American population. Wolverines are subject to poaching everywhere and are still legally hunted in Norway to minimize sheep and Reindeer predation. The North American population is managed via hunting, but there is protection for female den sites, and mitigation for habitat loss in the United States and Canada.[42]

It has been shown that winter recreation activities by people are quite disruptive to Wolverine routines, and they avoid areas with roads, logging, and helicopter and backcountry skiing.[43] In the winter, when food is scarce and temperatures are cold, they can ill afford the extra energy costs associated with this, particularly given their low reproductive rate.[44] Still, as of 2015, the IUCN listed them as of Least Concern, although declining. They are thought to be stable based on their wide distribution, although that plus their solitary nature makes this hard to quantify. In addition, their apparent dependence on significant levels of snow would seem to put them at risk from climate change.

ADDITIONAL TOP PREDATORS

SPOTTED HYENA
(Crocuta crocuta)

Hyenas get a bad rap. They are often viewed as strictly thieves and scavengers that will eat anything because they are incapable of hunting. In Disney movies they are inevitably portrayed as whiny sycophants or dim-witted thugs, bereft of the regal status of Lions or the threatening countenance of Tigers. They are a bit weird looking, with longer front legs than rear, and a short back, giving them a strange sloped and rather hunched posture. African lore has it that when God created animals,

he used the leftover pieces to make the hyena.

And they do not fit nicely into the classification schemes most people are familiar with. Hyenas look something like dogs and are part of the Carnivora, but they are *feliforms*—that is, catlike (but not cats). In her fascinating book *The Truth about Animals: Stoned Sloths, Lovelorn Hippos, and Other Tales from the Wild Side of Wildlife*, author and conservationist Lucy Cooke describes Spotted Hyenas as "souped-up mongooses."[45] They are the Hyaenidae group in our Carnivora phylogeny (see fig. 6.1), sharing a common ancestor with the mongooses, or Herpestidae, about 28 mya.

There are four extant hyena species: Spotted, Brown (*Hyaena brunnea*), and Striped Hyenas (*Hyaena hyaena*), and Aardwolves (*Proteles cristata*). Aardwolves are solitary termite eaters. We will focus on the biggest, most common, and arguably most predatory of the group, the Spotted Hyenas (fig. 9.16). These are the most common predator in sub-Saharan Africa. Spotted Hyenas are found in savannas, deserts, swamps, woodlands, and montane forests, at densities dependent on prey densities. In a desert this may work out to be 1 hyena/38 mi^2 (100 km^2), while on a prey-rich savanna it may be 3/mi^2 (2.6 km^2), arranged in clans of 4–5 to 90 individuals, averaging around 30. Home ranges are 1–400 mi^2 (2.6–1,036 km^2), averaging about 65 mi^2 (168 km^2), but these are not defended territories.[46] Hyenas essentially forage as far as they need to.

Spotted Hyenas weigh 88–190 lb (40–89 kg) and are 2.5–2.6 ft (~0.8 m) tall. Oddly, for a mammal, females are 10% bigger than males.[47] Perhaps not surprisingly, given the size difference, Spotted Hyenas have matriarchal societies; females rule the clans. There are permanent male and female members, with one or more matrilines of adult females and their offspring. Hyenas may live more than 18 years and are sexually mature at 3.5 years, so a single clan may contain five generations of relatives, in addition to males that have come in from elsewhere. It is rare that the whole clan gets together at once, instead traveling, resting, and foraging in subgroups whose membership continuously changes, something called *frisson-fusion*.

Within the clan, an individual's social rank determines priority access to food, and that rank depends entirely on who your relatives are. If you are the offspring of the alpha female, you are set. If you are her sister or niece or nephew, that is also good. The more removed the relationship, the lower the rank.[48] The males that stay with their natal clan retain their natal ranking, but generally males disperse, and immigrant males are below everybody, queued up in order of their arrival to the group. Except when it comes to mating, because females prefer to mate with immigrant males, such that natal males sire only about 3% of cubs.[49]

Why does this hierarchy matter? Females do not share cub-raising duties, as we saw with Lions, although cubs are raised in a communal den. Hyenas do not bring food back from a kill for the cubs (probably because there is none left, as you'll see!). But relatives cooperate in hunts and in defending food resources, and the highest-ranked individuals, including high-ranking cubs, get to eat first.[50] This is particularly important for hyenas because they are what is known as durophages (they eat absolutely everything from a kill). A clan of hyenas can make an entire zebra completely vanish in half an hour—bones, hair, hooves, and all (fig. 9.17). And they can eat up to a third of their body weight in one meal at a rate of almost 3 lb/min (1.3 kg/min).[51] You cannot come late to this buffet.

This feeding strategy is possible because Spotted Hyenas have, literally, bone-crushing jaws

FIGURE 9.16 Spotted Hyena (*Crocuta crocuta*) looks up into a tree where a Leopard stashed a kill. Spotted Hyenas will certainly steal a kill if possible (it might be easier than making their own kill), but contrary to common belief, they are quite capable hunters and hunt as much, if not more, than certain Lion populations.

FIGURE 9.17 Spotted Hyena (*Crocuta crocuta*) consuming an Impala rib cage. A common feeding strategy for Spotted Hyenas, if they can't consume an entire kill in one sitting, is to hide the remainder of the kill in a *pan* (a deep puddle of collected rainwater). The water hides the scent of the carcass and allows the hyena to return when hungry to finish a well-earned, if waterlogged meal.

BOX 9.1   A HYENA WALKS INTO A BAR

Like our story about the Black Bear in South Carolina, this sounds like the beginning of a joke but does occasionally happen. Author Rob reminisces about leading photographic safaris to South Africa. At one camp, the bar/lounge had a stuffed Pangolin (a scaly anteater) as a decoration sitting on one of the tables. Because Pangolins are incredibly rare and even more difficult to spot, Rob would pose with the stuffed Pangolin for a photo, pretending he found it in the wild. It was a yearly tradition.

Several years ago, Rob returned to the camp for another safari and saw that the Pangolin was no longer on the table. Rob asked the bartender, "What happened to the Pangolin?" The bartender replied that a month before, after guests had moved from the bar to the *boma* (a protective structure to keep out Lions) for dinner, a hyena walked in through the open doors of the bar and stole the Pangolin off the table. Rob was disappointed that his yearly tradition had come to an end, but probably not as disappointed as the hyena was after the first bite.

species, which we discuss below. Sea Otters are closely related to Giant River Otters, having diverged about 10 mya, a short time geologically speaking. Sea Otters are the largest of the otters, with an adult male averaging 64 lb (29 kg) but reaching as much as 100 lb (46 kg), with females averaging 20 lb (9 kg) less. They inhabit the rocky coasts of the northern Pacific Ocean.

While the Giant River Otter spends considerable time on land, the Sea Otter has evolved a few unique features to adapt to the marine environment and spends very little time on land, and none in freshwater. In fact, it is generally considered a marine mammal. These features include webbed hind feet, highly efficient kidneys to deal with the salt levels in the diet, higher volumes of oxygen in the blood and lungs to allow for diving and floatation, dense bones for diving, a high metabolism to generate heat in the chilly North Pacific, and the densest fur of any mammal.[15]

A Sea Otter has up to 1 million hairs/in^2 (100,000/cm^2) on the legs and chest, and up to four times as much on the sides and rump. Even the furry Chinchilla pales in comparison, with just 127,000 hairs/in^2 (50,000 hairs/cm^2). Fine blond human hair, the color that tends to be most dense, has just 480 hairs/in^2 (190 hairs/cm^2).[16] Otters need dense fur because they entirely lack the blubber other marine mammals use for insulation. They groom meticulously to keep that fur in top shape, including sometimes blowing bubbles into it, trapping extra air for insulation.[17] They also have shortened forelimbs with a shape adapted for manipulating tools and food, but also for grooming.[18]

Sea Otters lack the anal glands of the other mustelids and have reduced olfactory capabilities relative to other Carnivora, and like those of seals and sea lions. Sea Otters have advanced tactile perception. This helps them find and handle their food (fig. 9.9).

Like other otters, Sea Otters can be highly social and can be found floating together in what are called *rafts* containing up to 100 females and

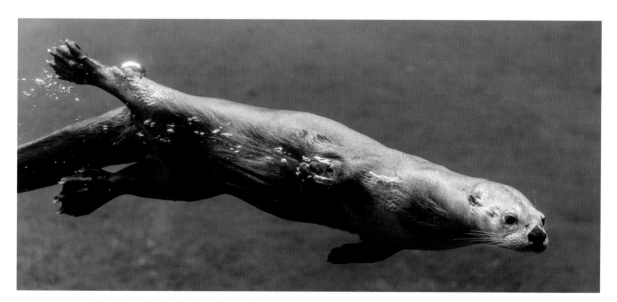

FIGURE 9.8 Sea Otter (*Enhydra lutris*). Sea Otters are the largest of the Mustelidae and a top predator in their environment. Recognized as a *keystone* species, Sea Otters are responsible for keeping sea urchin populations in check, a necessity for healthy kelp forests on which many species depend. Kris Wiktor / Shutterstock

FIGURE 9.9 Sea Otter (*Enhydra lutris*) eating a crab in Morro Bay, California. Increased tactile sensitivity allows Sea Otters to easily manipulate food with their forepaws, an important feature when your food can bite (or pinch) back. Jean-Edouard Rozey / Shutterstock

pups, or as many as 2,000 males. They hold paws or wrap up in kelp to stay in place and, like Giant River Otters, are quite talkative. They come and go from these groups, though, and are normally alone or in mother-offspring pairs.

Females reach sexual maturity at four to five years old. They copulate in the water, and the act is rough enough that females sometimes drown or are severely injured. A single pup is born in the water and is tended to and protected by the mother for four to nine months in California, to over a year in Alaska, during which time they must learn how to catch fish, a considerably bigger challenge than the stationary clams preferred farther south. Mothers are very attentive and often act as rafts for their learning young (fig. 9.10). Pups are vulnerable mainly while the mother dives, and orphaned pups are sometimes adopted by other females, or even males.

Sea Otters eat as many as 100 different types of mostly shellfish, urchins, clams, abalones, and so forth. They typically dive up to four minutes and no more than 130 ft (40 m) down to the bottom and carry the food up to the surface in their armpits. Their hard-shelled prey can be tough to get into, so with a thick-shelled clam or abalone, an otter will also bring a rock up to the surface, roll over on its back with the rock balanced on its chest, and bash the shellfish against the rock until it can access the meat. They consume from 20% to 30% of their body weight daily to produce enough metabolic heat to stay warm in the chilly North Pacific waters where they live, so if you are near a group of otters, you often hear them banging away.

and teeth. Their bite strength is stronger than that of a Lion, more than 1,000 lb (4,448 N), with enlarged premolars and a robust skull and teeth.[52] As a result of this diet (bones, specifically), their feces are bright white.

Spotted Hyenas prey mostly on whatever ungulates (grazers such as Wildebeests, zebras, Impalas, and others) are locally most abundant as different species migrate through, although they also eat hares, birds, and sometimes insects. Much of the time hyenas hunt alone, except when going after larger prey. Then they will team up in groups of as many as 11 animals. High-ranking hyenas tend to hunt in larger groups than low-ranking hyenas. Low-ranking hyenas must hunt more often, presumably because they do not get as much to eat each time, getting pushed out of the way by the higher-ranking individuals.

Adult hyenas are successful about 20%–35% of the time. Juveniles are generally over a year old before they are anywhere close to competent hunters, and prior to about age five they are simply not as proficient as adults. Hunting strategy, alone or together, is to rush at a herd of ungulates and then stop and watch them, apparently looking for weakness. They then pursue the chosen individual, sometimes for as much as several miles. Once they bring down the prey, they consume it immediately, and any hyenas from the clan who are nearby will come and eat, regardless of whether they were involved in the hunt. Any nearby hyenas will also defend a kill from Lions.

This sociality requires a communication system. We talked about territorial scent marking. In addition, hyenas exhibit some of the expressions common in social canines, for example bared teeth and laid-back ears signifying a threat (fig. 9.18). The tail is normally carried straight down, but straight up can indicate either a threat or just excitement. As you might predict, a subordinate hyena will approach a dominant one in a crouched posture.

These animals are also quite noisy. In fact, Spotted Hyenas are also called Laughing Hyenas, an apt description of one of their main sounds, although in his exhaustive study of the species in the 1970s, Hans Kruuk identified the following:

FIGURE 9.18 **Spotted Hyena** (*Crocuta crocuta*) showing its teeth. Hyena snarling is a form of social aggression, similar to that of many canine species. With a relatively domed and heavy skull, strong musculature, and large, sharp carnassial teeth, hyenas can chew through bone that most other predators cannot consume.

whoop, fast whoop, grunt, groan, low giggle, yell, growl, soft and loud grunt-laugh, whine, and soft squeal. Generally, high-pitched sounds are submissive, and loud, low-pitched sounds call the group together to attack or defend or just keep in touch. The whoop sound is one of these; it is somewhat comparable to a Wolf howl, and one of the most common sounds in wild Africa. The giggle and grunt-laugh are associated with excitement. Hyenas yell when getting bitten, and the soft squeal comes from the cubs.[53]

Finally, hyenas have a display that relies on some truly peculiar anatomy. Females are not only larger than males; they resemble males. Females possess what appear to be male external genitalia. Their clitoris is essentially extended into a *pseudopenis* with fused vaginal labia looking much like a scrotum. This appendage can get erect, and male and female hyenas greet each other by standing side by side, head to tail, checking out each other's erect penis and sniffing anal glands. The more erect the penis, the lower the social standing, which makes sense when you remember that the lowest of the low will always be an immigrant male. In fact, male hyenas have oddly low levels of androgen hormones like testosterone, so they do not aggressively compete with one another. They also cannot mate with a female without her full cooperation, so courting is much more important than fighting for male hyenas. And female hyenas do not need to fight either because of the established hierarchy, maintained by peaceful displays. The downside to this is that females lack a vagina, so they both urinate and give birth through the narrow pseudopenis canal. It does not always go well. About 10% of females die giving birth.[54] And you thought just human moms-to-be had it tough.

To complicate matters, cubs are *precocial* compared to the cubs of other big predators, meaning they are born relatively large, well developed, and ready for action. They weigh 2–3.5 lb (1–1.65 kg) at birth (compared to a Lion cub at 0.5 lb, or 0.27 kg, at birth). Their eyes are open and they have a complete row of deciduous ("baby") upper and lower incisors and canines.[55] They are fed exclusively milk for their first 5–8 months and then join the clan for kills. By 12–14 months their adult teeth are fully in and they eat exclusively meat.[56]

Females rarely have more than two cubs because they have only two nipples, and as we said, those cubs are born ready to rumble. High levels of androgens at birth make cubs quite aggressive, and siblicide is common among twins, particularly when both are the same sex. Since this happens whether resources are limited or not, it is thought that a female kills her sister to minimize competition with an equally ranked female for the rest of her days. For males, it helps to be bigger when you disperse to another clan, so maybe the stronger male cub kills his brother to monopolize the mother's resources.[57]

Whether it is one cub or two, the offspring of higher-ranking females get better access to food, grow more quickly, and can be weaned more quickly than low-ranking offspring. High-ranking cubs are more likely to survive to adulthood, and high-ranking adults live longer. So, it is better to be high ranking in this society. Pretty much the way of the world, no matter who you are.

Much popular opinion about hyenas stems from what we have learned from film and television, in particular Disney's *The Lion King*. As you probably know, Lions are depicted as heroes and hyenas as stupid, thieving villains. Observations show that hyenas are in fact quite intelligent, possessing some of the problem-solving intelligence of primates.[58] As for stealing kills, a Lion is just as apt to steal a kill from another predator as a hyena is.

Author Rob visited a hyena den in South Africa with wildlife ranger James Moodie. After making sure the adult members of the clan were not in the vicinity, Rob and James exited the Land Rover, not to interact with the two subadult hyena cubs but to film them and sit in their presence.[59] While obviously a bit shy of humans, the two cubs were very inquisitive, and one approached and tried to sample a bit of Rob's jacket sleeve (fig. 9.19). The hyenas showed no aggression, but rather a keen interest in discovering why two hairless apes had decided to visit their territory. Limiting the visit to 10 minutes to disrupt the den as little as possible, Rob and James returned to the vehicle. Before departing, they deployed a couple of infrared camera traps to monitor nocturnal activity at the den site. Upon retrieving the cameras the following morning, Rob discovered a couple of things. First, some amazing footage of the hyena clan returning to the den site. Second, *never* mount a camera trap at a hyena's face level if you want to get your camera back in one piece.

As of 2015, the IUCN listed Spotted Hyenas as of Least Concern, although the population was decreasing. A study at the time in the Masai Mara National Reserve in the Mara-Serengeti ecosystem found the largest Spotted Hyena clan ever observed, with 113 individuals. While increasing

FIGURE 9.19 Author Rob meets an inquisitive young Spotted Hyena (*Crocuta crocuta*) at a den site in South Africa. Hyenas are quite intelligent and inquisitive, despite their common depictions.

human activity on the edges of the preserve is resulting in increasing predation of livestock by Lions and hyenas, Lions have so far borne the brunt of retaliatory killings, perhaps because, at least initially, Lions were more likely to kill during the day when they could be seen and thus blamed. With the consequent decrease in Lions, there was an increase in hyena cub survival, so the hyena population has grown. This is leading to more livestock predation by them and more retaliatory killing. Even living in a protected park does not ensure population stability.[60]

FOSSA
(Cryptoprocta ferox)

The Fossa (fig. 9.20) is the largest predator on the island of Madagascar. It is the only member of the genus *Cryptoprocta* within the Eupleridae group on our Carnivora family tree, so it is a close relative of mongooses and civets (fig. 6.1). Fossas inhabit forested areas up to 6,500 ft (2,000 m) in altitude and avoid treeless habitats. An adult weighs 7–21 lb (3–9 kg), with males a little bigger than females. They get close to 6 ft (2 m) long, with a 3 ft (1 m) body and a 3 ft tail. They have retractable claws like a cat and flexible ankle joints like the Margay (a small cat; fig. 5.12), but they run on the ground flat footed like a bear, have round ears like a mongoose, and then there is that long monkey tail.[61] You can see why scientists needed to analyze this animal's genome in order to figure out what the heck it was!

Fossas were thought to be nocturnal, but that was just because they are so reclusive and inaccessible that nobody saw them during the day. In fact, they hunt day and night, both in the trees and on the ground. They eat animals as big

FIGURE 9.20 Fossa (*Cryptoprocta ferox*). Fossas are the largest predator of Madagascar and are endemic to this one island. A large part of their diet consists of various species of lemurs that also live only on Madagascar.
Dudarev Mikhail / Shutterstock

as the Diademed Sifaka, a medium-sized lemur weighing 6.5 lb (~3 kg). For comparison, the Ring-Tailed Lemur, commonly seen in zoos, is just shy of 5 lb (2.3 kg), so a Fossa could eat those as well as rodents and reptiles, and in higher mountain zones, insects, and in lowlands, crabs.[62] According to the San Diego Zoo website, they can eat any of the lemurs. Fossas are a quick ambush predator, grabbing prey with forelimbs and claws and dispatching it with a quick bite.

Breeding season is October to December, and other than this the animals are solitary except for a mother with cubs, which stay with her for at least a year or maybe longer, since it can take 20 months before the adult teeth are fully in. Fossas reach sexual maturity at three to four years and have two to four cubs every other year.[63]

As of 2015, the IUCN listed Fossas as Vulnerable and declining, with fewer than 9,000 individuals left on Madagascar. Author Rob spent some time with the Antandroy tribe in southern Madagascar in 2018. He discovered that several of the surrounding forests are considered sacred because the tribal dead are buried there. Hunting or the taking of plants or even firewood from these forests is strictly forbidden. As a result, the Ringtail, Sifaka, and Mouse Lemur populations are flourishing there. So too are the Fossas that rely on these prey animals. While these human traditions are helping Fossas in these particular locations, the same is not true throughout the rest of Madagascar. Fossas are hunted for food and parts for traditional medicine, are subject to predation by feral dogs, and are at risk from habitat loss as the human population on Madagascar expands. Like other predators, they have been persecuted out of suspicion and fear. For example, legend has it that they steal babies from cribs and lick sleeping people into a trance before disemboweling them.[64] Nope. As usual, they have much more to fear from people than people have to fear from them.

TASMANIAN DEVIL
(Sarcophilus harrisii)

Tasmanian Devils (fig. 9.21) are the world's largest carnivorous marsupial, so we would be remiss not to include them here. Marsupial mammals are those that carry their offspring in a pouch. There are a few species of marsupials in South America, and the single Opossum in most of North America, but they are the dominant type of mammal in Australia and the nearby islands, New Zealand and Tasmania. Marsupials give birth after a short gestation (just three weeks for a Tasmanian Devil) to tiny, hairless, blind babies with strong forelimbs that allow them to climb to their mother's pouch. Once there, the baby, called a *joey* regardless of the marsupial species, latches onto a nipple and continues to develop until it can survive out of the pouch. The isolation of marsupials in Australia seems to have been a quirk of who was where when the continents moved apart.

In addition to Devils, there were once Coyote-sized marsupial Thylacines, also variously known as Tasmanian Tigers, Lions, or Wolves. Obviously nobody knew quite what they were. The last known individual went extinct in captivity in 1936.[65] Fossils tell us that Devils were once abundant on mainland Australia but went extinct there about 400 years ago and now live only on Tasmania.

Devils vaguely resemble weird small dogs with tiny, round ears. They stand about 1 ft (30 cm) high at the shoulder and weigh 13–22 lb (6–10 kg). They scavenge but are also effective predators, eating Wallabies, small mammals, birds, and really any other small creatures. They are durophages like hyenas, with strong teeth and jaws allowing them to devour a carcass, bones, fur, and all. They have a threatening gaping display that shows off their teeth, and they famously make fierce noises, but generally these are only for competition for food among their fellow Devils. The animals are

FIGURE 9.21 Tasmanian Devil (*Sarcophilus harrisii*). Devils are the largest extant predatory marsupial, although larger marsupial predators (the Thylacine, or Tasmanian Tiger, for example) have only recently become extinct.

secretive rather than aggressive, and generally solitary except when feeding.[66] Young Devils can climb trees.

A female Devil usually breeds by age two but continues only to about age five. Females give birth to multiple young at a time, but with only four nipples, no more than four survive. The survivors live exclusively in the pouch for about four months. They are weaned at five to six months. Devils in the wild live five to six years if they don't succumb to Devil facial tumor disease, discussed below.[67]

Tasmanian Devils historically have suffered a population decline because of climate, even before their isolation from mainland Australia.[68] This population decline resulted in a loss of genetic diversity early on for the Devils. In addition, in the last two decades, the Tasmanian Devil has fallen victim to a contagious facial cancer, Devil facial tumor disease, transmitted through biting, which happens frequently during both feeding and mating. Possibly Devils are so susceptible to this in part because of that lack of genetic diversity, and it has resulted in about an 80% population decline and even further loss of genetic diversity.[69] As a result, Devils are listed as Endangered. There is some hope, though.

First, in 2005 an "insurance population" of Tasmanian Devils was assembled. Thirty healthy juveniles were collected from the wild, and this group has now grown to 10 times that many, held in captivity throughout Australia, Tasmania, and Maria Island, off Tasmania.[70] In addition, genetic studies have shown changes in Devil gene sequences that are generally associated with immune function and cancer risk in humans, suggesting that Devils may be evolving some level of resistance to the disease.[71] It remains to be seen whether this will happen in time to allow recovery of the wild population, but if not, the captive population could be used for reintroduction.

10
HUMANS

Why are we including humans in a book about top predators? You might be tempted to believe that humans are clearly the top predators on the planet, but perhaps not. We do not possess the physical characteristics of an ultimate predator: no claws, no fangs, no talons. Without tools, we would be hard pressed to take down a large prey animal. And no predators treat their food as cruelly as humans treat cows, pigs, chickens, and other animals in our industrialized animal factory system (fig. 10.1). Plus there are more of us than our planet can sustainably support. Normally there are far fewer predators than their prey.

Humans are not at the top of the food chain, at least as nature existed until a few thousand years ago. For most of human history, we have been prey for many of the true queens and kings of beasts. But we have been predators as well. We have developed technology that allows us to exploit the natural world, including its top predators. Humans are unique among species for extreme tool use. This ability, combined with overpopulation, has an impact on every environment on this planet. At this point, there is likely no place on Earth that has escaped a human "fingerprint" of some sort. This chapter explores our role in nature, our environmental impact, and some things we could do to ameliorate our negative impact.

HUMANS AND THE ENVIRONMENT

This chapter is about you. You know what you look like, so we have not included a photo of your species in its natural environment. If we had, we might have been so cynical as to feature a human watching TV, shopping, or driving. Or we could have highlighted humans living in squalor, since perhaps a billion or more live that way (fig. 10.2). Humans in nature, sadly, are mostly anecdotal. As we write, you are one of about 8 billion *Homo sapiens*, a number projected to grow to nearly 10 billion by 2050. And we humans do not have a good track record for sharing the planet with our fellow creatures, especially our fellow predators.

As early humans moved into new places around the planet, a spate of megafauna (big animal)

FIGURE 10.1 (A) Pigs and (B) chickens in animal factories. Do you think that animals such as these in large-scale, industrialized concentrated animal feeding operations, or CAFOs, are treated inhumanely?

(A), Mark Agnor; (B), Guitar photographer / Shutterstock

FIGURE 10.2 The Rocinha favela, Rio de Janeiro, Brazil. This neighborhood has become an urbanized slum, in most cases lacking sanitation and water. Rio de Janeiro has one of the world's highest human population densities.

Jefferson Bernardes / Shutterstock

extinctions typically followed them. Humans arrived in Australia around 65,000 years ago. About 9,000 years later, Australian megafauna, including Horned Turtles nearly the size of a Volkswagen Beetle, a flightless bird twice the height of a human, marsupial herbivores weighing about 4,400 lb (2,000 kg), and an 18 ft (5.5 m) long monitor lizard were driven to extinction. We are so sorry we missed those!

Humans moved into North America around 14,000 years ago. The continent then was populated with saber-tooths, camels, Giant Ground Sloths, Mastodons, and mammoths. By 10,000–12,000 years ago, 73% of the genera (remember, *genus* is the taxonomic category just above *species*) of big mammals were gone.

Extinctions on islands follow the same pattern worldwide. On islands in the Mediterranean and the Arctic Ocean in Russia, and on New Zealand, megafauna mostly disappeared a few thousand years after the arrival of humans. On Madagascar, we lost giant lemurs and a 9 ft (3 m) tall elephant bird! Even now, Hawai'i, the Galapagos, and the islands of Indonesia have high rates of extinction. Correlation, of course, is not causation. Climate was changing naturally throughout these times, and we may have introduced disease rather than purposely killed these creatures, but various studies lay blame at our feet.[1]

Since the eighteenth century in North America, we have indeed directly killed off the Dodo, Great Auk, Steller's Sea Cow, and a population of possibly two to three *billion* Passenger Pigeons. We very nearly did in all several million Bison. The Caribbean Monk Seal was gone in the 1950s. The Baiji, or Chinese River Dolphin, was declared extinct in 2006 (although the IUCN waits 25 years after the last sighting to truly give up hope). The last Vaquita porpoise may well be gone as you read this. No question about what happened in these cases.

And what is particularly pernicious about humans as predators is that we do not tend to take the weak and sick animals, leaving the strong to enhance the gene pool. At least not since we invented guns, poison, and suburbanization (see below). No, we take the strong. No safari hunter wants the head of a Lion cub mounted on the wall. A Gazelle calf? An old Cheetah with broken teeth? No. Hunters want a big, healthy, impressive adult, the very same animals that are needed to keep the population robust.

Even if we do not kill animals directly, we take up space and resources. In some ways this inadvertently helps predators. By shifting from hunting and gathering to farming, we laid out a buffet of grazers for predators. Our cities attract rodents and pigeons, easy picking for raptors and Coyotes brave enough to settle around us. Our roads provide predators with passage and access to prey. And even our best intentions can backfire. Fish ladders to allow salmon around hydroelectric dams? Sea lions descend to take advantage of fish trapped in the ensuing traffic jam. Tunnels under roadways? Dingoes learn to wait on one side for Wallabies to come through. Predators become our competitors. Often we do not really intend to kill predators, but there are so many of us now, we have not left them space or resources.

Because humans must eat, we require land for agriculture, and much of that land is irrigated, taking up and/or contaminating water that might otherwise be flowing downstream. In most instances the land is also fertilized, usually with industrially produced fertilizers, sometimes with animal waste. Fertilizers contribute to nutrient pollution. How can one have too many nutrients? This is in fact one of the most pervasive forms of water pollution (fig. 10.3). Excess fertilizer runs off from farmlands and turfgrass into waterways and ultimately to the ocean. An additional pollution load is generated by animal waste during meat production at so-called factory farms (also known as concentrated animal feeding operations, or CAFOs) (fig. 10.1).

So what is the problem? Fertilizers and animal waste promote the growth of tiny phytoplankton like algae. These then die and decompose, in the process depleting the water of oxygen. Sometimes the phytoplankton themselves are toxic, such as those that have sickened and caused die-offs of Bottlenose Dolphins off the East Coast of the United States. And perhaps you have heard of a toxic *red tide*. The problem is made worse by increased ocean temperatures caused by climate change. Globally, nutrient pollution leads to huge expanses of ocean where life cannot survive. These areas are known as dead zones, which may be permanent or seasonal. The number and area are growing. In 2019, the dead zone in the Gulf of Mexico was 6,952 mi^2 (18,000 km^2).

People require transportation. Roads generate polluted runoff (even including toxic dust from tires wearing down!). We cover permeable land with impervious paving that prevents water from soaking into groundwater. Cars on roads take an enormous toll on plants and animals (including humans). Roads divide habitats, which can accelerate species loss. Roads also provide human access to wilderness and forestland, frequently resulting in trampling, hunting, forest fires, and invasion of exotic plant species (by seeds affixing to shoes or tires, for example) (fig. 10.4).

Human society differs from natural ecosystems in that we process resources in a linear fashion, like this: find → produce → use → discard. In nature, resource use is generally cyclic; one

FIGURE 10.3 Graphic depicting effects of commercial fertilizer. Rain and overspray result in runoff of synthetic fertilizer, which often has major downstream consequences, including development of dead zones and accompanying fish kills.

FIGURE 10.4 Aerial view of trucks transporting soy on a road cut through the Amazon rain forest near Novo Progresso, Brazil. Beyond the clear-cut forest, roads create physical boundaries that separate populations of wildlife, preventing migration and potential for isolated populations to breed. PARALAXIS/Shutterstock

creature's waste is another's food. There is no concept known as "waste." A good example of this is the relationship between plants and animals. Plant "waste" (oxygen) is essential for animals, and animal "wastes" (nitrogen and phosphorus compounds, and carbon dioxide) are essential for plants. (Although, as we just saw, not at artificially high levels.) Because humans do not process resources cyclically, the larger and more complex our societies become, the more waste is generated. Handling that waste in a way that does not degrade the environment becomes increasingly challenging. And humans produce chemicals that are not found in nature (dioxins, furans, PCBs, etc.) and for which there is no "natural" means of disposal. We

discussed the effects of PCBs on the top predators among marine mammals, for example.

Wastes may be collected and reused, recycled, buried in landfills, incinerated, exported, or dumped illegally (figs. 10.5 and 10.6). Laws, incentives, and higher costs have increased the amount of some waste recycling since 1980. Many states and federal government agencies have set ambitious goals for further waste reduction. This does not mean the problem has "gone away," though. Even waste that is recycled must be collected, transported, and processed.

Availability of water for domestic, agricultural, and industrial uses is also affected by population growth. Both surface and groundwater sources

FIGURE 10.5 (A) Landfill, or dump. Solid waste and the regular consumption of disposable goods is a major obstacle for sustainability. (B) Outdated televisions and other e-waste in a dumpster in Murrells Inlet, South Carolina, USA, a few days after Thanksgiving sales. (A), vchal/Shutterstock

FIGURE 10.6 Polluted beach. Illegal disposal of trash is likely to have negative long-term environmental impacts, particularly with the introduction of synthetic, nonbiodegradable materials like plastics. These images, taken by author Rob, show pollution on Komodo Island's "Pink Beach," home to the famous Komodo Dragons discussed in chapter 3. Among the trash and debris, he came across a vial of blood!

FIGURE 10.7 Polluted waterway in Ghana. Two-fifths of American waterways are contaminated to the point of being classified as "impaired" by environmental agencies, though few are as polluted as this water body adjacent to the Korle Lagoon at the Agbogbloshie dump in Accra, Ghana.

are becoming endangered by pollution and excessive withdrawals. Mainly because of population increase, global per capita water supplies are more than one-third lower than they were in 1970, and water quality is declining in many areas. Water scarcity has become a major obstacle to sustainable development. By United Nations estimates, two-thirds of humanity will face shortages of clean fresh water by 2025. At least two-fifths of the waterways of the United States are still classified as "impaired"— that is, contaminated by one or more pollutants including sediment, microorganisms, nutrients (nitrogen and phosphorus), and industrial discharges (fig. 10.7).

In addition, less-regulated world trade and a more integrated global economy mean citizens of one country can have an increasing adverse impact on the environment in another. A country can import raw materials such as tropical hardwoods (fig. 10.8) or ores, and it can ship toxins, in the form of solid waste and air and water pollution, across political boundaries, intentionally or not.

Humans also require energy. For the near term, that still includes fossil fuels. These impose a cost on the environment when they are extracted (land degradation, water pollution), refined (air pollution), transported (tanker and pipeline spills), used (air and water pollution), and even disposed of (ash and mine waste). Much, indeed most, of this cost is not reflected in the retail prices we pay for this energy. This represents a substantial subsidy encouraging the wasteful use of fossil fuels. And many countries still directly subsidize fossil fuel use, keeping prices artificially low.

FIGURE 10.8 Hardwood logging site. Exportation of tropical hardwood, like that from this logging operation in Malaysia, is often unregulated and unsustainable. Rich Carey / Shutterstock

Thus, although each additional human being contributes something unique to the planet and may be the source of problem-solving ingenuity, we also place physical demands on the Earth. The more humans, the greater the demand.

The damage our species does to the natural world has not gone unnoticed, at least by the scientific community. In 1992, 1,700 scientists, including most of the living Nobel laureates in the sciences, issued the "World Scientists' Warning to Humanity":

> Human beings and the natural world are on a collision course. Human activities inflict harsh and often irreversible damage on the environment and on critical resources. If not checked, many of our current practices put at serious risk the future that we wish

for human society and the plant and animal kingdoms, and may so alter the living world that it will be unable to sustain life in the manner that we know. Fundamental changes are urgent if we are to avoid the collision our present course will bring about.[2]

Twenty-five years later more than 15,000 scientists from 184 countries revisited the warning in a paper titled "World Scientists' Warning to Humanity: A Second Notice,"[3] assessing whether humans had changed the trajectory of the "collision course" of the original warning. It is a sad statement of the times that you likely already know the answer. Overwhelmingly, we have not only failed to correct the causes of the damage but have largely accelerated them. Scientists are urgently imploring humanity to work toward a sustainable planet for all inhabitants, predators included. What do we need to do?

The textbook *Environmental Science: The Way the World Works* lists five "Principles of Ecosystem Sustainability."[4] These represent a good starting point to assess how humans fit into the natural world. For humans to begin to live sustainably on the planet, as a species we must strive to obey the same ecological laws that other organisms do. Read through the list and keep these ideas in mind as you read the rest of the chapter.

1. Ecosystems use sunlight as their energy source.
2. Ecosystems dispose of wastes and replenish nutrients by recycling all elements.
3. The size of consumer populations is maintained such that overgrazing and other forms of overuse do not occur.
4. Ecosystems show resilience when subject to disturbance.
5. Biodiversity is maintained.

The problems we face are immense. Virtually every indicator of planetary environmental health is in decline, from productivity of the natural world to clean air and water. We are nearing the tipping point for a drastically diminished quality of life, and perhaps unparalleled human suffering and mortality, in a world where perhaps a billion already live in abject poverty.

The planet's most pressing global environmental problems are declines in biodiversity and climate destabilization. You doubtless know about climate destabilization, although you might call it "global warming" or "climate change." You may be less aware of the loss of biodiversity, both the variety of species and their population sizes either in particular ecosystems or on the planet as a whole.

Reliable scientific assessments indicate entire ecosystems will be irrevocably altered,

or even lost. The list of threatened ecosystems is long and growing. As it does with individual species, the IUCN maintains a Red List of Ecosystems that classifies ecosystems as Critically Endangered (CR), Endangered (EN), Vulnerable, Near Threatened, Least Concern, Data Deficient, Not Evaluated, and Collapsed. The Red List of Ecosystems is far from complete and is not expected to conclude its first comprehensive assessment before 2025. Ecosystems classified as CR or EN include the Mesoamerican barrier reef system, Caribbean coral reefs, Alaskan kelp forest (United States), Murray-Darling basin wetlands (Australia), Sydney coastal wetlands (Australia), south karst springs (Australia), Coorong lagoon and Murray River estuary (Australia), mountain fynbos near Cape Town (South Africa), Rhineland raised bogs (Germany), acacia forests in the Senegal River basin (Senegal, Mali, and Mauritania), and the Aral Sea (Kazakhstan and Uzbekistan). The list is certain to expand and will include forests of all types (tropical, temperate, and boreal), ocean ecosystems (coral reefs, salt marshes, mangrove systems, estuaries, kelp forests, seagrass communities, polar seas), grasslands, freshwater wetlands, rivers, shrublands, and deserts.

Some of the numbers from the United States are staggering: total loss of the dry prairies of Florida to cattle pasture and agriculture; 90% loss of ancient (old-growth) forests; 97% of Connecticut's coastline developed.[5] Globally, the situation is no better. In 2018, about 30 million ac (12 million ha) of tropical forest were lost, 9 million of which were old growth.[6] A comprehensive 2019 United Nations report on biodiversity put it like this: "Much of nature has already been lost, and what remains is continuing to decline."[7]

Individual species are faring no better. We are currently experiencing the Earth's sixth mass extinction, akin to what happened to the dinosaurs

66 mya. There is no compelling evidence that humans have a chance of surviving this, at least with any quality of life. Recent extinction rates among organisms range as high as 1,000 times the normal background extinction rate that occurs naturally. All groups of organisms are threatened, especially amphibians and birds, where over 40% of species among both groups are declining in numbers. Predators as a group are also imperiled. As of 2009, ecologists had assessed the health of nearly 48,000 species of animals and plants and concluded that an astonishing 36% were threatened with extinction. That number is now substantially larger. About three-quarters of the world's marine fish stocks are fully exploited or overexploited. Even about 20% of livestock breeds may go extinct.

What specifically is causing these declines? Habitat change related to land use (e.g., alteration and loss of forests, mangroves, coral reefs, etc.), overexploitation (e.g., overfishing, poaching), pollution, invasive species, and of course climate change. Let's consider a few examples of the issues related to these.

Land use refers to alterations of landscapes by and for human activities. Before the arrival of Europeans, most of the eastern United States was

FIGURE 10.9 (A) Housing development. A poster child for suburbanization: Atlanta, Georgia, USA, where farms and forest habitats have been converted into residential subdivisions with miles and miles of McMansion-sized tract homes serviced by highways and power lines. (B) Sod truck delivering turfgrass from a sod farm that will be used to create an "instant lawn." Although sod is clearly better at erosion control than bare dirt, much turf has replaced forest and agricultural land. Most lawns receive inappropriate doses of fertilizer, insecticides, and herbicides and require fossil fuel–powered lawn-care devices.

(A), trekandshoot;
(B), Noel V. Baebler / Shutterstock

hardwood forest. Changes in land use during the ensuing 400 years have been profound. The first major change involved clearing old-growth forest for agriculture, which was substantially complete by 1920. Land-use changes continue, most notably the suburbanization of much of the United States, which was a hallmark of the twentieth century (fig. 10.9A). Accompanying this profound shift in land use has been the conversion of large areas of the United States to turfgrass (fig. 10.9B). We owe to the twentieth century the idea of a "smooth, green carpet"—that is, a lawn—as a necessary adjunct to the perfect home.

Natural ecosystems develop because of adaptations by organisms to climate and geological change. These adaptations generally require centuries or millennia for ecosystems to become fully established. Human land-use changes differ in that they occur quickly. Thus, the clearing of the American West (west of the Allegheny Mountains) began near the end of the eighteenth century and was essentially complete by the early part of the twentieth.

Rapid land-use changes usually impose significant environmental costs. The clearing of eastern old-growth forests increased sedimentation rates into rivers manyfold. The Chesapeake Bay erosion rate, for example, increased over 12 times from 1700 to 1750. This process literally changed the geography of many areas of the Atlantic coast. Deforestation contributes to global climate change in many ways, including increasing carbon dioxide release by soils. Burning vegetation in the process also releases carbon dioxide to the atmosphere. We also lose the service of the forest as a major sink for carbon (a place for uptake and storage).

Most people know that we are rapidly losing tropical forests. A study by the World Resources Institute (WRI) concluded that over half the original planetary forest cover, or about 7.4 *billion* ac

(3 billion ha), has been lost because of human activity. We continue to lose forest at a rate of about approximately 39.5 million ac (16 million ha) a year. For comparison, the total forest area of Canada is about 1 billion ac (418 million ha). During the 2020 coronavirus pandemic, tropical forest loss accelerated, which is ironic since humans invading forests may have contributed to the problem in the first place (box 10.1).

Less than 40% of the planet's remaining forests are the especially valuable frontier forests. Frontier forests have large contiguous areas with limited human influence and can sustain biodiversity without human interference. Other kinds of modern forests, such as palm tree monocultures (fig. 10.10), have undergone so much human alteration that their diversity is much reduced from prehistoric times. Large predators, for example, require extensive tracts of contiguous habitat and cannot survive in fragmented forests. And a monoculture forest lacks the diversity of plants that can support a diversity of animals. Almost half the world's forest has been replaced by agriculture, pasture, or settlement over the past 8,000 years.

According to the Nature Conservancy, "[As of 2005] every second, a slice of rainforest the size of a football field is mowed down. That's more than 56,000 square miles of natural forest lost each year." In Brazil alone between 2000 and 2005, 7.7 million ac (3.1 million ha) of forest were cut down per year. In 2018, about 30 million ac (12.1 million ha) of tropical forest were lost. *We MUST stop this.* But there are just so many of us!

As he developed the theory of evolution by natural selection, Charles Darwin observed that for any species, more individuals are born than will survive to reproduce. If resources are unlimited and environmental conditions are ideal, the number of offspring reaches a maximum. This state of highest reproductive power, known as a population's *biotic potential*, varies widely among different organisms.

FIGURE 10.10 Palm oil plantation. Although visually very lush, palm oil plantations, like many farmed monocultures, support very little biodiversity compared to the land's original, predeveloped state. In Indonesia, native forests are being clear-cut to grow this lucrative crop. In the process, the habitats of Tigers, Orangutans, rhinos, elephants, bears, and many other species are being destroyed. muhd fuad abd Rahim / Shutterstock

For example, we noted that there are normally far fewer predators than prey in an ecosystem.

Several factors collectively known as *environmental resistance* normally prevent populations from reaching their full biotic potential and thus growing explosively. Environmental resistance includes disease, predation, drought, temperature extremes, lack of food, and other adverse physical or chemical conditions. Most of these factors are density dependent; their effects are most pronounced when the population density—the number of individuals in a given area—gets very high. Thus, the interplay between biotic potential and density-dependent environmental resistance keeps a population in balance (see box 10.1 for an example of a density-dependent disease).

An important ecological concept related to this balancing act is *carrying capacity*. Carrying capacity is the maximum number of individuals of a given species that an area's resources can support in the long term without significantly depleting or degrading those resources. For humans, this definition includes (1) not degrading our cultural and social environment and (2) not harming our physical environment in ways that would adversely affect future generations.[8]

Determining carrying capacities for most organisms throughout the animal and plant kingdoms is, at least in theory, a reasonably direct calculation. However, when carrying capacity is used in a human context, the discussion may become contentious and the resultant calculations subject to dispute. Some would argue that the term *human carrying capacity* is meaningless because human ingenuity and technological advances will continue to keep pace with population increases. Nevertheless, the human carrying capacity of the Earth, or a portion thereof, can be estimated, granted that different populations have different impacts based on their technology, consumption, and ethics, as well as the simple number of individuals.

When estimating global or regional human carrying capacity, scientists study environmental changes as well as the rate at which these changes occur. Some useful indicators of environmental change include the rates of topsoil loss, species loss, water quality degradation, and changes in atmospheric composition. Carrying capacity for humans may ultimately have less to do with population density and more to do with a society's technology, resource demand, and waste.

Another way of examining the impact of humans is to estimate the individual impact each of us has. This is known as our *ecological footprint*. Every individual has an ecological footprint that extends well beyond the geographic area in which that person lives. The Global Footprint Network estimates how much of the Earth's area we appropriate for our "needs." The average American in 2010, for example, used 20 ac (8 ha) to support his or her lifestyle. This includes farmland, forests, mines, dumps, schools, hospitals, roads, playgrounds, malls, and so on. You can find websites that will help you calculate your own carbon footprint. Given that information, you can estimate how many folks living like you could sustainably "fit" on the planet.

Suffice it to say there are more than that many people already. When all factors are considered, many environmental scientists have concluded that the carrying capacity of planet Earth for humans is about two billion. We're closing in on eight billion.

Our very survival as a species could ultimately rely on maintaining the integrity of ecosystems we barely understand. Biodiversity is particularly critical for sustainability because of the specialized and often little understood roles each species plays in maintaining the dynamic state of ecological balance. Moreover, surprisingly little is known about key ecosystems like soils and the deep ocean.

Esthetics and ethics must also play a part since humans can survive, after a fashion, on an Earth with drastically reduced species diversity. The question is, do we wish to eradicate species and ecosystems for future generations without the input of those descendants? That our ancestors did not know any better is no excuse for us to do the same thing.

HOW ARE WE DOING?

Humans had negligible impact on the planet thousands of years ago, as hunter-gatherers. Of course there were far fewer of us then. Human impact was minimized because our nomadic lifestyle was governed largely by Darwinian natural selection. Early humans hunted where they settled. Like other top predators in any ecosystem, they selected the less-fit animals and gathered the "low-hanging" (i.e., easy to pick) fruit and vegetables. When it became progressively more difficult to locate and kill animals—because the ones remaining had superior predator-avoidance adaptations—these nomads packed up and moved on. This pattern allowed ecosystems to assimilate the direct human impact (wastes and landscape changes). Organisms were able to repopulate and recover relatively quickly. In other words, ecosystems had the resiliency to naturally restore what minimal damage human presence caused.

It is an understatement to say that those days are long gone. According to environmental

BOX 10.1 UNINTENDED CONSEQUENCES

Wet market in Tomohon, Indonesia. A wet market sells meat from a variety of wild or nontraditional meat animals. Shoppers can purchase meat from bats, dogs, snakes, turtles, and a wide variety of species that could be the genesis of the next great pandemic.

Sony Herdiana / Shutterstock

In the movie *The Matrix*, Agent Smith, an artificial intelligence computer program, likens humans to a virus. We eat, reproduce, and ultimately make our host (Earth) sick. While this may be a bit of Hollywood hyperbole, the similarities are too close to the truth to ignore. Humans are rapidly converting nature's biodiversity and biomass into *human mass*.

Nature typically restores balance by reducing overpopulated species through emergence of competing species, environmental change, and the spread of disease, allowing only the fittest to survive. Humans have been very successful at using technology to overcome these challenges. We tend to kill any species that we see as a threat or competition. We have been able to survive in even the harshest of environments. We have been able to treat and prevent many of the most lethal diseases that have "plagued" our species.

As your authors are writing this chapter about humans, we find ourselves amid a global pandemic caused by COVID-19. Although the origin of the virus is not yet clear, leading experts believe that it may have originated in wildlife, with the likely suspect being a bat.[9] The first diagnosed case of COVID-19 occurred in Wuhan, China, a city where bats, pangolins, and a large variety of other species are available to purchase for human consumption in "wet markets." We mentioned zoonotic diseases in our discussion of invasive Burmese Pythons. These diseases

philosopher and provocateur Garrett Hardin, the post-hunter-gatherer human approach became "exploit, ruin, and move on." This was based on the mistaken assumption that the Earth was too big and its resources too

abundant for humans to cause irreversible harm. There are now nearly eight billion of us, and virtually no ecosystem—land or sea—has escaped our footprint. But we have no place else to "move on" to.

originating in animals do not generally pass from animals to humans, largely because of physiological differences between species. "Not likely" should not, however, be confused with "not possible."

The rapid evolution of viruses and bacteria can result in a disease spreading from the original host to an unrelated species, sometimes with runaway transmission between its members who have no immunity to the novel disease. Zoonotic diseases can occur in humans after we consume or have extended contact with an infected animal (which is a prime suspect for the origin of COVID-19). Other known zoonotic diseases include bird flu, swine flu, parrot flu, anthrax, mad cow disease, Ebola, cat scratch fever, malaria, bovine tuberculosis, hepatitis E, toxoplasmosis, rabies, brucellosis, and many more. Millions of people die annually from zoonotic diseases. Although it is impossible to prevent the spread of all these diseases, our actions could greatly reduce their transmission. We should at least be careful to not help create a new strain of disease.

As we write this chapter, there are still many unanswered questions about COVID-19. The global impact of this virus is yet to be realized, but we hope that the event brings a greater appreciation of the consequences our actions could have. We would be well advised to tread with a bit more caution.

Fortunately, there is action to undo some of the damage humans have caused on the planet, a transition to sustainable development and ecological restoration. The Society for Ecological Restoration defines ecological restoration as "the process of assisting the recovery of an ecosystem that has been degraded, damaged, or destroyed." Ecological restoration can be conducted on a small scale (like a backyard or a park) or a large scale (e.g., Great Plains, Everglades, San Francisco Bay delta, Great Lakes).

The year 2021 marked the beginning of the UN Decade on Ecosystem Restoration (2021–2030). Its goal, to set "the world on a course to massively scale up the restoration of degraded ecosystems and halt further degradation to fight the climate crisis, enhance food security and water supply, and protect biodiversity," is a noble and virtuous one that will require the support and effort of all of us.

For example, arising out of the growing need to support principles of sustainability, many if not most governments now encourage some sort of alternative to sprawl development. This is sometimes called *smart growth*, which means sustainable development. The components of sustainable development must be applied together. Sustainable development should, among other things, do the following:

- Provide realistic transportation alternatives to the automobile.
- Be powered by renewable energy.
- Plan for, rather than react to, development.
- Create pedestrian- and bike-friendly neighborhoods.
- Encourage mixed-use development within existing communities and use urban growth boundaries to control sprawl.
- Preserve open space to protect critical environmental areas and services.
- Conserve water.
- Encourage a variety of affordable housing choices, all built to green standards.

We could all live with these! But we are not yet.

FIGURE 10.11 Visitors exploring Yellowstone National Park. Taking a break from digital entertainment and going into nature is an important step in remedying our collective disconnect from the natural world. With 61 national parks, 8,565 state parks, and countless county, city, and other public parks in the United States, finding a bit of nature is not that difficult.

FINAL THOUGHTS

A human population that is still growing, and that continues to practice a resource-consumptive lifestyle, is the single biggest roadblock to planetary sustainability. World trade has enormous potential to foster the objectives of development but can also be the source of massive environmental degradation without multinational agreements. It remains to be seen whether the Paris Agreement or some alternative can stem the seemingly inexorable rise of greenhouse gas emissions. Major failures in the United States are top-down governmental intransigence and the lack of action to regulate greenhouse gas emissions. The system is clearly broken and requires a transition to a more sustainable economic model. Do not ask us which one; we are biologists, not economists.

But there are also hopeful signs. Nature is resilient. Human spirit and determination can have a positive rather than negative effect. Effective management of wildlife (by which we mean managing human impact on wildlife) is necessary. We are already seeing signs of recovery for many species, including Gray Wolves, Bald Eagles, American Alligators, and some sharks. We see hope in you, who we suspect are reading this book because of your interest and concern for the fate of predators in general. Nature and the planet have the potential to heal themselves.

Many may put at least some faith in the ability of larger institutions to begin to turn the supertanker of human insults to the planet in general, and to predators in particular. In the interim or even as an alternative, there are many actions you can take.

First, get out into nature and away from the addicting screens of your phone and laptop (fig. 10.11). Richard Louv, in his groundbreaking book *Last Child in the Woods*, asserted that a new disorder has arisen in an entire generation as a result of replacing outdoor experiences with staring at screens: nature-deficit disorder (NDD). NDD leads to a lack of interest in and concern about environmental issues, and addiction to entertainment and electronics. Sound scholarship demonstrates that time spent watching screens leads to shorter attention spans, diminished vocabularies, and impaired cognitive skills; it also reinforces overconsumption, a key environmental problem. Children and young adults can name more corporate logos than the trees and birds they see every day. To value nature, you must first experience it. Seeing a shark swimming or a Lion walking in its environment is a life-transforming experience. It cannot be replicated on a screen.

Other actions are more intuitive. Actively participate in a movement to make the planet more sustainable.[10] If possible, reduce your ecological footprint by eating responsibly, being a more responsible consumer, and conserving energy. Reduce the amount of, or even eliminate, meat and dairy in your diet (fig. 10.12). Buy local and, if possible, organic vegetables or grow your own. Avoid processed food in wasteful packaging. Educate yourself about the social and environmental impacts of what you buy, and purchase less. Use durable goods instead of disposables. Skip the next computer and smartphone upgrades. Use mass transit. Have a smaller family. These choices may seem inconvenient to some, but our choices are largely a function of habit. Better habits result in a better chance of survival for humans and the myriad species we share the planet with.

Our last advice: make conservation biology and education a career, or at least a central part of your life. Remember that humans and predators are both stakeholders in achieving and sustaining a livable, thriving planet.

Gallons of water required to produce:

Beef (1 lb)   = 1799 gal
Cheese (1 lb) = 700 gal
Bread (1 lb)   = 22 gal

FIGURE 10.12 Water requirement for fast-food production. Personal choices, such as diet (A), can have massive environmental (and social) impacts. According to the US EPA, the production of a single 1/3 lb cheeseburger requires approximately 660 gallons of water. (B) Waiting at the drive-through. Little things make a difference: use mass transit, carpool, or walk instead of driving and, when you must drive, avoid idling in long lines.
(A), Noel V Baebler; (B), Michael O'Keene / Shutterstock

NOTES

CHAPTER 1. INTRODUCTION TO TOP PREDATORS

1. Prugh, Laura L., et al. 2009. The rise of the mesopredator. BioSci. 59, no. 9: 779–791.

2. Moore, John A. *Science as a Way of Knowing: The Foundations of Modern Biology* (Cambridge, MA: Harvard University Press, 1993).

3. Bolen, Eric G., and Robinson, William. *Wildlife Ecology and Management*, 4th ed. (New York: Pearson Education, 1999).

4. "How Many People Can Earth Actually Support?" Australian Academy of Science. https://www.science.org.au/curious/earth-environment/how-many-people-can-earth-actually-support, accessed 6/14/20.

5. Paine, R. 1966. Food web complexity and species diversity. Amer. Naturalist 100, no. 910: 65–75.

6. Ripple, W. J., and Beschta, R. L. 2003. Wolf reintroduction, predation risk, and cottonwood recovery in Yellowstone National Park. Forest Ecol. Manag. 184: 2.

7. Henke, S. E., and Bryant, F. C. 1999. Effects of coyote removal on the faunal community in western Texas. J. Wildl. Manag. 63: 1066–1081.

8. Hildebrand, Grant V., et al. 1999. Role of brown bears (*Ursus arctos*) in the flow of marine nitrogen into a terrestrial ecosystem. Oecologia 121, no. 4: 546–550.

9. Ripple, W. J., and Beschta, R. L. 2008. Trophic cascades involving cougar, mule deer, and black oaks in Yosemite National Park. Biol. Cons. 141: 1249–1256; Rabinowitz, Alan. *An Indomitable Beast: The Remarkable Journey of the Jaguar* (Washington, DC: Island Press, 2014).

10. Bondavalli, C., and Ulanowicz, R. E. 1999. Unexpected effects of predators upon their prey: The case of the American alligator. Ecosyst. 2: 49.

11. Daily, Gretchen, et al. 1997. Ecosystem services: Benefits supplied to human societies by natural ecosystems. Issues in Ecol. 2. https://www.esa.org/esa/wp-content/uploads/2013/03/issue2.pdf, accessed 6/10/20.

12. "Hunters and Anglers: Fueling Our Nation's Economy and Paying for Conservation." 2012. National Wildlife Federation. https://www.nwf.org/~/media/PDFs/Water/WOTUS%20Econ%20fact%20sheet%203252014.pdf, accessed 6/10/20.

13. "2016 National Survey of Fishing, Hunting, and Wildlife-Associated Recreation (FHWAR)." US Census Bureau. https://www.census.gov/programs-surveys/fhwar.html, accessed 6/10/20.

14. Lindsay, P. A., et al. 2007. Economic and conservation significance of the trophy hunting industry in sub-Saharan Africa. Biol. Cons. 134: 455–469.

15. "Africa Tourism Report." 2013. World Bank. http://www.worldbank.org/en/region/afr/publication/africa-tourism-report-2013, accessed 6/10/20.

16. Butler, Declan. 2013. Fungus threatens top banana: Fears rise for Latin American industry as devastating disease hits leading variety in Africa and Middle East. Nature 504: 195–196.

17. Roberson, Emily. 2008. "Nature's Pharmacy, Our Treasure Chest." Center for Biological Diversity. https://www.biologicaldiversity.org/publications/papers/Medicinal_Plants_042008_lores.pdf, accessed 6/10/20.

18. Costanza, Robert, et al. 2014. Changes in the global value of ecosystem services. Global Env. Change 26: 151–158.

19. Muir, John. *My First Summer in the Sierra* (Boston: Houghton-Mifflin, 1911).

20. Louv, Richard. *Last Child in the Woods* (Chapel Hill, NC: Algonquin Press, 2005).

21. Wilson, E. O. *Biophilia* (Cambridge, MA: Harvard University Press, 1984).

CHAPTER 2. SHARKS

1. The accepted common name for this species is Sand Tiger, not Sand Tiger Shark.

2. Neff, C., and Hueter, R. 2013. Science, policy, and the public discourse of shark "attack": A proposal for reclassifying human-shark interactions. J. Envir. Stud. and Sciences 3: 65–73.

3. Nosal, A. P., et al. 2016. The effect of background music in shark documentaries on viewers' perceptions of sharks. PloS ONE 11, no. 8: e0159279.

4. "Lesley Rochat: The Shark Warrior." http://www.lesleyrochat.com, accessed 3/21/22.

5. Oeffner, J., and Lauder, G. 2012. The hydrodynamic function of shark skin and two biomimetic applications. J. Exp. Biol. 215: 785–795.

6. "Shark Kills Man during Rescue Attempt by Coastguard in Caribbean." December 22, 2015. *Guardian*. https://www.theguardian.com/environment/2015/dec/22/man-killed-by-shark-in-caribbean-while-being-rescued-by-coast-guard, accessed 3/21/22.

7. We recognize that hand-feeding wild animals is a controversial activity, and there are good reasons to oppose it. We consider this a valid educational activity, and it is done infrequently enough that the likelihood of long-term impact is low.

8. Theisen, B., et al. 1986. Functional morphology of the olfactory organs in the Spiny Dogfish (*Squalus acanthias* L.) and the Small-spotted Catshark (*Scyliorhinus canicula* (L.)). Acta Zool. 67: 73–86.

9. "International Shark Finning Bans and Policies." Animal Welfare Institute. https://awionline.org/content/international-shark-finning-bans-and-policies, accessed 3/21/22.

10. Bangley, C. W., et al. 2018. Increased abundance and nursery habitat use of the bull shark (*Carcharhinus leucas*) in response to a changing environment in a warm-temperate estuary. Sci. Rep. 8: 6018.

11. Brown, C. J., et al. 2019. Life-history traits inform population trends when assessing the conservation status of a declining tiger shark population. Biol. Conserv. 239. https://doi.org/10.1016/j.biocon.2019.108230.

12. Gulak, S.J.B., et al. 2015. Hooking mortality of scalloped hammerhead *Sphyrna lewini* and great hammerhead *Sphyrna mokarran* sharks caught on bottom longlines. Afr. J. Mar. Sci. 37: 267–273.

CHAPTER 3. NONAVIAN REPTILES

1. Pough, F. Harvey, and Janis, Christine M. *Vertebrate Life*, 10th ed. (New York: Oxford University Press, 2019).

2. Hsiang, Allison Y., et al. 2015. BMC Evol. Biol. https://doi.org/10.1186/s12862-015-0358-5.

3. Head, Jason J., et al. 2009. Giant boid snake from the Palaeocene neotropics reveals hotter past equatorial temperatures. Nature 457, no. 5: 715–717.

4. Coleman, Loren. 1979. Alligators-in-the-sewers: A journalistic origin. J. Amer. Folklore 92, no. 365: 335–338; "The Truth about Alligators in the Sewers of New York." February 26, 2020. *New York Times*. https://www.nytimes.com/2020/02/26/nyregion/alligators-sewers-new-york.html, accessed 5/11/20.

5. Harmon, Beau. 2020. "What Are the Top 10 Most Famous Lizards?" Pets on Mom. https://animals.mom.me/top-10-famous-lizards-8985.html, accessed 5/11/20.

6. "Native American Lizard Mythology." 2015. Native Languages of the Americas Website. http://www.native-languages.org/legends-lizard.htm, accessed 5/11/20.

7. "Primate Facts." 2022. New England Primate Conservancy. https://neprimateconservancy.org/primate-facts/, accessed 6/29/22.

8. Hockenbury, Sandra E., and Nolan, Susan A. 2019. *Discovering Psychology*, 8th ed. (New York: Worth, 2019).

9. "Turtles All the Way Down." 2020. Wikipedia. https://en.wikipedia.org/wiki/Turtles_all_the_way_down, accessed 5/7/20.

10. Laufer, Peter. *Dreaming in Turtle* (New York: St. Martin's Press, 2018).

11. The Reptile Database. 2021. http://reptile-database.reptarium.cz/, accessed 5/7/20.

12. Pough, F. Harvey, and Janis, Christine M. *Vertebrate Life*, 10th ed. (New York: Oxford University Press, 2019).

13. Dinets, V. 2015. Play behavior in crocodilians. Anim. Behav. Cog. 2, no. 1: 40–55.

14. Pough, F. Harvey, et al. *Herpetology*, 4th ed. (Sunderland, MA: Sinauer Associates, 2016).

15. Pough, F. Harvey, and Janis, Christine M. *Vertebrate Life*, 10th ed. (New York: Oxford University Press, 2019).

16. Ibid.

17. Sillman, A. J., Ronan, S. J., and Loew, E. R. 1991. Histology and microspectrophotometry of the photoreceptors of a crocodilian, *Alligator mississippiensis*. Proc. Biol. Sci. 243, no. 1306: 93–98.

18. Erickson, Gregory M., et al. 2012. Insights into the ecology and evolutionary success of crocodilians revealed through bite-force and tooth-pressure experimentation. PLoS ONE 7, no. 3: e31781.

19. "Battle at Kruger." 2004. YouTube. https://www.youtube.com/watch?v=LU8DDYz68kM, accessed 6/29/22.

20. Pough, F. Harvey, and Janis, Christine M. *Vertebrate Life*, 10th ed. (New York: Oxford University Press, 2019).

21. Dinets, V. 2015. Play behavior in crocodilians. Anim. Behav. Cog. 2, no. 1: 40–55.

22. "Major Conservation Initiatives." 2020. IUCN Crocodile Specialist Group. http://www.iucncsg.org/pages/Major-Conservation-Initiatives.html, accessed 5/11/20.

23. Pough, F. Harvey, et al. *Herpetology*, 4th ed. (Sunderland, MA: Sinauer Associates, 2016).

24. Davis, Jon R., and DeNardo, Dale F. 2010. Seasonal patterns of body condition, hydration state, and activity of Gila monsters (*Heloderma suspectum*) at a Sonoran Desert site. J. Herp. 44, no. 1: 83–93.

25. Herrel, Anthony, et al. 2009. Aggressive behavior and performance in the tegu lizard *Tupinambis merianae*. Physiol. Biochem. Zool. 82, no. 6: 680–685.

26. Tattersall, Glenn J., et al. 2016. Seasonal reproductive endothermy in tegu lizards. Sci. Adv. 2(1): e1500951.

27. Pernas, Tony, et al. 2012. First observations of nesting by the Argentine black and white tegu, *Tupinambis merianae*, in South Florida. Southeas. Nat. 11, no. 4: 765–770.

28. Manrod, Jennifer D., Hartdegen, Ruston, and Burghardt, Gordon M. 2008. Rapid solving of a problem apparatus

by juvenile black-throated monitor lizards (*Varanus albigularis albigularis*). Anim. Cog. 11: 267–273.

29. Fry, Bryan G., et al. 2009. A central role for venom in predation by *Varanus komodoensis* (Komodo dragon) and the extinct giant *Varanus* (*Megalania*) *priscus*. Proc. Natl. Acad. Sci. U.S. 106, no. 22: 8969–8974.

30. Ibid.

31. Vitt, Laurie J., and Caldwell, Janalee P. *Herpetology*, 3rd ed. (Burlington, MA: Academic Press, 2019).

32. Adriantiono, Tim S., et al. 2018. Effects of human activities on Komodo dragons in Komodo National Park. Biodiv. Conserv. 27: 3329–3347.

33. Pough, F. Harvey, and Janis, Christine M. *Vertebrate Life*, 10th ed. (New York: Oxford University Press, 2019).

34. "Sonic Hedgehog and Beethoven." 2018. StatNews. https://www.statnews.com/2018/07/16/gene-names-oral-history/, accessed 5/11/20.

35. Kvon, Evgeny, et al. 2016. Progressive loss of function in a limb enhancer during snake evolution. Cell 167: 633–642.

36. Pough, F. Harvey, and Janis, Christine M. *Vertebrate Life*, 10th ed. (New York: Oxford University Press, 2019).

37. Ibid.

38. Miller, A. K., et al. 2015. An ambusher's arsenal: Chemical crypsis in the puff adder (*Bitis arietans*). Proc. Roy. Soc. Lon. B. 282. https://doi.org/10.1098/rspb.2015.2182.

39. "Misplaced Fears: Rattlesnakes Are Not as Dangerous as Ladders, Trees, Dogs, or Large TVs." 2020. Los Angeles County Natural History Museum. https://nhm.org/stories/misplaced-fears-rattlesnakes-are-not-dangerous-ladders-trees-dogs-or-large-tvs, accessed 5/11/20.

40. "Deforestation and Forest Degradation." 2020. IUCN. https://www.iucn.org/resources/issues-briefs/deforestation-and-forest-degradation, accessed 5/11/20.

41. McCleery, Robert A., et al. 2015. Marsh rabbit mortalities tie pythons to the precipitous decline of mammals in the Everglades. Proc. Roy. Soc. Lon. B. 282. http://dx.doi.org/10.1098/rspb.2015.0120.

42. Hoyer, Isaiah J., et al. 2017. Mammal decline, linked to invasive Burmese python, shifts host use of vector mosquito towards reservoir hosts of a zoonotic disease. Biol. Let. 13. http://dx.doi.org/10.1098/rsbl.2017.0353.

43. Johnson, Robert D., and Nielson, Cynthia L. 2016. Traumatic amputation of finger from an alligator snapping turtle bite. Wildern. Env. Med. 27: 277–281.

44. Conant, Roger, and Collins, Joseph T. *A Field Guide to Reptiles and Amphibians: Eastern and Central North America* (New York: Houghton-Mifflin, 1978).

45. Laufer, Peter. *Dreaming in Turtle* (New York: St. Martin's Press, 2018).

CHAPTER 4. THE RAPTOR CHAPTER

1. Mindel, David P., et al. Phylogeny, taxonomy, and geographic diversity of diurnal raptors: Falconiformes, Accipitriformes, and Cathartiformes. In *Birds of Prey*, Sarasola, J., et al., eds. (New York: Springer, Cham, 2018), 3–32. https://link.springer.com/chapter/10.1007/978-3-319-73745-4_1, accessed 4/14/20.

2. *Homeotherms* refers to animals that maintain a stable internal temperature, while *endotherms* refers to animals that are themselves the source of heat, more than the environment.

3. Listen to it yourself at "Red-Tailed Hawk." The Cornell Lab. https://www.allaboutbirds.org/guide/Red-tailed_Hawk, accessed 1/11/20.

4. Bildstein, Keith L. *Raptors* (Ithaca, NY: Cornell University Press, 2017).

5. Simmons, R. E. *Harriers* (Oxford: Oxford University Press, 2000).

6. "In Good Years, Snowy Owls Build Nests out of Dead Lemmings." 2014. *The World*. https://www.pri.org/stories/2014-03-23/good-years-snowy-owls-build-nests-out-dead-lemmings, accessed 5/6/20.

7. Parker, J. W. 1999. Raptor attacks on people. J. Raptor Res. 23, no. 1: 63–66.

8. Berger, L. R., and McGraw, W. S. 2007. Further evidence for eagle predation of, and feeding damage on, the Taung child. Soc. Afr. J. Sci. 103: 496–498.

9. Scofield, R. P., and Ashwell, K.W.S. 2009. Rapid somatic expansion causes the brain to lag behind: The case of the brain and behavior of New Zealand's Haast's Eagle (*Harpagornis moorei*). J. Vert. Paleo. 29, no. 3: 637–649.

10. "Eagle Tumbles down Cliff Face Clutching Prey in Hair-Raising Hunt." 2016. Earth Touch News. https://www.earthtouchnews.com/natural-world/predator-vs-prey/eagle-tumbles-down-cliff-face-clutching-prey-in-hair-raising-hunt/, accessed 1/13/20.

11. Penteriani, V., et al. 2006. The importance of visual cues for nocturnal species: Eagle owls signal by badge brightness. Behav. Ecol. https://doi.org/10.1093/beheco/arl060.

12. Tucker, V. A. 2000. The deep fovea, sideways vision and spiral flight paths in raptors. J. Exp. Biol. 203: 3745–3754.

13. Unwin, Mike, and Tipling, David. *Enigma of the Owl* (New Haven, CT: Yale University Press, 2016).

14. Konishi, Masakazu. 1973. How the owl tracks its prey: Experiments with trained barn owls reveal how their acute sense of hearing enables them to catch prey in the dark. Amer. Sci. 61, no. 4: 414–424.

15. Rice, W. R. 1982. Acoustical location of prey by the marsh hawk: Adaptation to concealed prey. Auk 99: 403–413.

16. Bildstein, Keith L. *Raptors* (Ithaca, NY: Cornell University Press, 2017).

17. Stager, K. E. 1964. The role of olfaction in food location by the turkey vulture (*Cathartes aura*). LA County Museum Contrib. Sci. 81: 1–63.

18. Smith, Steven A., and Paselk, Richard A. 1986. Olfactory sensitivity of the turkey vulture (*Cathartes aura*) to three carrion-associated odorants. Auk 103, no. 3: 586–592.

19. Bildstein, Keith L. *Raptors* (Ithaca, NY: Cornell University Press, 2017).

20. Einoder, Luke, and Richardson, Alastair. 2006. An ecomorphological study of the raptorial digital tendon locking mechanism. Ibis 148: 515–525.

21. Unwin, Mike, and Tipling, David. *Enigma of the Owl* (New Haven, CT: Yale University Press, 2016).

22. Ibid.

23. Authors Gilman and Abel think he faked the injury so he would be excused from typing this manuscript!

24. Rüppell, G. 1981. Analyse des Beautefanges des Fischadlers (*Pandion haliaëtus*). J. Ornith. 122, no. 3: 285–305.

25. "Fastest Bird (Diving)." 2005. Guinness World Records. https://www.guinnessworldrecords.com/world-records/70929-fastest-bird-diving, accessed 6/30/22.

26. Ponitz, Benjamin, et al. 2014. Diving-flight aerodynamics of a peregrine falcon (*Falco peregrinus*). PLoS ONE 9, no. 2: e86506.

27. DeCandido, Robert, and Allen, Deborah. 2006. Nocturnal hunting by peregrine falcons at the Empire State Building, New York City. Wilson J. Ornith. 118, no. 1: 53–58.

28. Rafferty, John P., ed. *Meat-Eaters* (Chicago: Britannica Educational Publishing, 2011).

29. McClure, Christopher J., et al. 2018. State of the world's raptors: Distributions, threats, and conservation Recommendations. Biol. Cons. 227: 390–402.

30. "Birds and Climate Change." 2009. Audubon. https://www.audubon.org/news/birds-climate-change, accessed 6/30/22.

31. Ims, Rolf A., and Fuglei, Eva. 2005. Trophic interaction cycles in tundra ecosystems and the impact of climate change. Biosci. 55, no. 4: 311–322.

32. Lehman, Robert N., et al. 2007. The state of the art in raptor electrocution research: A global review. Biol. Cons. 136: 159–174.

33. De Lucas, Manuela, et al. 2008. Collision fatality of raptors in wind farms does not depend on raptor abundance. J. App. Ecol. 45, no. 6: 1695–1703.

34. Krijgsveld, K. L., et al. 2009. Collision risk of birds with modern large wind turbines. BioOne 97, no. 3: 357–366.

35. McClure, J. W., et al. 2018. Automated monitoring for birds in flight: Proof of concept with eagles at a wind power facility. Biol. Cons. 234: 26–33.

36. Hager, Stephen B. 2009. Human-related threats to urban raptors. J. Raptor Res. 43, no. 3: 210–226.

37. McClure, J. W., et al. 2018. Automated monitoring for birds in flight: Proof of concept with eagles at a wind power facility. Biol. Cons. 234: 26–33.

38. Swan, Gerry, et al. 2006. Removing the threat of diclofenac to critically endangered Asian vultures. PLoS Biol. 4, no. 3: e66.

39. McClure, J. W., et al. 2018. Automated monitoring for birds in flight: Proof of concept with eagles at a wind power facility. Biol. Cons. 234: 26–33.

40. O'Bryan, Christopher J., et al. 2018. The contribution of predators and scavengers to human well-being. Nature Ecol. Evol. 2: 229–236.

41. Snyder, Noel F. R., and Meretsky, Vicky J. 2003. California condors and DDE: A re-evaluation. Ibis 45, no. 1: 136–151.

42. Finkelstein, Myra E., et al. 2012. Lead poisoning and the deceptive recovery of the critically endangered California condor. Proc. Nat. Acad. Sci. U.S. 109, no. 28: 11449–11454.

43. "Steller's Sea-Eagle." 2020. San Diego Zoo Wildlife Alliance. https://animals.sandiegozoo.org/animals/stellers-sea-eagle, accessed 4/14/20.

44. "Steller's Sea-Eagle." 2016. IUCN Red List. https://www.iucnredlist.org/species/22695147/93492859, accessed 4/14/20.

45. "Andean Condor." 2018. Birdforum. https://www.birdforum.net/opus/Andean_Condor, accessed 4/14/20.

46. "Gyrfalcon." The Peregrine Fund. https://peregrinefund.org/explore-raptors-species/falcons/gyrfalcon, accessed 4/14/20.

47. "Eurasian Eagle-Owl." 2018. Animalia. http://animalia.bio/eurasian-eagle-owl, accessed 4/14/20.

48. "An Owl Visiting a Football Game." 2008. Dailymotion. https://www.dailymotion.com/video/x273f5, accessed 4/14/20.

CHAPTER 5. CATS

1. Tseng, Z. J., et al., 2014. Himalayan fossils of the oldest known pantherine establish ancient origin of big cats. Proc. Roy. Soc. B. https://royalsocietypublishing.org/doi/full/10.1098/rspb.2013.2686.

2. Li, G., Davis, B., Eizirik, E., and Murphy, W. 2016. Phylogenomic evidence for ancient hybridization in the genomes of living cats (Felidae). Genome Res. 26: 1–11.

3. Tucker, Abigail. 2009. "The Most Ferocious Man-Eating Lions." Smithsonian.com. https://www.smithsonianmag.com/science-nature/the-most-ferocious-man-eating-lions-2577288/, accessed 6/30/22.

4. Tilson, Ronald, and Nyhus, Philip. *Tigers of the World*, 2nd ed. (New York: Academic Press, 2010).

5. "Why Are India's Tigers Killing Humans?" 2014. BBC News. https://www.bbc.com/news/world-asia-india-25755104, accessed 6/30/22.

6. Unless we're talking about Florida Panthers, we're going to call *Puma concolor* "Puma." You might also know it as Cougar, Mountain Lion, Catamount, or Panther.

7. Shankar, Malini. 2008. "ENVIRONMENT-INDIA: Illegal Trade Decimating Wildlife." Inter Press Service News Agency. http://www.ipsnews.net/2008/10/environment-india-illegal-trade-decimating-wildlife/, accessed 6/30/22.

8. Ibid.

9. "2016 Wild Feline Census." 2016. Feline Conservation Foundation. https://www.felineconservation.org/2016-wild-feline-census/, accessed 4/13/20.

10. Wielebnowski, N. 2003. Stress and distress: Evaluating their impact for the well-being of zoo animals. J. Amer. Vet. Med. Assoc. 223, no. 7: 973–977.

11. Szokalski, M. S., et al. 2012. Enrichment for captive tigers (*Panthera tigris*): Current knowledge and future directions. App. Anim. Behav. Sci. 139, no. 1–2: 1–9.

12. Li, G., et al. 2016. Phylogenomic evidence for ancient hybridization in the genomes of living cats (Felidae). Genome Res. 26: 1–11; Zhang, W. Q., and Zhang, M. H. 2013. Complete mitochondrial genomes reveal phylogeny relationship and evolutionary history of the family Felidae. Gen. Molec. Res. 12, no. 3: 3256–3262.

13. Sanderson, James G., and Watson, Patrick. *Small Wild Cats* (Baltimore: Johns Hopkins University Press, 2014).

14. Brehm, Denise. 2010. "The Surprising Physics of Cats' Drinking." MIT News. http://news.mit.edu/2010/cat-lapping-1112, accessed 4/13/20; "Cutta Cutta." 2010. YouTube. https://www.youtube.com/watch?time_continue=4&v=NTCxZWYlWC0, accessed 12/29/17.

15. Eizirik, E., et al. 2003. Molecular genetics and evolution of melanism in the cat family. Curr. Biol. 13, no. 5: 448–453.

16. Fennel, J. G., et al. 2019. Optimizing colour for camouflage and visibility using deep learning: The effects of the environment and the observer's visual system. J. Royal Soc. 16. https://royalsocietypublishing.org/doi/10.1098/rsif.2019.0183, accessed 4/13/20.

17. Xu, X., et al. 2013. The genetic basis of white tigers. Curr. Biol. 23: 1031–1035.

18. "Have You Seen These Images of India's Only Golden Tiger Spotted in Assam." 2020. Indian Express. https://indianexpress.com/article/trending/trending-in-india/look-at-this-beauty-picture-of-golden-tiger-in-assam-goes-viral-6502244/, accessed 6/30/22.

19. Xu, X., et al. 2017. The genetics of tiger pelage color variations. Cell Research 27: 954–957.

20. Luo, S.-J., and Xu, X. 2014. "Save the White Tigers." Scientific American. https://www.scientificamerican.com/article/save-the-white-tigers/, accessed 6/30/22.

21. "Serval vs. Guinea Fowl Smithsonian." YouTube. https://www.youtube.com/watch?v=pK9-jsgbDus, accessed 6/30/22.

22. Turner, Alan, and Anton, Maurico. *The Big Cats and Their Fossil Relatives* (New York: Columbia University Press, 1997).

23. Stoyanov, G. S., et al. 2018. The human vomeronasal (Jacobson's) organ: A short review of current conceptions, with an English translation of Potiquet's original text. Cureus 10(5): e2643.

24. Sanderson, James G., and Watson, Patrick. *Small Wild Cats* (Baltimore: Johns Hopkins University Press, 2014).

25. McLean, C. Y., et al. 2011. Human-specific loss of regulatory DNA and the evolution of human-specific traits. Nature 471: 216–219.

26. Sanderson, James G., and Watson, Patrick. *Small Wild Cats* (Baltimore: Johns Hopkins University Press, 2014).

27. Hunter, Luke. *Wild Cats of the World* (London: Bloomsbury, 2015).

28. Heffner, H. E., and Heffner, R. S. 2007. Hearing ranges of laboratory animals. J. Amer. Assoc. Lab. Anim. Sci. 46, no. 1: 20–22.

29. Leyhausen, P., and Tonkin, B. A. *Cat Behavior* (New York: Garland STPM Press, 1979).

30. Krausman, Paul R., and Morales, Susana. 2005. *Acinonyx jubatus*. BioOne 771: 1–6.

31. Christianson, Per. 2007. Canine morphology in the larger Felidae: Implications for feeding ecology. Biol. J. Linnean Soc. 91: 573–592.

32. Sicuro, F. L., and Oliveira, L.F.B. 2011. Skull morphology and functionality of extant Felidae (Mammalia: Carnivora): A phylogenetic and evolutionary perspective. Zool. J. Linnean Soc. 161: 414–462.

33. Sakamoto, M., et al. 2010. Phylogenetically structured variance in felid bite force: The role of phylogeny in the evolution of biting performance. J. Evol. Biol. 22: 463–478.

34. Sicuro, F. L., and Oliveira, L.F.B. 2011. Skull morphology and functionality of extant Felidae (Mammalia: Carnivora): A phylogenetic and evolutionary perspective. Zool. J. Linnean Soc. 161: 414–462.

35. "IUCN Cat Specialist Group." http://www.catsg.org/index.php?id=61, accessed 4/13/20.

36. "Save China's Tigers." http://www.savechinastigers.org/, accessed 1/6/18.

37. Bale, Rachel. 2016. "The World Is Finally Serious about Tiger Farms." National Geographic. https://www.nationalgeographic.com/animals/article/wildlife-watch-tiger-farms-cites-protections, accessed 6/20/22.

38. Curry, C. J. 2020. "Spatiotemporal Genetic Diversity

of Lions." bioRxiv. https://www.biorxiv.org/content/10.1101/2020.01.07.896431v1.abstract.

39. Hall, Jani. 2018. "Cecil the Lion Died amid Controversy—Here's What Happened Since." National Geographic. https://www.nationalgeographic.com/news/2016/06/cecil-african-lion-anniversary-death-trophy-hunting-zimbabwe/, accessed 6/30/22.

40. Gross, Josh. 2015. "Jaguars in the United States: Part 1." The Jaguar and Its Allies. https://thejaguarandallies.com/2015/10/22/jaguars-in-the-united-states-part-1/, accessed 1/15/18.

41. Davis, Tony. 2013. "Major Jaguar Biologist Opposes Plan." *Arizona Daily Star.* http://tucson.com/news/science/environment/major-jaguar-biologist-opposes-plan/article_9a62c14a-02ab-565f-9eda-b766873f1134.html, accessed 1/3/2017.

42. Wilkinson, Allie. 2015. "In Brazil, Cattle Industry Begins to Fight Deforestation." Science Magazine. http://www.sciencemag.org/news/2015/05/brazil-cattle-industry-begins-help-fight-deforestation, accessed 1/3/18.

43. "Journey of the Jaguar." 2022. Panthera. https://panthera.org/journey-jaguar, accessed 6/30/22.

44. Coniff, S. D. and Winter, S. 2015. Learning to live with leopards. National Geographic, December 2015. https://www.nationalgeographic.com/magazine/2015/12/leopards-moving-to-cities/, accessed 6/5/20.

45. "Livestock Guarding Dogs with Cheetah Outreach." YouTube. https://www.youtube.com/watch?v=c6k5kX0zpmA, accessed 7/1/22.

46. "Common Questions about Mountain Lions." 2018. California Department of Fish and Wildlife. https://wildlife.ca.gov/Conservation/Mammals/Mountain-Lion/FAQ#359951244-why-cant-mountain-lions-be-hunted-in-california, accessed 4/13/20.

47. "Cougar Kills Jogger in Rare Human Attack." 1991. *Journal Times.* https://journaltimes.com/news/national/cougar-kills-jogger-in-rare-human-attack/article_936648f6-fe50-561d-846c-e5096262df6b.html, accessed 4/13/20; Hall, Bennett. 2019. "Oregon Cougar Encounter." *Corvallis Gazette-Times.* https://www.oregonlive.com/pacific-northwest-news/2019/09/oregon-cougar-encounter-renews-debate-about-threats-from-animal.html, accessed 4/13/20.

48. Bansal, Agam, et al. 2018. Selfies: A boon or bane? J. Family Med. Prim. Care 7, no. 4: 828–831. https://www.ncbi.nlm.nih.gov/pmc/articles/PMC6131996/, accessed 4/13/20.

49. "Florida Panther Population." 2017. US Fish and Wildlife Service. https://www.fws.gov/southeast/news/2017/02/florida-panther-population-estimate-updated/, accessed 4/13/20.

50. "Puma." 2015. IUCN Redlist. https://www.iucnredlist.org/species/18868/97216466#population, accessed 4/13/20.

51. "Jaguar (*Panthera onca*) Fact Sheet: Population & Conservation Status." San Diego Zoo Wildlife Alliance Library. https://ielc.libguides.com/sdzg/factsheets/jaguar/population, accessed 6/30/22.

52. "Snow Leopard Facts." Snow Leopard Trust. https://snowleopard.org/snow-leopard-facts/habitat/, accessed 6/30/22.

53. "Leopard (*Panthera pardus*) Fact Sheet: Population & Conservation Status." San Diego Zoo Wildlife Alliance Library. https://ielc.libguides.com/sdzg/factsheets/leopard/population, accessed 6/30/22.

54. Wiley, Hannah. 2018. "Leopard Grabs and Eats Toddler." *USA Today.* https://www.usatoday.com/story/news/world/2018/05/08/leopard-grabs-kills-toddler-uganda-national-park/590022002/, accessed 4/13/20.

CHAPTER 6. CANIDS

1. Bralower, Timothy, and Bice, David. 2020. "Ancient Climate Events: Paleocene Eocene Thermal Maximum." Pennsylvania State University. https://www.e-education.psu.edu/earth103/node/639, accessed 6/30/22.

2. Sole, F., et al. 2014. Dental and tarsal anatomy of *"Miacis" latouri* and a phylogenetic analysis of the earliest carnivoraforms (Mammalia, Carnivoramorpha). J. Vert. Paleo. 34, no. 1: 1–21.

3. Castelló, José R. *Canids of the World* (Princeton, NJ: Princeton University Press, 2018).

4. Ibid.

5. Botigue, Laura R., et al. 2017. Ancient European dog genomes reveal continuity since the Early Neolithic. Nat. Comm. https://doi.org/10.1038/ncomms16082.

6. Dietz, James M. Ecology and social organization of the maned wolf (*Chrysocyon brachyurus*). Smithsonian Contr. Zool. 392 (Washington, DC: Smithsonian Institution Press, 1984).

7. "Coyote." Vocabulary.com. https://vocabulary.com/dictionary/coyote, accessed 6/15/20.

8. Smith, Bradley, and Litchfield, Carla A. 2009. A review of the relationship between indigenous Australians, dingoes (*Canis dingo*) and domestic dogs (*Canis familiaris*). Anthrozoos 22, no. 2: 111–128.

9. Peace, Adrian. 2002. The cull of the wild: Dingoes, development and death in an Australian tourist location. Anthrop. Today 18, no. 5: 14–19.

10. Linnell, J.D.C., et al. 2002. The fear of wolves: A review of wolf attacks on humans. Norsk Inst. Naturforskning Oppdragsmelding 731: 1–65.

11. Barclay, Shelly. 2011. "The History of the Werewolf Legend." Historic Mysteries. https://www.historicmysteries.com/history-of-the-werewolf-legend/, accessed 6/15/20.

12. "Hypertrichosis (Werewolf Syndrome)." 2018. Healthline. https://www.healthline.com/health/hypertrichosis, accessed 6/15/20.

13. Tripp, Drew. 2017. "Chupacabra in the Carolinas?" ABC 4 News. https://abcnews4.com/news/local/is-this-a-chupacabra-in-the-lowcountry-or-simply-a-coyote-with-mange, accessed 6/15/20.

14. Drouilly, Marine, Nattrass, Nicoli, and O'Riain, M. Justin. 2018. Dietary niche relationships among predators on farmland and a protected area. J. Wildl. Manag. 82, no. 3: 507–518.

15. Shaffer, H. Bradley, and McKnight, Mark L. 1996. The polytypic species revisited: Genetic differentiation and molecular phylogenetics of the tiger salamander *Ambystoma tigrinum* (Amphibia: Caudata) complex. Evol. 50, no. 1: 417–433.

16. Leathlobhair, Maire, et al. 2018. The evolutionary history of dogs in the Americas. Science 361: 81–85.

17. "Gray Wolf." 2020. Yellowstone National Park. https://www.nps.gov/yell/learn/nature/wolves.htm, accessed 6/15/20.

18. Allen, B. J. et al. 2020. The latitudinal diversity gradient of tetrapods across the Permo-Triassic mass extinction and recovery interval. Proc. Roy. Soc. B. https://royalsocietypublishing.org/doi/full/10.1098/rspb.2020.0690.

19. Miao, Benpeng, et al. 2020. Genomic analysis reveals hypoxia adaptation in the Tibetan mastiff by introgression of the gray wolf from the Tibetan Plateau. Molec. Biol. Evol. 34, no. 3: 734–743. https://academic.oup.com/mbe/article/34/3/734/2843179.

20. "Webinar: Understanding Species Hybridization." 2018. National Academies of Sciences, Engineering, and Medicine. http://nas-sites.org/dels/studies/wolf-taxonomy-study/webinar-hybridization/, accessed 6/30/22.

21. Hersteinsson, Pal, and Macdonald, David W. 1992. Interspecific competition and the geographical distribution of red and Arctic foxes *Vulpes vulpes* and *Alopex lagopus*. Oikos 64: 501–515; Prestrud, Pal. 1991. Adaptations by the Arctic fox (*Alopex lagopus*) to the polar winter. Arctic 44, no. 2: 132–138; Pough, F. Harvey, and Janis, Christine M. *Vertebrate Life*, 10th ed. (New York: Oxford University Press, 2018).

22. Asa, Cheryl S., and Valdespino, Carolina. 1998. Canid reproductive biology: An integration of proximate mechanisms and ultimate causes. Amer. Zool. 38, no. 1: 251–259.

23. Hayward, Matt W., et al. 2017. Factors affecting the prey preferences of jackals (Canidae). Mamm. Biol. 85: 70–82.

24. Psova, Klara, et al. 2016. Golden jackal (*Canis aureus*) in the Czech Republic: The first record of a live animal and its long-term persistence in the colonized habitat. ZooKeys 641: 151–163. https://doi.org/10.3897/zookeys.641.10946.

25. Smith, Bradley, and Litchfield, Carla A. 2009. A review of the relationship between indigenous Australians, dingoes (*Canis dingo*) and domestic dogs (*Canis familiaris*). Anthrozoos 22, no. 2: 111–128.

26. "Dogs 101 New Guinea Singing Dog." YouTube. https://youtu.be/ttwt6xDO0M0, accessed 6/30/22.

27. Wallach, Adrian D., et al. 2015. Promoting predators and compassionate conservation. Cons. Biol. 29, no. 5: 1481–1484.

28. Parker, Merryl. 2007. The cunning dingo. Soc. Anim. 15: 69–78.

29. Letnic, Mike, et al. 2012. Top predators as biodiversity regulators: The dingo *Canis lupus dingo* as a case study. Biol. Rev. 87: 390–413.

30. Mech, L. David, and Boitani, Luigi. Wolf social ecology. In *Wolves: Behavior, Ecology, and Conservation*, Mech, L. David, and Boitani, Luigi, eds. (Chicago: University of Chicago Press, 2003), 1–34.

31. Castelló, José R. *Canids of the World* (Princeton, NJ: Princeton University Press, 2018); Creel, Scott, and Creel, Nancy Marusha. 1995. Communal hunting and pack size in African wild dogs, *Lycaon pictus*. Anim. Behav. 50: 1325–1339.

32. Creel, Scott, and Creel, Nancy Marusha. 1995. Communal hunting and pack size in African wild dogs, *Lycaon pictus*. Anim. Behav. 50: 1325–1339.

33. Courchamp, Franck, and Macdonald, David W. 2001. Crucial importance of pack size in the African wild dog *Lycaon pictus*. Anim. Cons. 4: 169–174.

34. Harrington, Fred H., and Asa, Cheryl S. Wolf communication. In *Wolves: Behavior, Ecology, and Conservation*, Mech, L. David, and Boitani, Luigi, eds. (Chicago: University of Chicago Press, 2003), 66–103.

35. Voldina, Elena, et al. 2006. Biphonation may function to enhance individual recognition in the dhole, *Cuon alpinus*. Ethol. 112, no. 8: 815–825.

36. Estes, Richard D., and Goddard, John. 1967. Prey selection and hunting behavior of the African wild dog. J. Wildl. Manag. 31, no. 1: 52–70.

37. Walker, Reena H., et al. 2017. Sneeze to leave: African wild dogs (*Lycaon pictus*) use variable quorum thresholds facilitated by sneezes in collective decisions. Proc. Roy. Soc. B 284: 20170347. http://dx.doi.org/10.1098/rspb.2017.0347.

38. Montague, Michael, et al. 2014. Comparative analysis of the domestic cat genome reveals genetic signatures underlying feline biology and domestication. PNAS 48. https://www.pnas.org/content/111/48/17230/tab-figures-data.

39. Sillero-Zubiri, Claudio, et al. 1996. Male philopatry, extra-pack copulations and inbreeding avoidance in Ethiopian wolves (*Canis simensis*). Behav. Ecol. Sociob. 38, no. 5: 331–340.

40. Castelló, José R. *Canids of the World* (Princeton, NJ: Princeton University Press, 2018); Asa, Cheryl S., and Valdespino, Carolina. 1998. Canid reproductive biology: An integration of proximate mechanisms and ultimate causes. Amer. Zool. 38, no. 1: 251–259.

41. Packard, Jane. Wolf behavior. In *Wolves: Behavior, Ecology, and Conservation*, Mech, L. David, and Boitani, Luigi, eds. (Chicago: University of Chicago Press, 2003), 35–65.

42. Acharya, Bhaskar. 2007. The ecology of the dhole or Asiatic wild dog (*Cuon alpinus*) in Pench Tiger Reserve, Madhya Pradesh. PhD diss., Saurashtra University, Rajkot; Girman, Derek J., et al. 1997. A molecular genetic analysis of social structure, dispersal, and interpack relationships of the African wild dog (*Lycaon pictus*). Behav. Ecol. Sociob. 40, no. 3: 187–198.

43. Erlandsson, R., et al. 2017. Indirect effects of prey fluctuation on survival of juvenile Arctic fox (*Vulpes lagopus*): A matter of maternal experience and litter attendance. Can. J. Zool. 95: 239–246.

44. Walker, Reena H., et al. 2017. Sneeze to leave: African wild dogs (*Lycaon pictus*) use variable quorum thresholds facilitated by sneezes in collective decisions. Proc. Roy. Soc. B 284: 20170347. http://dx.doi.org/10.1098/rspb.2017.0347.

45. Craven, Brent A., et al. 2010. The fluid dynamics of canine olfaction: Unique nasal airflow patterns as an explanation of macrosmia. J. Roy. Soc. Interface 7: 933–943.

46. "Bat-Eared Fox." 2020. San Diego Zoo Wildlife Alliance, Animals and Plants. https://animals.sandiegozoo.org/animals/bat-eared-fox, accessed 7/1/22.

47. Coren, Stanley. *How Dogs Think* (New York: Free Press, 2004).

48. Rogers, Lesley J., and Kaplan, Gisela. *Spirit of the Wild Dog* (Crows Nest, Australia: Allen and Unwin, 2003).

49. Stahler, Daniel, et al. 2002. Common ravens, *Corvus corax*, preferentially associate with grey wolves, *Canis lupus*, as a foraging strategy in winter. Anim. Behav. 64, no. 2: 283–290.

50. "Flying Foxes: Did You Know That Gray Foxes Can Climb Trees?" YouTube. https://youtu.be/aa27i1dkS9w, accessed 7/1/22.

51. Estes, Richard D., and Goddard, John. 1967. Prey selection and hunting behavior of the African wild dog. J. Wildl. Manag. 31, no. 1: 52–70.

52. Creel, Scott, and Creel, Nancy Marusha. 1995. Communal hunting and pack size in African wild dogs, *Lycaon pictus*. Anim. Behav. 50: 1325–1339.

53. "Wolf Restoration." 2020. Yellowstone National Park. https://www.nps.gov/yell/learn/nature/wolf-restoration.htm, accessed 7/1/22.

54. "How Wolves Change Rivers." YouTube. https://www.youtube.com/watch?v=ysa5OBhXz-Q, accessed 7/1/22.

55. Ripple, William J., and Beschta, Robert L. 2003. Wolf reintroduction, predation risk, and cottonwood recovery in Yellowstone National Park. Forest Ecol. Manag. 184: 299–313.

56. Fortin, Daniel, et al. 2005. Wolves influence elk movements: Behavior shapes a trophic cascade in Yellowstone National Park. Ecol. 86, no. 5: 1320–1330.

57. Ripple, William J., and Beschta, Robert L. 2012. Trophic cascades in Yellowstone: The first 15 years after wolf reintroduction. Biol. Cons. 145: 205–213.

58. "Yellowstone Science: Celebrating 20 Years of Wolves." 2016. Yellowstone Science 24, no. 1. https://www.nps.gov/yell/learn/upload/YELLOWSTONE-SCIENCE-24-1-WOLVES.pdf, accessed 7/1/22.

59. Kays, Roland, et al. 2009. Rapid adaptive evolution of northeastern coyotes via hybridization with wolves. Evol. Biol. 6, no. 1. https://doi.org/10.1098/rsbl.2009.0575.

60. vonHoldt, Bridgett, et al. 2016. Whole-genome sequence analysis shows that two endemic species of North American wolf are admixtures of the coyote and gray wolf. Sci. Adv. 2, no. 7: e1501714. https://doi.org/10.1126/sciadv.1501714.

61. Heppenheimer, Elizabeth, et al. 2018. Rediscovery of red wolf ghost alleles in a canid population along the American Gulf Coast. Genes 9, no. 12. https://www.mdpi.com/2073-4425/9/12/618.

62. Nagy, Christopher, et al. 2017. Initial colonization of Long Island, New York by the eastern coyote, *Canis latrans* (Carnivora, Canidae), including first record of breeding. Check List 13: 901–907. https://doi.org/10.15560/13.6.901.

63. Arnold, Janosch, et al. 2012. Current status and distribution of golden jackals *Canis aureus* in Europe. Mamm. Rev. 42, no. 1: 1–11.

64. Nasowitz, Dan. 2013. "Can I Have a Pet Fox?" Popular Science. https://www.popsci.com/science/article/2012-10/fyi-domesticated-foxes#page-21, accessed 7/1/22.

65. Castelló, José R. *Canids of the World* (Princeton, NJ: Princeton University Press, 2018).

66. "Decline of the Island Fox." 2020. Channel Island National Park, National Park Service. https://www.nps.gov/chis/learn/nature/fox-decline.htm, accessed 7/1/22.

67. Castelló, José R. *Canids of the World* (Princeton, NJ: Princeton University Press, 2018).

68. Botigue, Laura R., et al. 2017. Ancient European dog genomes reveal continuity since the Early Neolithic. Nat. Comm. https://doi.org/10.1038/ncomms16082.

69. "How Many Dog Breeds Are There?" 2020. Breeding Business. https://breedingbusiness.com/how-many-dog-breeds-are-there/, accessed 7/1/22.

70. Mardsen, Clare D., et al. 2016. Bottlenecks and selective sweeps during domestication have increased deleterious genetic variation in dogs. PNAS 113, no. 1: 152–157. https://www.pnas.org/content/113/1/152.short.

71. "Grey Wolf." Canid Specialist Group. https://www.canids.org/species/view/PREKLD895731, accessed 7/1/22.

CHAPTER 7. BEARS

1. Brown, Gary. *The Bear Almanac: A Comprehensive Guide to the Bears of the World* (Guilford, CT: Lyons Press, 2009).

2. Ritland, Kermit, Newton, Craig, and Marshall, H. Dawn. 2001. Inheritance and population structure of the white-phased "Kermode" black bear. Curr. Biol. 11: 1468–1472.

3. Amstrup, Steven C. Polar bear (*Ursus maritimus*). In *Wild Mammals of North America: Biology, Management, and Conservation*, Feldhamer, George A., et al., eds. (Baltimore: Johns Hopkins University Press, 2003), 587–610.

4. Caro, Tim, et al. 2017. Why is the giant panda black and white? Behav. Ecol. 28, no. 3: 657–667.

5. Not sure what that sounds like? Neither was one of your authors. Try this: "Original Chewbacca Scream." YouTube. https://www.youtube.com/watch?v=HswrFHRfcns, accessed 1/20/20.

6. Henry, J. D., and Herrero, S. M. 1974. Social play in the American black bear: Its similarity to canid social play and an examination of its identifying characteristics. Amer. Zool. 14, no. 1: 371–389.

7. Pough, F. Harvey, and Janis, Christine M. *Vertebrate Life*, 10th ed. (New York: Oxford University Press, 2019).

8. "Bears in Our Pool." YouTube. https://www.youtube.com/watch?v=PjRnPFl00FA, accessed 1/23/20.

9. Pagano, A, M. et al. 2013. Long-distance swimming by polar bears (*Ursus maritimus*) of the southern Beaufort Sea during years of extensive open water. Can. J. Zool. 90: 663–676.

10. "Do Bears Really Hibernate?" National Forest Foundation. https://www.nationalforests.org/blog/do-bears-really-hibernate, accessed 6/15/20.

11. "Fat Bear Week." National Park Service. https://www.nps.gov/katm/learn/fat-bear-week.htm, accessed 3/27/22.

12. Linzey, Donald W. *Vertebrate Zoology*, 2nd ed. (Baltimore: Johns Hopkins University Press, 2012).

13. Hellgren, Eric C., Physiology of Hibernation in Bears. 1998. Ursus 10: 467–477.

14. McGee-Lawrence, Meghan E., et al. 2009. Grizzly bears (*Ursus arctos horribilis*) and black bears (*Ursus americanus*) prevent trabecular bone loss during disuse (hibernation). Bone 45: 1186–1191; Hellgren, Eric C. 1998. Physiology of hibernation in bears. Ursus 10: 467–477.

15. Hailer, F. 2015. Introgressive hybridization: Brown bears as vectors for polar bear alleles. Molec. Ecol. 24: 1161–1163.

16. Cahill, J. A., et al. 2015. Genomic evidence of geographically widespread effect of gene flow from polar bears into brown bears. Molec. Ecol. 24: 1205–1217.

17. Kumar, V., et al. 2017. The evolutionary history of bears is characterized by gene flow across species. Sci. Rep. 7: 46487.

18. Peyton, Bernard. 1980. Ecology, distribution, and food habits of spectacled bears, *Tremarctos ornatus*, in Peru. J. Mamm. 61, no. 4: 639–652.

19. Rahming, Deevon. 2019. "Not Your Average Customer: Black Bear Walks into Murrells Inlet Restaurant." WPDE. com. https://wpde.com/news/local/not-your-average-customer-black-bear-walks-into-murrells-inlet-restaurant, accessed 5/28/20.

20. Kasbohm, John W., et al. 1996. Effects of gypsy moth infestation on black bear reproduction and survival. J. Wildl. Manag. 60, no. 2: 408–416.

21. French, Steven P., et al. 1994. Grizzly bear use of army cutworm moths in the Yellowstone ecosystem. Int. Conf. Bear Res. and Manage. 9, no. 1: 389–399.

22. Oka, Teruki, et al. 2004. Relationship between changes in beechnut production and Asiatic black bears in northern Japan. J. Wildl. Manag. 68, no. 4: 979–986.

23. Peyton, Barnard. 1980. Ecology, distribution, and food habits of spectacled bears, *Tremarctos ornatus*, in Peru. J. Mamm. 61, no. 4: 639–652.

24. Bergali, H. S., et al. 2004. Feeding ecology of sloth bears in a disturbed area in central India. Ursus 15, no. 2: 212–217.

25. Wong, Siew Te, et al. 2004. Home range, movement and activity patterns, and bedding sites of Malayan sun bears *Helarctos malayanus* in the rainforest of Borneo. Biol. Cons. 119: 169–181.

26. Freeman, Scott, and Herron, Jon C. *Evolutionary Analysis*, 5th ed. (New York: Pearson/Prentice Hall, 2014).

27. Peichl, L., et al. 2005. Retinal cone types in brown bears and the polar bear indicate dichromatic color vision. Invest. Ophthal. Vis. Sci. 46: 4539.

28. Klinka, D. R., and Reimchen, T. E. 2002. Nocturnal and diurnal foraging behaviour of brown bears (*Ursus arctos*) on a salmon stream in coastal British Columbia. Can. J. Zool. 80: 1317–1322.

29. Gende, S. M., et al. 2001. Consumption choice by bears feeding on salmon. Oecologia 127, no. 3: 372–382.

30. Hilderbrand, Grant V., et al. 1999. Role of brown bears (*Ursus arctos*) in the flow of marine nitrogen into a terrestrial ecosystem. Oecologia 121, no. 4: 546–550.

31. Rauset, Geir Rune, et al. 2012. Modeling female brown bear kill rates on moose calves using global positioning satellite data. J. Wildl. Manag. 76, no. 8: 1597–1606.

32. French, Steven P., and French, Marilynn G. 1990. Predatory behavior of grizzly bears feeding on elk calves in Yellowstone National Park, 1986–88. Int. Conf. Bear Res. and Manage. 8: 335–341.

33. Peyton, Bernard. 1980. Ecology, distribution, and food habits of spectacled bears, *Tremarctos ornatus*, in Peru. J. Mamm. 61, no. 4: 639–652.

34. Bowen W. D., and Siniff, D. B. Distribution, population biology, and feeding ecology of marine mammals. In *Biology of Marine Mammals*, Reynolds, John E., III, and Rommel, Sentiel A., eds. (Washington, DC: Smithsonian Institution Press, 1999), 423–484.

35. Stempniewicz, Lech. 2006. Polar bear predatory behaviour toward molting barnacle geese and nesting glaucous gulls on Spitsbergen. Arctic 59, no. 3: 247–251.

36. Bowen W. D., and Siniff, D. B. Distribution, population biology, and feeding ecology of marine mammals. In *Biology of Marine Mammals*, Reynolds, John E., III, and Rommel, Sentiel A., eds. (Washington, DC: Smithsonian Institution Press, 1999), 423–484.

37. Amstrup, Steven C. Polar bear (*Ursus maritimus*). In *Wild Mammals of North America: Biology, Management, and Conservation*, Feldhamer, George A., et al., eds. (Baltimore: Johns Hopkins University Press, 2003), 587–610.

38. Derocher, Andrew F., et al. 2000. Predation of Svalbard reindeer by polar bears. Polar Biol. 23: 675–678.

39. Gjertz, Ian, and Lydersen, Christian. 1986. Polar bear predation on ringed seals in the fast-ice of Hornsund, Svalbard. Polar Res. 4, no. 1: 65–68.

40. IUCN Red List. 2021–2023. https://www.iucnredlist.org, accessed 7/2/22.

41. Zhang, Baowei, et al. 2007. Genetic viability and population history of the giant panda, putting an end to the "evolutionary dead end"? Molec. Biol. Evol. 24, no. 8: 1801–1810.

42. Quammen, David. *Monster of God* (New York: W. W. Norton, 2003).

43. Brown, Gary. *The Bear Almanac: A Comprehensive Guide to the Bears of the World* (Guilford, CT: Lyons Press, 2009).

44. Ibid.

45. "Brown Bears." 2020. National Park Service. https://www.nps.gov/subjects/bears/brown-bears.htm, accessed 6/7/20.

46. Brown, Gary. *The Bear Almanac: A Comprehensive Guide to the Bears of the World* (Guilford, CT: Lyons Press, 2009).

47. "Brown Bears." 2020. IUCN. https://www.iucnredlist.org/species/41688/121229971#conservation-actions, accessed 6/7/20.

48. Pagano, A. M., et al. 2013. Long-distance swimming by polar bears (*Ursus maritimus*) of the southern Beaufort Sea during years of extensive open water. Can. J. Zool. 90: 663–676.

49. Amstrup, Steven C. Polar bear (*Ursus maritimus*). In *Wild Mammals of North America: Biology, Management, and Conservation*, Feldhamer, George A., et al., eds. (Baltimore: Johns Hopkins University Press, 2003), 587–610.

50. "Polar Bear Range States Circumpolar Action Plan." 2020. https://polarbearagreement.org/circumpolar-action-plan, accessed 6/6/20.

CHAPTER 8. MARINE MAMMALS

1. "Humpback Whale Almost Swallows Kayakers of California Coast." 2020. YouTube. https://youtu.be/UIuWBNgLqZY, accessed 7/2/22.

2. "The History of Animals." http://classics.mit.edu/Aristotle/history_anim.6.vi.html, accessed 3/3/20.

3. Pantidou, Georgia. Dolphins in secondary education. In *Dolphins*, Samuels, Joshua B., ed. (Hauppauge, NY: Nova Science Publishers, 2014), 83–128.

4. Leibowitz, Elissa. 2010. "Five Myths about Amazon River Dolphins." World Wildlife Fund. https://www.worldwildlife.org/blogs/good-nature-travel/posts/five-myths-about-amazon-river-dolphins, accessed 7/2/22.

5. Delfour, F., and Marten, K. 2001. Mirror image processing in three marine mammal species: Killer whales (*Orcinus orca*), false killer whales (*Pseudorca crassidens*) and California sea lions (*Zalophus californianus*). Behav. Proc. 53, no. 3: 181–190.

6. Lee, Meredith. 2014. "Blackfish." Center for Journalism Ethics, University of Wisconsin-Madison. https://ethics.journalism.wisc.edu/tag/blackfish, accessed 7/2/22.

7. Mason, Georgia J. 2010. Species differences in responses to captivity: Stress, welfare and the comparative method. Trends in Ecol. Evol. 25, no. 12: 713–721.

8. Jiang, Yixing, et al. 2007. Public awareness, education, and marine mammals in captivity. Tour. Rev. Int. 11, no. 3: 237–249. https://www.ingentaconnect.com/content/cog/tri/2007/00000011/00000003/art00006.

9. Arendt, Florian, and Matthes, Jorg. 2016. Nature documentaries, connectedness to nature, and pro-environmental behavior. Env. Comm. 10, no. 4: 453–472. https://www.tandfonline.com/doi/abs/10.1080/17524032.2014.993415.

10. Smith, Brian D., et al. 2009. Catch composition and conservation management of a human-dolphin cooperative cast-net fishery in the Ayeyarwady River, Myanmar. Biol. Cons. 142: 1042–1049.

11. Fahlman, A., et al. 2018. Lung mechanics and pulmonary function testing in cetaceans. Front. in Physiol. 9: 886. https://doi.org/10.3389/fphys.2018.00886.

12. Ridgeway, S. H., et al. 1969. Respiration and deep diving in the bottlenose porpoise. Science 166, no. 3913: 1651–1654.

13. Ridgway, S. H., and Johnston, D. G. 1966. Blood oxygen and ecology of porpoises of three genera. Science 151, no. 3709: 456–458.

14. Elsner, R. Living in water: Solutions to physiological problems. In *Biology of Marine Mammals*, Reynolds, J. E., and Rommel, S. A., eds. (Washington, DC: Smithsonian Press, 1999), 73–116.

15. Noren, S. R., and Williams, T. M. 2000. Body size and skeletal muscle myoglobin of cetaceans: Adaptations for maximizing dive duration. Comp. Biochem. Physiol. Part A: Molec. and Integrat. Physiol. 126, no. 22: 181–191.

16. Some consider the "dive response" to include all the anatomical, physiological, and behavioral adaptations that work together to allow prolonged, often deep dives.

17. Hanke, F. D., et al. 2009. Basic mechanisms in pinniped vision. Exp. Brain. Res. 199: 299–311.

18. Mauck, B., et al. 2005. How a harbor seal sees the night sky. Mar. Mam. Sci. 21, no. 4: 646–656.

19. Dehnhardt, Guido, et al. 1998. Seal whiskers detect water movements. Nature 394: 235–236.

20. Wieskotten, Sven, et al. 2011. Hydrodynamic discrimination of wakes caused by objects of different size or shape in a harbour seal (*Phoca vitulina*). J. Exp. Biol. 214: 1922–1930.

21. Oliver, Guy W. 1978. Navigation in mazes by a grey seal, *Halichoerus grypus* (Fabricius). Behav. 67, no. 1: 97–114.

22. Kilian, M., et al. 2015. How harbor seals (*Phoca vitulina*) pursue schooling herring. Mam. Biol. 80: 385–389.

23. Hocking, David. 2015. Prey capture and processing in otariid pinnipeds with implications for understanding the evolution of aquatic foraging in marine mammals. PhD diss., Monash University, Australia.

24. Adam, Peter J., and Berta, Annalisa. 2001. Evolution of prey capture strategies and diet in the Pinnipedimorpha (Mammalia, Carnivora). Oryktos 4: 3–27.

25. Ibid.

26. Marshall, Christopher D., et al. 2015. Feeding kinematics and performance of basal otariid pinnipeds, Steller sea lions and northern fur seals: Implications for the evolution of mammalian feeding. J. Exp. Biol. 218: 3229–3240.

27. Keinle, S. S., et al. 2018. Comparative feeding strategies and kinematics in phocid seals: Suction without specialized skull morphology. J. Exp. Biol. 221. https://doi.org/10.1242/jeb.179424.

28. Wells, Randall S., et al. Behavior. In *Biology of Marine Mammals*, Reynolds, J. E., and Rommel, S. A., eds. (Washington, DC: Smithsonian Press, 1999), 324–422.

29. Panagiotopoulou, Olga, et al. 2016. Architecture of the sperm whale forehead facilitates ramming combat. Peer Journals, Zool. Sci. https://www.ncbi.nlm.nih.gov/pubmed/27069822, accessed 5/22/2019.

30. Barrett-Lennard, Lance Godfrey. 2000. Population structure and mating patterns of killer whales (*Orcinus orca*) as revealed by DNA analysis. PhD diss., University of British Columbia.

31. Boyd, Ian L., et al. Reproduction in marine mammals. In *Biology of Marine Mammals*, Reynolds, J. E., and Rommel, S.A., eds. (Washington, DC: Smithsonian Press, 1999), 218–286.

32. "Orca Mother Drops Calf, after Unprecedented 17 Days of Mourning." 2018. National Geographic. https://www.nationalgeographic.com/animals/2018/08/orca-mourning-calf-killer-whale-northwest-news/, accessed 7/2/22.

33. Wartzok, D., and Ketten, D. R. Marine mammal sensory systems. In *Biology of Marine Mammals*, Reynolds, J. E., and Rommel, S. A., eds. (Washington, DC: Smithsonian Press, 1999), 118–175.

34. Wells, Randall S., et al. Behavior. In *Biology of Marine Mammals*, Reynolds, J. E., and Rommel, S. A., eds. (Washington, DC: Smithsonian Press, 1999), 324–422.

35. Fox, A., and Young, R. 2012. Foraging interactions between wading birds and strand-feeding bottlenose dolphins (*Tursiops truncatus*) in a coastal salt marsh. Can. J. Zool. 90: 744–752.

36. "Orcas' Upside-Down Hunting Moves May Be a Clever Way to 'Zombify' Stingrays." 2017. Earthtouch News Network. https://www.earthtouchnews.com/oceans/whales-and-dolphins/watch-orcas-upside-down-hunting-moves-may-be-a-clever-way-to-zombify-stingrays/, accessed 7/2/22.

37. Duignan, Padraig J., et al. 2000. Stingray spines: A potential cause of killer whale mortality in New Zealand. Aquat. Mam. 26, no. 2: 143–147.

38. Saulitis, Eva, et al. 2000. Foraging strategies of sympatric killer whale (*Orcinus orca*) populations in Prince William Sound, Alaska. Mar. Mam. Sci. 16, no. 1: 94–109.

39. Visser, Ingrid N., et al. 2007. Antarctic peninsula killer whales (*Orcinus orca*) hunt seals and a penguin on floating ice. Mar. Mam. Sci. 24, no. 1: 225–234.

40. Jefferson, Thomas A., et al. 1991. A review of killer whale interactions with other marine mammals: Predation to co-existence. Mam. Rev. 21, no. 4: 151–180.

41. Hoelzel, A. Rus. 1991. Killer whale predation on marine mammals at Punta Norte, Argentina: Food sharing, provisioning and foraging strategy. Behav. Ecol. Sociobiol. 29, no. 3: 197–204.

42. Fleming, P. A., and Bateman, P. W. 2018. Novel predation opportunities in anthropogenic landscapes. Anim. Behav. 138: 145–155.

43. Lauriano, G., et al. 2009. An overview of dolphin depredation in Italian artisanal fisheries. J. Mar. Biol. Assoc. U.K. 89, no. 5: 921–929.

44. "Marine Mammal Protection." 2020. NOAA Fisheries. https://www.fisheries.noaa.gov/topic/marine-mammal-protection, accessed 7/2/22.

45. Whitehead, Hal. 2002. Estimates of the current global population size and historical trajectory for sperm whales. Mar. Ecol. Prog. Ser. 242: 295–304.

46. IUCN Red List, November 2018. https://iucn-csg.org/5-updated-cetacean-red-list-assessments-published-in-march-2019/, accessed 7/2/22.

47. Reynolds, John E., et al. 2009. Marine mammal conservation. Endan. Spec. Res. 7: 23–28.

48. Kovaks, Kit M., et al. 2012. Global threats to pinnipeds. Mar. Biodiv. 41, no. 1: 414–436.

49. Reynolds, John E., et al. 2009. Marine mammal conservation. Endan. Spec. Res. 7: 23–28.

50. "Vaquita." Marine Mammal Commission. https://www.mmc.gov/priority-topics/species-of-concern/vaquita/, accessed 7/2/22.

51. Goldstein, Tracey, et al. 2009. Phocine distemper virus in northern sea otters in the Pacific Ocean, Alaska, USA. Emerging Infect. Dis. 15, no. 6: 925–927.

52. "Gulf of Mexico 'Dead Zone' Is the Largest Ever Measured." 2017. NOAA. https://www.noaa.gov/media-release/gulf-of-mexico-dead-zone-is-largest-ever-measured, accessed 7/2/22.

53. Ylitalo, Gina M., et al. 2001. Influence of life-history parameters on organochlorine concentrations in free-ranging killer whales (*Orcinus orca*) from Prince William Sound, AK. Sci. Total Env. 281: 183–203.

54. "Dead Whale Found with 40 Kilograms of Plastic in its Stomach." 2018. CNN. https://www.cnn.com/2019/03/18/asia/dead-whale-philippines-40kg-plastic-stomach-intl-scli/index.html, accessed 7/2/22; "Young Sperm Whale Found Full of Plastic." 2019. CNN. https://www.cnn.com/2019/05/20/europe/sperm-whale-sicily-plastic-stomach-intl-scli/index.html, accessed 7/2/22.

55. Weilgart, Linda S. 2007. A brief review of known effects of noise on marine mammals. Int. J. Comp. Phys. 20: 159–168.

56. Kovacs, Kit M., et al. 2011. Global threats to pinnipeds. Mar. Biodiv. 41, no. 1: 181–194.

57. Flores, H., et al. 2012. Impact of climate change on Antarctic krill. Mar. Ecol. Prog. Ser. 459: 1–19.

58. "Leopard Seals." 2018. Australian Antarctic Program. http://www.antarctica.gov.au/about-antarctica/wildlife/animals/seals-and-sea-lions/leopard-seals, accessed 7/2/22.

59. "Danger beneath the Water: 10 Facts about Leopard Seals." 2020. Oceanwide Expeditions. https://oceanwide-expeditions.com/blog/danger-beneath-the-water-10-facts-about-leopard-seals, accessed 7/2/22.

CHAPTER 9. SOME OTHER TOP PREDATORS

1. Fromentin, Jean-Marc, and Powers, Joseph E. 2005. Atlantic bluefin tuna: Population dynamics, ecology, fisheries, and management. Fish and Fisheries 6: 281–306.

2. Ibid.

3. Specia, Megan. 2019. "Japan's 'King of Tuna' Pays Record $3 Million for Bluefin at New Tokyo Fish Market." *New York Times*. https://www.nytimes.com/2019/01/05/world/asia/record-tuna-price-japan.html, accessed 7/3/22.

4. "Atlantic Bluefin Tuna." 2020. IUCN. https://www.iucnredlist.org/species/21860/9331546, accessed 7/3/22.

5. Bozek, Michael A., et al. 1999. Diets of muskellunge in northern Wisconsin lakes. N. Amer. J. Fish. Manag. 19: 258–270.

6. "Muskellunge." 2020. Lake Scientist. https://www.lakescientist.com/lake-facts/fish/muskellunge/, accessed 6/5/20.

7. Richards, K. Trends in muskellunge populations and fisheries in Wisconsin during 1980–2015. In *Muskellunge Management*, Kapuscinski, K. L., et al., eds., American Fisheries Society Symposium 85 (Bethesda, MD, 2017), 33–36.

8. Valenciano, Alberto, et al. 2016. *Megalictis*, the bone-crushing giant mustelid from the early Miocene of North America. PLoS ONE 11, no. 4: https://doi.org/10.1371/journal.pone.0152430.

9. Hutchings, Michael R., and White, Piran C. L. 2000. Mustelid scent-marking in managed ecosystems: Implications for population management. Mamm. Rev. 30, no. 3–4: 157–169.

10. Kruuk, Hans. *Otters: Ecology, Behavior and Conservation* (Oxford: Oxford University Press, 2007).

11. Londono, G. Corredor, and Munoz, Tigreros. 2006. Reproduction, behaviour and biology of the giant river otter. Int. Zoo Yearbook 40: 360–371.

12. dos Santos Lima, Danielle, et al. 2012. Site and refuge use by giant river otters (*Pteronura brasiliensis*) in western Brazilian Amazonia. J. Natl. Hist. 46, no. 11–12: 729–739.

13. Kruuk, Hans. *Otters: Ecology, Behavior and Conservation* (Oxford: Oxford University Press, 2007).

14. dos Santos Lima, Danielle, et al. 2012. Site and refuge use by giant river otters (*Pteronura brasiliensis*) in western Brazilian Amazonia. J. Natl. Hist. 46, no. 11–12: 729–739.

15. Kruuk, Hans. *Otters: Ecology, Behavior and Conservation* (Oxford: Oxford University Press, 2007).

16. Ward, Simon. 2015. "Amazing Facts about Fur: Nature's Densest Furs." Truth about Fur. https://www.truthaboutfur.com/blog/amazing-facts-about-fur-natures-densest-furs/, accessed 7/3/22.

17. Kruuk, Hans. *Otters: Ecology, Behavior and Conservation* (Oxford: Oxford University Press, 2007).

18. Beichman, Annabel C., et al. 2019. Aquatic adaptation and depleted diversity: A deep dive into the genomes of the sea otter and giant otter. Molec. Biol. Evol. 36, no. 12: 2631–2655.

19. Bell, Tom W., et al. 2015. Geographical variability in the controls of giant kelp biomass dynamics. J. Biogeog. 42: 2010–2021.

20. Shelton, Andrew O., et al. 2018. From the predictable to the unexpected: Kelp forest and benthic invertebrate community dynamics following decades of sea otter expansion. Oecologia 188: 1105–1119; Bell, Tom W., et al. 2015. Geographical variability in the controls of giant kelp biomass dynamics. J. Biogeog. 42: 2010–2021.

21. Beichman, Annabel C., et al. 2019. Aquatic adaptation and depleted diversity: A deep dive into the genomes of the sea otter and giant otter. Molec. Biol. Evol. 36, no. 12: 2631–2655.

22. White, LeAnn, et al. 2018. Mortality trends in northern sea otters (*Enhydra lutris kenyoni*) collected from the coasts of Washington and Oregon, USA (2002–15). J. Wildl. Diseas. 54, no. 2: 238–247.

23. Waxman, O. 2014. "A Brief History of the Honey Badger Meme." Time. https://newsfeed.time.com/2014/02/19/a-brief-history-of-the-honey-badger-meme/, accessed 7/3/22.

24. Begg, C. M., et al. 2003. Scent-marking behavior of the honey badger, *Mellivora capensis* (Mustelidae), in the southern Kalahari. Anim. Behav. 66: 917–929; Carter, Shenrae, et al. 2017. The honey badger in South Africa: Biology and conservation. Int. J. Avian and Wildl. Biol. 2, no. 2. https://doi.org/10.15406/ijawb.2017.02.00016.

25. Begg, C. M., et al. 2005. Life history variables of an atypical mustelid, the honey badger, *Mellivora capensis*. J. Zool. Soc. Lon. 265: 17–22.

26. Begg, C. M., et al. 2003. Sexual and seasonal variation in the diet and foraging behavior of a sexually dimorphic carnivore, the honey badger (*Mellivora capensis*). J. Zool. Soc. Lon. 260: 301–316.

27. Allen, Maximilian R., et al. 2018. No respect for apex carnivores: Distribution and activity patterns of honey badgers in the Serengeti. Mamm. Biol. 89: 90–94.

28. Drabek, Danielle H., et al. 2015. Why the honey badger don't care: Convergent evolution of venom-targeted nicotinic acetylcholine receptors in mammals that survive venomous snake bites. Toxicon 99: 68–72.

29. Carter, Shenrae, et al. 2017. The honey badger in South Africa: Biology and conservation. Int. J. Avian and Wildl. Biol. 2, no. 2. https://doi.org/10.15406/ijawb.2017.02.00016.

30. Banci, Vivian. Wolverine. In *The Scientific Basis for Conserving Forest Carnivores: American Marten, Fisher, Lynx, and Wolverine in the Western United States.* Gen. Tech. Rep. RM-254, Ruggiero, Leonard F., et al., eds. (Fort Collins, CO: US Dept. of Agriculture, Forest Service, Rocky Mountain Forest and Range Experiment Station, 1994), 99–127.

31. Lundberg, Murray. 2009. "The Wolverine." Explore North. http://www.explorenorth.com/library/weekly/aa121500a.htm, accessed 7/3/22.

32. Magoun, Audrey, et al. 2018. Characteristics of wolverine reproductive den sites. Can. Field-Natur. 132, no. 4. https://doi.org/10.22621/cfn.v132i4.2050.

33. Van Dijk, Jiska, et al. 2008. Foraging strategies of wolverines within a predator guild. Can. J. Zool. 86: 966–975.

34. Krebs, John, et al. 2007. Multiscale habitat use by wolverines in British Columbia, Canada. J. Wildl. Manag. 71, no. 7: 2180–2192.

35. van der Veen, Bert, et al. 2020. Refrigeration or anti-theft? Food caching behavior of wolverines (*Gulo gulo*) in Scandinavia. Behav. Ecol. Sociol. 74. https://doi.org/10.1007/s00265-020-2823-4.

36. Inman, Robert M., et al. 2013. Developing priorities for metapopulation conservation at the landscape scale: Wolverines in the western United States. Biol. Cons. 166: 276–286.

37. Persson, J., et al. 2010. Space use and territoriality of wolverines (*Gulo gulo*) in northern Fennoscandia. Europ. J. Wildl. Res. 56, no. 1: 49–57.

38. Higgins, Stephanie. 2019. Pedigree reconstruction reveals large scale movement patterns and population dynamics of wolverines (*Gulo gulo*) across Fennoscandia. Master's thesis, Swedish University of Agricultural Sciences.

39. Copeland, Jeffrey P., and Kucera, Thomas E. Wolverine. In *Mesocarnivores of Northern California: Biology, Management, and Survey Techniques*, Harris, John E., and Ogan, Chester V., eds. (Arcata, CA: Wildlife Society, California North Coast Chapter, 1997), 23–34.

40. Magoun, Audrey, and Copeland, Jeffrey P. 1998. Characteristics of wolverine reproductive den sites. J. Wildl. Manag. 62, no. 4: 1313–1320.

41. Higgins, Stephanie. 2019. Pedigree reconstruction reveals large scale movement patterns and population dynamics of wolverines (*Gulo gulo*) across Fennoscandia. Master's thesis, Swedish University of Agricultural Sciences.

42. Persson, Jens, et al. 2006. Reproductive characteristics of female wolverines (*Gulo gulo*) in Scandinavia. J. Mamm. 87, no. 1: 75–79.

43. Krebs, John, et al. 2007. Multiscale habitat use by wolverines in British Columbia, Canada. J. Wildl. Manag. 71, no. 7: 2180–2192.

44. Thiel, Alexandra, et al. 2019. Effects of reproduction and environmental factors on body temperature and activity patterns of wolverines. Front. Zool. 16. https://doi.org/10.1186/s12983-019-0319-8.

45. Cooke, Lucy. *The Truth about Animals: Stoned Sloths, Lovelorn Hippos, and Other Tales from the Wild Side of Wildlife* (London: Transworld, 2018).

46. Holecamp, K. E., et al. 2012. Society, demography and genetic structure in the spotted hyena. Molec. Ecol. 21: 613–632.

47. Bradford, Alina. 2016. "Facts about Hyenas." LiveScience. https://www.livescience.com/55037-hyenas.html, accessed 4/24/20.

48. Holecamp, Kay E., et al. 1997. Patterns of association among female spotted hyenas (*Crocuta crocuta*). J. Mamm. 78, no. 1: 55–64.

49. East, Marion L., and Hofer, Heribert. 2001. Male spotted hyenas (*Crocuta crocuta*) queue for status in social groups dominated by females. Behav. Ecol. 12, no. 5: 558–568.

50. Holekamp, Kay E., et al. 1997. Patterns of association among female spotted hyenas (*Crocuta crocuta*). J. Mamm. 78, no. 1: 55–64.

51. Kruuk, Hans. *The Spotted Hyena* (Chicago: University of Chicago Press, 1972).

52. Binder, Wendy, and van Valkenburgh, Blaire. 2000. Development of bite strength and feeding behavior of spotted hyenas (*Crocuta crocuta*). J. Zool. Soc. Lon. 252: 273–283.

53. Kruuk, Hans. *The Spotted Hyena* (Chicago: University of Chicago Press, 1972).

54. Cooke, Lucy. *The Truth about Animals: Stoned Sloths, Lovelorn Hippos, and Other Tales from the Wild Side of Wildlife* (London: Transworld, 2018).

55. Frank, Laurence G., et al. 1991. Fatal sibling aggression, precocial development, and androgens in neonatal spotted hyenas. Science 252, no. 5006: 702–704; East, Marion L., et al. 1993. The erect "penis" is a flag of submission in a female-dominated society: Greetings in Serengeti spotted hyenas. Behav. Ecol. Soc. 33: 355–370.

56. Binder, Wendy, and van Valkenburgh, Blaire. 2000. Development of bite strength and feeding behavior of spotted hyenas (*Crocuta crocuta*). J. Zool. Soc. Lon. 252: 273–283.

57. Frank, Laurence G., et al. 1991. Fatal sibling aggression, precocial development, and androgens in neonatal spotted hyenas. Science 252, no. 5006: 702–704.

58. Hambrick, David Z. 2017. "What Hyenas Can Tell Us about the Origins of Intelligence." Scientific American. https://www.scientificamerican.com/article/what-hyenas-can-tell-us-about-the-origins-of-intelligence/, accessed 7/3/22.

59. "Hyena Den with Dr. Robert Johnson in South Africa." 2021. YouTube. https://www.youtube.com/watch?v=uHzizZ__4EA, accessed 7/3/22.

60. Green, D. S., et al. 2018. Anthropogenic disturbance induces opposing population trends in spotted hyenas and African lions. Biodiv. Cons. 27: 871–889.

61. "Fossa." 2020. San Diego Zoo Animals and Plants. https://animals.sandiegozoo.org/animals/fossa, accessed 7/3/22.

62. Goodman, S. M., et al. 1997. Food habits of *Cryptoprocta ferox* in the high mountain zone of the Andringitra Massif, Madagascar. Mammalia 61, no. 2. https://doi.org/10.1515/mamm.1997.61.2.185.

63. Köhncke, M., and Leonhardt, K. 1986. *Cryptoprocta ferox*. Amer. Soc. of Mammalogists, Mammalian Species, no. 254: 1–5.

64. "Fossa." 2020. San Diego Zoo Animals and Plants. https://animals.sandiegozoo.org/animals/fossa, accessed 7/3/22.

65. Pough, F. Harvey, and Janis, Christine M. *Vertebrate Life*, 10th ed. (New York: Oxford University Press, 2019).

66. Keely, T., et al. 2017. Seasonality and breeding success of captive and wild Tasmanian devils (*Sarcophilus harrisii*). Theriogenology 95: 33–41.

67. Minton, Ella. 2019. "Tasmanian Devil." Australian Museum. https://australianmuseum.net.au/learn/animals/mammals/tasmanian-devil/, accessed 7/3/22.

68. Bruniche-Olsen, Anna, et al. 2014. Extensive population decline in the Tasmanian devil predates European settlement and devil facial tumor disease. Biol. Letters 10. http://dx.doi.org/10.1098/rsbl.2014.0619.

69. Hendricks, Sarah, et al. 2017. Conservation implications of limited genetic diversity and population structure in Tasmanian devils (*Sarcophilus harrisii*). Cons. Genet. 18: 977–982.

70. Keeley, T., et al. 2017. Seasonality and breeding success of captive and wild Tasmanian devils (*Sarcophilus harrisii*). Theriogenology 95: 33–41.

71. Epstein, Brendan, et al. 2016. Rapid evolutionary response to a transmissible cancer in Tasmanian devils. Nat. Comm. https://doi.org/10.1038/ncomms12684.

CHAPTER 10. HUMANS

1. Linzey, Donald W. *Vertebrate Life*, 2nd ed. (Baltimore: Johns Hopkins University Press, 2012).

2. "1992 World Scientists' Warning to Humanity." 1992, updated 2022. Union of Concerned Scientists. http://www.ucsusa.org/about/1992-world-scientists.html, accessed 6/23/2022,

3. Ripple, W. J., et al. 2017. World scientists' warning to humanity: A second notice. Bioscience 67, no. 12: 1026–1028.

4. Nebel, B. J., and Wright, R. T. *Environmental Science: The Way the World Works* (Upper Saddle River, NJ: Pearson, 1999).

5. Noss, R. F., and Scott, J. M. 1995. *Endangered Ecosystems of the United States: A Preliminary Assessment of Loss and Degradation.* Vol. 28. US Department of the Interior, National Biological Service.

6. World Resources Institute. https://www.wri.org./, accessed 7/2/2022.

7. Intergovernmental Science-Policy Platform on Biodiverstiy and Ecosystem Services (IPBES). *The Global Assessment Report on Biodiversity and Ecosystem Services*, Brondizio, E. S., Settele, J., Díaz, S., and Ngo, H. T., et al., eds. (Bonn, Germany: IPBES Secretariat, 2019).

8. Bouvier, L., and L. Grant. *How Many Americans?* (San Francisco: Sierra Club Books, 1994).

9. "Genomic Study Points to Natural Origin of COVID-19." 2020. NIH Director's Blog. https://directorsblog.nih.gov/2020/03/26/genomic-research-points-to-natural-origin-of-covid-19/, accessed 6-23-2022

10. Here are two recommendations: support Shark Advocates International (for sharks), and Dogwood Alliance (for climate change and forest protection).

INDEX